Arrow Logic and
Multi-Modal Logic

Studies in Logic, Language and Information

Studies in Logic, Language and Information is the official book series of the European Association for Logic, Language and Information (FoLLI).

The scope of the series is the logical and computational foundations of natural, formal, and programming languages, as well as the different forms of human and mechanized inference and information processing. It covers the logical, linguistic, psychological and information-theoretic parts of the cognitive sciences as well as mathematical tools for them. The emphasis is on the theoretical and interdisciplinary aspects of these areas.

The series aims at the rapid dissemination of research monographs, lecture notes and edited volumes at an affordable price.

Managing editor: Robin Cooper, University of Gothenburg

Executive editor: Maarten de Rijke, University of Warwick

Editorial board:

Peter Aczel, Manchester University
Nicholas Asher, The University of Austin, Texas
Jon Barwise, Indiana University, Bloomington
John Etchemendy, CSLI, Stanford University
Dov Gabbay, Imperial College, London
Hans Kamp, Universität Stuttgart
Godehard Link, Universität München
Fernando Pereira, AT&T Bell Laboratories, Murray Hill
Dag Westerståhl, Stockholm University

Arrow Logic and Multi-Modal Logic

edited by
Maarten Marx,
László Pólos,
and Michael Masuch

CSLI Publications
Center for the Study of Language and Information
Stanford, California
&
FoLLI
The European Association for
Logic, Language and Information

Copyright ©1996
Center for the Study of Language and Information
Leland Stanford Junior University
Printed in the United States
00 99 98 97 96   5 4 3 2 1

Library of Congress Cataloging-in-Publication Data

Arrow logic and multi-modal logic / edited by Maarten Marx,
  László Pólos, and Michael Masuch.
      p. cm.
Includes bibliographical references and index.
ISBN 1-57586-025-2(alk. paper). – ISBN 1-57586-024-4 (pbk.)
1. Logic. 2. Modality (Logic) I. Marx, Maarten. I. Pólos, László.
III. Masuch, Michael, 1949–   .
   BC71.A77
   160–dc20   96–24385
                                          CIP

∞ The acid-free paper used in this book meets the minimum requirements of the American National Standard for Information Sciences—Permanence of Paper for Printed Library Materials, ANSI Z39.48-1984.

Contents

Contributors vii
Preface xi

Part I Arrow Logic 1

1 **A Crash Course in Arrow Logic** 3
   YDE VENEMA

2 **Investigations in Arrow Logic** 35
   MAARTEN MARX, SZABOLCS MIKULÁS, ISTVÁN NÉMETI AND ILDIKÓ SAIN

3 **Causes and Remedies for Undecidability in Arrow Logics and in Multi-Modal Logics** 63
   HAJNAL ANDRÉKA, ÁGNES KURUCZ, ISTVÁN NÉMETI, ILDIKÓ SAIN AND ANDRÁS SIMON

4 **Associativity Does Not Imply Undecidability without the Axiom of Modal Distribution** 101
   VIKTOR GYURIS

5 **Dynamic Arrow Logic** 109
   MAARTEN MARX

6 **Complete Calculus for Conjugated Arrow Logic** 125
   SZABOLCS MIKULÁS

7 **Many-Dimensional Arrow Structures: Arrow Logics II** 141
   DIMITER VAKARELOV

Part II Multi-Modal Logic 189

8 **What is Modal Logic?** 191
   MAARTEN DE RIJKE

9   Content versus Wrapping: an Essay in Semantic
    Complexity   203
    JOHAN VAN BENTHEM

10 A Fine-Structure Analysis of First-Order Logic   221
    ISTVÁN NÉMETI

Contributors

HAJNAL ANDRÉKA is a research scientist in the Mathematical Institute of the Hungarian Academy of Science. Her research interests include most aspects of logic and many aspects of computer science.
Current address: Mathematical Institute, P.O.Box 127, Budapest, H-1364 Hungary.
E-mail: andreka@math-inst.hu.

JOHAN VAN BENTHEM is a professor of Logic at the University of Amsterdam and Stanford University. His research interests include Logic and its applications in Linguistics and Computer Science.
Current address: Institute for Logic, Language and Computation, Plantage Muidergracht 24, 1018 TV Amsterdam, The Netherlands.
E-mail: johan@fwi.uva.nl.

VIKTOR GYURIS is a Ph.D. student at the University of Illinois at Chicago. His research interests are in Logic, Algebraic Logic, Universal Algebra and Computer Science.
Current address: Department of Mathematics, University of Illinois at Chicago, 851 S. Morgan (m/c 249) Chicago, IL 60607 USA.
E-mail: viktor@math.uic.edu.
WWW: http://zariski.math.uic.edu/~viktor/.

ÁGNES KURUCZ is a Ph.D. student in the Symbolic Logic Department of Eötvös Loránd University, Budapest. Her research interests include mathematical logic, algebraic logic, decision problems.
Current address: Mathematical Institute of the Hungarian Academy of Sciences, P.O. Box 127, H-1364 Budapest, Hungary.
E-mail: kuag@math-inst.hu.

MAARTEN MARX is a Research Associate at the Department of Computing, Imperial College, London. His research interests are in modal and algebraic logic.

Current address: Department of Computing, Imperial College, 180 Queen's Gate, London SW7 2BZ, United Kingdom.
E-mail: m.marx@doc.ic.ac.uk.

MICHAEL MASUCH is the scientific director of the Center for Computer Science in Organization and Management (CCSOM) of the University of Amsterdam. His research interests include the application of logic in the social sciences. *Current address:* CCSOM, University of Amsterdam, Sarphatistraat 143, 1018 GD, Amsterdam, The Netherlands.
E-mail: michael@ccsom.uva.nl.

SZABOLCS MIKULÁS is a Research Associate at King's College London. His research interests include algebraic and modal logic.
Current address: Department of Computer Science, King's College London, Strand, London WC2R 2LS, United Kingdom.
E-mail: szabolcs@dcs.kcl.ac.uk.

ISTVÁN NÉMETI is a research scientist in the Mathematical Institute of the Hungarian Academy of Science. His research interests include most aspects of logic and many aspects of computer science.
Current address: Mathematical Institute, P.O.Box 127, Budapest, H-1364 Hungary.
E-mail: nemeti@math-inst.hu.

LÁSZLÓ PÓLOS is a senior researcher at the Center for Computer Science in Organization and Management (CCSOM) of the University of Amsterdam. His research interests include formal semantics, situation theory, non-monotonic reasoning, and the application of logic in the social sciences.
Current address: CCSOM, University of Amsterdam, Sarphatistraat 143, 1018 GD, Amsterdam, The Netherlands.
E-mail: laszlo@ccsom.uva.nl.

MAARTEN DE RIJKE is a Warwick Research Fellow at the University of Warwick. His research interests are in logic, process theory, and complex systems.
Current address: Department of Computer Science, University of Warwick, Coventry CV4 7AL, England.
E-mail: mdr@dcs.warwick.ac.uk.

ILDIKÓ SAIN is a research mathematician in the Mathematical Institute of the Hungarian Academy of Science. Her research interests include most aspects of logic and many aspects of theoretical computer science and universal algebra. Currently she is most active in algebraic logic.
Current address: Mathematical Institute, P.O.Box 127, Budapest, H-1364 Hungary.
E-mail: sain@math-inst.hu.

ANDRÁS SIMON is an assistant research fellow in the Mathematical Institute of Hungarian Academy of Sciences. His research interests include algebraic logic, universal algebra and model theory.
Current address: Mathematical Institute of the Hungarian Academy of Sciences 1364 Budapest, P.O. Box 127, Hungary.
E-mail: andras@math-inst.hu.

DIMITER VAKARELOV is a professor in the Faculty of Mathematics and Computer Science, Sofia University. His research interests are in Non-Classical Logic with applications in Computer Science and AI.
Current address: Department of Mathematical Logic with Laboratory of Applied Logic, Faculty of Mathematics and Computer Science, Sofia University, blvd James Bouchier 5, 1126 Sofia , Bulgaria.
E-mail: dvak@fmi.uni-sofia.bg.

YDE VENEMA is currently a research fellow of the Royal Dutch Academy of Arts and Sciences. His research interests include modal and algebraic logic, and logical aspects of dynamics.
Current address: Department of Mathematics and Computer Science, Free University, De Boelelaan 1081, 1081 HV Amsterdam, The Netherlands.
E-mail: yde@cs.vu.nl.

Preface

In the last decade the trend towards dynamics is one of the most important developments in formal semantics (of natural language). In this school, sentence meaning is characterized in terms of information state potential. To know what a (declarative) statement means is to know what changes it can bring about in the information state of the receiving agent. Understanding a sentence results in a transition from one information state to another. From this perspective the logic of information processing is the logic of transitions.

Conceived by Johan van Benthem and Yde Venema (1991), arrow logic was an attempt to give an account of the logic of transitions in general. This general approach to transitions led to new applications in areas ranging from philosophy to computer science. Transitions are modeled by arrows; the meaning of a proposition is given by a set of arrows. The connection between arrow logic and (multi-)modal logic is very close, since arrow logic can be seen as the modal logic of transitions.

The language of arrow logic is a propositional language enriched with three modalities: a binary modality "∘" for composing arrows, a unary one "⊗" for taking the inverse of an arrow and a constant which denotes the identity arrows. Thus the language of arrow logic coincides with the one of relation algebras. In the form of relation algebras, arrow logic is studied for about 150 years, starting with de Morgan, Peirce and Schröder. The main interest of the relation algebra community was and is one particular semantics for the above given language. In this semantics arrows are equated with ordered pairs and the models are full Cartesian products $U \times U$ ("square semantics"). This is the study of the class of representable relation algebras (RRA). This particular semantics has the disadvantage that it is highly undecidable and not finitely axiomatizable.

What is new in the study of arrow logic is that one looks for natural semantics which do have such desirable meta-properties as for instance *decidability, finite axiomatizability, Craig interpolation* and *Beth definability*.

Van Benthem argues in his contribution to this volume that these weaker but nice-behaving logics are in many cases better suited for applications than the well-known classical arrow logic with the square semantics.

The arrow logic day at the Logic@Work conference in December 1992 was the first formal meeting fully devoted to this subject. At that conference it became apparent that arrow logic was just an example of a more general scientific program — sketched explicitly in van Benthem's paper — which can be described as *finding new versions of well-known and widely applied undecidable logics which have desirable meta-logical properties (in particular decidability) and which are still strong enough for applications.* We call this program the "arrow logic analysis". A prime candidate for such an analysis is of course first order (FO) logic. The contribution of Németi contains several decidable but still quite strong versions of FO logic.

As a result of these developments the field of arrow logic was considerably broadened in the years following this conference. The arrow logic analysis covers the general enterprise of finding the undecidability/decidability border of widely applied logics and finding decidable, but still intuitive, versions of them.

A result of this analysis is a whole landscape of possible semantic modelings for a logic where the original (undecidable) logic turns out to be just one particular option. Such a landscape makes it easier to choose a specific version of a logic for a specific application. In chapters 2 and 10 interesting parts of this landscape are drawn for arrow logic and FO logic respectively.

The contents of this volume reflect the widening of the field. Besides chapters which deal with arrow logic proper and where one can find many techniques which can be used for an "arrow logic analysis" of other logics, we included three papers which are in the spirit of the general research program.

Short Description of the Content

This volume is divided into two parts. The first part is about arrow logic proper, the second about the arrow logic analysis.

Part I. This part starts with an extensive introduction and overview of the field of arrow logic written by Yde Venema. Motivations and applications are provided and the connection of arrow logic with modal logic and with relation algebras is established.

In chapter 2 (Marx et al.) the focus is on that particular semantics for arrow logic where arrows are equated with ordered pairs. It is shown that with this semantics arrow logic has the four above mentioned positive properties if and only if there are models with non-transitive universes.

In chapters 3 and 4 (Andréka et al. and Gyuris) a sharp border is provided between decidable and undecidable (modal) logics. A main result is that the Booleans plus an associative binary modality lead to undecidability. This gives a clear division between decidable and undecidable arrow logics. Several stronger versions are also given, together with several applications to logics other than arrow logic.

Chapter 5 (Marx) provides a comparison between arrow logic and another logic of transitions: Propositional Dynamic Logic (PDL). One of the results is that expansions with the Kleene star of the nice behaving versions of arrow logic studied in chapter 2 stay decidable and finitely axiomatizable.

In chapter 6 (Mikulás) the converse operator is dropped from the language and instead the two residuals of composition are taken as primitive operators. This version of arrow logic can be seen as an expansion of the Lambek Calculus with the Booleans.

Chapter 7 (Vakarelov) gives an account of the Sophia approach to arrow logic. The emphasis is on multi-graphs as models for arrow logic. The basic language differs from the relation algebra language and lies somewhere in between that language and the language of FO logic. In this paper the notion of an arrow as an object with a head and a tail is generalized to an object with n dimensions.

Part II. In this part it is the arrow logic analysis which is the binding factor of the papers.

The contribution of de Rijke (chapter 8) puts arrow logic in a wider perspective, using modal logic as a unifying framework. The sharp dividing line between (decidable) propositional logic and (undecidable) first order logic disappears when viewing these logics — and almost everything in between — as modal logics.

In chapter 9 van Benthem sketches the research program of the arrow logic analysis. He provides strong motivations from the perspective of applied logic for a quest for decidable versions of such widely applied logics as first order logic and the logic of binary relations. He emphasizes the distinction between the classical versions of these logics and their "logical core", which should have positive meta-properties.

Chapter 10 (Németi) provides several decidable versions of FO logic. This chapter is a good example of the research program outlined by van Benthem and shows that the program can indeed be successfully extended beyond arrow logic proper.

About the Authors

Arrow logic is studied mainly in three cities in Europe: Amsterdam, Budapest and Sophia. Each city correspond roughly with a particular view on arrow logic. In Amsterdam arrow logic is closely connected to modal logic and Kripke semantics. In Budapest arrow logic is mainly studied as a weakened version of Representable relation algebras. In Sophia the emphasis is on the arrow logic of multi-graphs. This volume contains contributions of authors from all these three cities; a short description of them can be found in the list of contributors.

We are indebted to Maarten de Rijke of SiLLI for his help during the preparation of this volume.

Amsterdam, March 1996
Maarten Marx

Part I
Arrow Logic

1

A Crash Course in Arrow Logic
YDE VENEMA

ABSTRACT. This contribution gives a short introduction to arrow logic. We start by explaining the basic idea underlying arrow logic and the motivation for studying it (sections 1 and 2). We discuss some elementary duality theory between arrow logic and the algebraic theory of binary relations (section 3). In the sections 4 and 5 we give a brief survey of the theory that has been developed on the semantics (definability), axiomatics and decidability of various systems of arrow logic. We briefly describe some closely related formalisms and some extensions and reducts of arrow logic in section 7. We end with mentioning some promising research lines and open problems, in section 8.

1 The Basic Idea

Summarizing in six words what arrow logic is about, one could say that

arrow logic is the basic modal logic of arrows,

a slogan calling for a few more words to discuss its key words: 'basic', 'modal' and 'arrow'.

To start with the latter, the language of arrow logic is designed to talk about all such objects as may be represented in a picture by arrows. As a concrete representation of an arrow the mathematically inclined reader might think of a vector, a function or a morphism in some category; the computer scientist of a program; the linguist of the dynamic meaning of a grammatically well-formed piece of text or discourse; the philosopher of some agent's (cognitive) action; etc. Note that in the last three examples (which will be discussed in some more detail later on) arrows are *transitions* related to some space of (information) states.

The essential characteristic of arrow logic is that the entities at which the truth of a formula is evaluated are arrows, i.e. that the basic statement made in arrow logic is of the form

$$\mathfrak{M}, a \models \phi.$$

Here $\mathfrak{M}$ is an arrow model, i.e. a structure of which the universe consists of arrows and arrows only, and a is an *arrow*. This is not to say that arrows need always be the primitive entities of the model. For instance, in the semantics of arrow logic an important role is played by two-dimensional models. Here an arrow a is seen as a *pair* (a_0, a_1) of which a_0 may be thought of as the starting point of a and a_1 as the endpoint of a.

Having defined the intended models of arrow logic as consisting of objects that are graphically representable as arrows, we should say something about the *structure* imposed on these arrows. Let us consider the question what the *basic* relations between arrows are. The obvious first candidate is *composition*: vector spaces have an additive structure, functions can be composed, language fragments can be concatenated, etc. Therefore, the central relation of arrows will be a ternary *composition relation* C, $Cabc$ denoting the fact that a can be seen as an outcome of the composition of b and c, or conversely, that a can be decomposed into b and c. Note that in many concrete examples, C is actually a (partial) function; for instance, in the two-dimensional framework we have

(1) $\qquad Cabc$ iff $a_0 = b_0$, $a_1 = c_1$ & $b_1 = c_0$.

Second, in all the examples listed, the composition function has a neutral element; think for instance of the identity function or the SKIP-program. So, arrow models will contain degenerate arrows, transitions that do not lead to a different state. Formally, there will be a designated subset I of *identity arrows*; in the pair-representation, I will be (a subset of) the diagonal:

(2) $\qquad Ia$ iff $a_0 = a_1$.

Slightly more debatable is the presence among the basic arrow relations of the third candidate: converse. For instance, in linguistic applications it seems to be difficult to imagine what the meaning of a reversed arrow could be. On the other hand, in some theories of cognitive science, reverse arrows appear with the clear meaning of 'undo-ing' an action; and in many fields of mathematics, arrow-like objects have converses (vectors) or inverses (bijective functions). Therefore, a generous viewpoint prevails: an arrow structure will have a binary *reverse relation* R. Again, in many cases this relation will be a function; for instance in the two-dimensional picture, the function f given by

(3) $\qquad fa = (a_1, a_0) \qquad$ (if fa is defined).

Now there is a question as to whether this triple of arrow relations really forms the basic set. A natural further candidate would be some manifestation of *iteration*, if only because of its fundamental importance in computer science. However, it is less obvious how to give a mathematically simple and transparent definition of a primitive 'iteration relation' between arrows than it is for the other relations; besides that, the notion of iteration can be captured nicely without extending the signature of arrow frames (cf. section 7). As there are no other plausible candidates[1], we can now give a formal definition of an arrow frame:

Definition 1.1 An arrow frame is a quadruple $\mathfrak{F} = (W, C, R, I)$ such that $C \subseteq W \times W \times W$, $R \subseteq W \times W$ and $I \subseteq W$.

A nice aspect of arrow logic is that one can draw quite perspicuous pictures, clarifying the meaning of the relations:

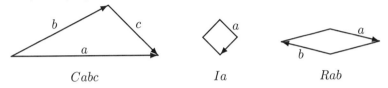

$Cabc$ Ia Rab

Now that we have explained two of the three key words of our slogan, we should discuss the *modal* nature of arrow logic. Let us first consider briefly what we understand with the notion 'modal logic'. The last decade has seen a development in modal logic towards a more abstract and technical approach. In this perspective of what we will call *abstract modal logic*, arbitrary relational structures can be seen as models for an (extended) modal language: any relation is a potential accessibility relation of some suitably defined modal operator. The essentially *modal* aspects of the framework include the following. First, in contrast to what happens in the semantics of first order logic, the quantification over the model is restricted to an accessible part of the structure. And second, modal logicians focus on both the structure of the model itself (like in first order logic) and (subalgebras of) the power set algebra of the structure (like in algebraic logic).

What this boils down to for arrow logic is that we define a modal operator-language such that the truth of its formulas can be evaluated at arrows. If one wants to use traditional terminology from modal logic, this means that the transitions are *not* links between the possible worlds of the model; we treat the transitions *themselves* as the possible worlds. The three modal operators of the language are taken such that the composition, converse and identity relation are their accessibility relations. For instance, the language has a binary operator ∘ for C; intuitively speaking, the truth

---

[1]This is not entirely true: for instance, one could study relations like R_{ll}: $R_{ll}ab$ holds between two arrows a and b iff a and b start at the same point (cf. the discussion on Bulgarian-style arrow logic in section 7). However, in many classes of arrow frames, these relations can be *defined* using our 'basic' signature, for instance $R_{ll}ab \equiv \exists x\, Cabx$.

definition of ∘ states that $\phi \circ \psi$ is true at an arrow iff it can be decomposed into two arrows at which ϕ and ψ hold, respectively.

Now we are ready to give a formal definition of the syntax and semantics of arrow logic:

Definition 1.2 The alphabet of arrow logic consists of an infinite set of propositional variables, the boolean connectives $\neg$ and $\vee$, and the modal operators $\circ$ (composition), $\otimes$ (converse) and $\iota\delta$ (identity).[2] Its set of formulas is defined as follows:
(i)   every propositional variable is a formula, and so is $\iota\delta$,
(ii) if ϕ and ψ are formulas, then so are $\neg\phi$, $\phi \vee \psi$, $\phi \circ \psi$ and $\otimes\psi$,
(iii) and nothing else is a formula.

We will freely use the standard boolean abbreviations $\wedge$, $\rightarrow$ and $\leftrightarrow$, and also $\underline{\circ}$ and $\underline{\otimes}$ as duals of $\circ$ and $\otimes$, i.e. $\phi \underline{\circ} \psi = \neg(\neg\phi \circ \neg\psi)$ and $\underline{\otimes}\phi = \neg\otimes\neg\phi$.

Definition 1.3 An arrow model is a pair $\mathfrak{M} = (\mathfrak{F}, V)$ such that $\mathfrak{F} = (W, C, R, I)$ is an arrow frame and V is a valuation, i.e. a function mapping propositional variables to subsets of W. Truth of a formula ϕ at an arrow a of $\mathfrak{M}$, notation: $\mathfrak{M}, a \Vdash \phi$, is inductively defined as follows:

$\mathfrak{M}, a \Vdash p$ if $a \in V(p)$
$\mathfrak{M}, a \Vdash \iota\delta$ if Ia
$\mathfrak{M}, a \Vdash \neg\phi$ if $\mathfrak{M}, a \nVdash \phi$
$\mathfrak{M}, a \Vdash \phi \vee \psi$ if $\mathfrak{M}, a \Vdash \phi$ or $\mathfrak{M}, a \Vdash \psi$
$\mathfrak{M}, a \Vdash \phi \circ \psi$ if there are b, c with $Cabc$, $\mathfrak{M}, b \Vdash \phi$ and $\mathfrak{M}, c \Vdash \psi$
$\mathfrak{M}, a \Vdash \otimes\phi$ if there is a b with Rab and $\mathfrak{M}, b \Vdash \phi$.

A formula ϕ is valid in a class K of frames, notation $\mathsf{K} \models \phi$, if for every frame $\mathfrak{F}$ in K, every valuation V on $\mathfrak{F}$ and every arrow a in $\mathfrak{F}$, we have $\mathfrak{F}, V, a \Vdash \phi$. A formula ϕ is valid in a frame $\mathfrak{F}$ if it is valid in $\{\mathfrak{F}\}$.

The reader will have noticed that while we fixed the syntax of arrow logic, we are more liberal concerning its semantics. For instance, any *group* $\mathfrak{G} = (G, \cdot, (\cdot)^{-1}, e)$ counts as an arrow frame, if we put

(4) $\quad\quad\quad\begin{aligned} C &= \{(x, y, z) \mid x = y \cdot z\}, \\ R &= \{(x, y) \mid y = x^{-1}\}, \\ I &= \{e\}. \end{aligned}$

We agree that such examples may stretch one's intuitions concerning 'arrows' to the limit[3]. Nevertheless, the generality of our semantics reflects the opinion that there is no such thing as *the* proper semantics for arrow logic: different applications invoke different kinds of model. It is the task of the logician to clarify the choices that are to be made, to investigate the

---
[2]There are other notations in the literature as well, e.g. • or ; for ∘, $\phi^\smile$ for $\otimes\phi$ and *id* for $\iota\delta$.

[3]Note that in the Cayley representation, the elements of any group do have a natural representation as (sets of) pairs.

relations between the various options, and to search for logical patterns in the landscape.

This is not to say that discussions concerning the ontology of arrows are not interesting or important. Consider for instance the problem whether arrows should be primitive entities or constructed from more basic material (for instance, as pairs of states). These matters touch upon the philosophical question whether dynamic or static aspects of knowledge and action should have primacy. Technically speaking however, the two-dimensional frames form just another special subclass of the class of arrow frames.

As a last remark, let us mention some other dimensions along which the semantics of arrow logic may be varied. As an example, one might consider arrow models in which the valuations are subjected to certain restrictions. For instance, in the two-dimensional case one might think of the so-called flat models, in which the truth of atomic propositions is not dependent on both coordinates, but perhaps only on the first one. Or, having some informational interpretation in mind, we may need a many-valued semantics to formalize reasoning with incomplete information. As these options have not yet been explored, we will not go into detail here.

2 Some Intuitive Examples

Let us now consider those examples of arrow frames that have been studied most intensively in the literature: squares, graphs (or relativized squares) and multigraphs.

To start with the latter, a *multigraph* is a quadruple (E, P, l, r) of which E is a set of directed edges, P is a set of points and l and r are total functions mapping edges to points. Intuitively we understand an edge a as an arrow leading from its left point $l(a)$ to its right point $r(a)$:

$$l(a) \xrightarrow{\quad a \quad} r(a)$$

In a most natural way, we can define the following relations on the set E of edges of a multigraph:

$Cabc$ iff a leads from $l(b)$ to $r(c)$ and $r(b) = l(c)$,
Rab iff a leads from $r(b)$ to $l(b)$,
Ia iff $l(a) = r(a)$,

thus creating a multigraph arrow frame with universe E.[4] Now an arrow

---

[4]Although we will not treat *category* frames in detail, it is easy to imagine how one may change the definition of a multigraph frame into that of a category frame. We may see a (small) category as a multigraph endowed with a binary *composition function* on arrows; this composition function will give the ternary accessibility relation C of the arrow frame. Note that we cannot define a reverse relation R on category frames, hence

model arises if we *label* the edges of the multigraph with sets of propositions, consider the example below:

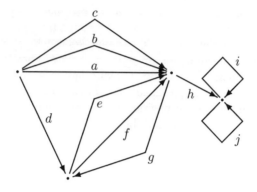

If we label the edges of this multigraph by the following valuation:
$$V(p) = \{a,b,e,i\}$$
$$V(q) = \{i,g\}$$
$$V(r) = \{h\},$$
we obtain that
$$d \Vdash p \circ q$$
$$g \Vdash \otimes p \wedge \otimes \neg p$$
$$h \Vdash r \circ (\iota\delta \wedge q)$$
etc.

Note that in a multigraph frame, there may be *various* arrows leading from one point to another, as in the picture above. This implies that neither C nor R needs to be functional in a multigraph frame.

This is different in *graph* frames: a graph is a multigraph with *at most one* edge between each pair of points. So we arrive at the two-dimensional semantics mentioned earlier on, as we may identify an arrow a with the pair $(l(a), r(a))$. From this perspective, a graph frame is a structure $\mathfrak{F} = (E, C, R, I)$ such that E is a subset of a cartesian square $P \times P$; C and I are defined as in (1) resp. (2); and R is the function f satisfying (3). This viewpoint also explains the other names in use for graph frames: *pair* frames, *two-dimensional* frames and *relativized squares* (in this paper, we will mainly use the latter name).

Note that in a graph model, the modal clauses of the truth definition

---

we can only interpret the ⊗-free arrow formulas in category frames. This is different for *allegories* (cf. Freyd and Scedrov 1990), where arrows do have reverses; the reader is invited to investigate the connections between arrow logic and allegories.

boil down to the following: for (x,y) in the model $\mathfrak{M}$,

$\mathfrak{M},(x,y) \Vdash \phi \circ \psi$ iff   there is a z s. t. $\mathfrak{M},(x,z) \Vdash \phi$ and $\mathfrak{M},(z,y) \Vdash \psi$,
$\mathfrak{M},(x,y) \Vdash \otimes \phi$ iff $\mathfrak{M},(y,x) \Vdash \phi$,
$\mathfrak{M},(x,y) \Vdash \iota\delta$ iff $x = y$.

Relativized squares can be depicted as graphs, but we can also draw them in a way that reflects their two-dimensional aspects more clearly:

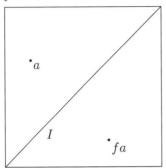 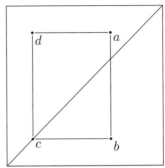

The identity relation I consists of the *diagonal* elements of the universe; f is the operation of *mirroring* in the diagonal. C consists of all triples (a, b, d) with the property that we can draw a *rectangle abcd* such that: b lies on the vertical line through a; d lies on the horizontal line through a; and c (which does not have to belong to the universe of the model) lies on the diagonal.

We can narrow down the class of graphs further by demanding that the universe E of 'available' arrows satisfies certain properties, like *reflexivity* (if (x, y) is available, then so are its starting arrow (x, x) and its ending arrow (y, y)) or *symmetry* (if (x, y) is available, then so is its converse (y, x)). A particularly interesting condition is that *all* pairs of points are present as arrows. In the setting of multigraphs, this means that there is *precisely one* arrow running between every pair of points. With this constraint we arrive at the class of *square* frames or *full* graphs having a full cartesian square as their universe.

Definition 2.1 An arrow frame is called a two-dimensional frame, a pair frame, a relativized square or a graph frame, if for some base set U, W is a subset of $U \times U$, C and I are given as in (1) and (2) and R is a partial function f satisfying (3). If W is a reflexive and symmetric binary relation, we call the arrow frame locally square; if W is equal to the full cartesian product over U, i.e. $W = U \times U$, the frame is called a square. The classes of squares, relativized squares and locally square arrow frames are denoted by SQ, RSQ and LSQ, respectively.

3 The Motivation

As we have already hinted in the introduction, arrow logic is a widely applicable system: many different notions from various disciplines like mathematics, computer science, linguistics and cognitive science can be formalized in it. Here we will briefly discuss the possible applications of arrow logic in some more detail, first focusing on the notion of *dynamic semantics*, which unifies insights from computer science, linguistics and philosophy, and then describing the position of arrow logic with respect to other parts of (mathematical) logic.

So let us start with a brief explanation of the dynamic perspective on semantics; for more detailed information we refer the reader to van Benthem 1991b. Traditionally, the central notion of semantics in logic is a static one: a proposition is interpreted as a declarative statement over a model. In recent years however, alternative viewpoints have been proposed in which not only the information content, but also the use of this information forms an object of study. In particular, the basic idea underlying dynamic semantics is that part of the meaning of a proposition is formed by its potential to *change* the information that a given agent has about the model. Let us give a few examples.

Consider for instance the meaning of anaphora in natural language. Comparing the texts *a man walks in the park; he whistles* with *he whistles; a man walks in the park*, the reader will notice that a purely truth-conditional approach to semantics cannot explain the difference in meaning. Clearly, the meaning of the pronoun *he* has an active component, in that it passes information concerning the anaphoric linkage. In Groenendijk and Stokhof 1991, the system of Dynamic Predicate Logic is introduced, with the intention of formalizing this dynamic view on the semantics of anaphora; an essential characteristic of this formalism is that the formulas of ordinary first order logic are interpreted as sets of transitions over a state space. To be more precise: the objects at which the evaluation of formulas takes place, are *pairs* consisting of an input- and an output-state.

Related to the anaphora problem in natural language is the philosophical notion of presupposition; the intuitive idea to see a presupposition as a condition to be fulfilled in order to enable the agent to process some piece of information, turns presupposition into an undeniably dynamic notion (cf. van Eijck 1993, Visser 1994).

Similar proposals have been made in the literature on Categorial Grammar, cf. van Benthem 1991b. The paradigm that parsing a sentence is performing a logical deduction, has been combined with the observation that these processes require an activity from the agent. A surprising insight resulting from this connection is that Lambek's Syntactic Calculus, in fact the main Categorial Grammar in use, allows a dynamic, procedural interpretation, and as such can be taken as a generalized reduct of arrow logic (we will give more details in section 7).

These ideas have also been investigated in the wider context of cognitive science, cf. for instance Gärdenfors 1988 or Veltman to appear. Here the old philosophical idea of cognition as activity is molded into a formal framework, in which information processing is modeled by epistemic operations for changing information states, e.g. by updating or revising them.

The central idea emerging from the literature is that propositions are interpreted as sets of *transitions*. Because of this, dynamic semantics forms a bridge from formalisms developed in linguistics and philosophy towards approaches followed in computer science, such as process algebra (cf. Milner 1980, Bergstra and Klop 1984) or dynamic logic (cf. Harel 1984). Note that in the denotational semantics of programming languages, the standard meaning of a program α is a set of input-output pairs.

It seems as if a research field of modal transition logics is arising, i.e. modal formalisms designed to talk about transition structures, and as we saw in the previous section, arrow logic occupies a central position in this landscape.

Finally, within the field of mathematics and logic, arrow logic is of interest because of its intimate and important ties with algebraic logic and with the predicate calculus.

The program of algebraic logic (cf. Németi 1991 for an overview) is to study logical formalisms with algebraic tools. A familiar example is the use of Boolean Algebras to study classical propositional logic. Several kinds of algebras of relations have been studied in the literature of algebraic logic, one of the most important ones is formed by the Relation Algebras defined in Tarski 1941. One of the motivations for developing arrow logics was to investigate the theory of Relation Algebras from an abstract modal point of view. In section 4 we will see how Relation Algebras emerge as the *modal* or *complex* algebras of arrow logic.

With respect to the predicate calculus, arrow logic is interesting from the viewpoint of correspondence theory (cf. van Benthem 1985, the field of modal logic in which the expressive power of modal logic is compared to that of first order logic and other formalisms. One of the motivations for breaking out of the traditional modal framework was to increase the expressive power of modal languages. An interesting aspect of arrow logic is that over the class of square models, arrow logic is equally expressive as the three-variable fragment of the predicate calculus: everything that can be said about binary relations using three variables (and first order quantifiers) only, is expressible in arrow logic as well, and vice versa. For more details we refer to Venema 1991.

4 The Algebras

As we already mentioned in the previous section, arrow logic has intimate connections with algebraic logic; the insights and tools obtained in the literature on Relation Algebras have played an important role in the development of arrow logic. Therefore, we feel it might be useful to provide the reader with a brief introduction to the algebraic theory of binary relations and its connections with arrow logic. As a starting point, consider the following picture, due to C. Brink (cf. Brink 1993):

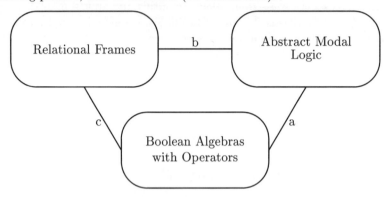

It has manifestations for a manifold of logical formalisms; for instance, in the case of classical propositional logic (which we may see as a degenerate kind of modal logic) the relational frames are just unstructured sets, and the algebras are boolean algebras without extra operators.

Let us now mention some key words to explain the connections in the picture. The relation (a) between abstract modal logic and boolean algebras with operators[5] (BAOs) is very tight; for instance, BAOs appear as the Lindenbaum-Tarski algebras of extended modal logics. As we saw in the introductory section, (b) relational frames are the central structures in the semantics of modal logic, and this is so since the work of Kripke. Note however that, years before the terminology of 'possible worlds' connected by 'accessibility relations' was introduced, (cf. Jónsson and Tarski 1951) had already investigated the relation (c) between BAOs and relational frames — in particular, they had already studied *arrow frames*. For a recent overview of the *duality theory* which was started by the Jónsson-Tarski paper we refer to Goldblatt 1989, Sambin and Vaccaro 1989. From a mathematical perspective, the development of arrow logic can be seen as *filling in* the modal part of the picture above, where the structural and algebraic sides already existed.

---

[5] A boolean algebra with operators is a structure $\mathfrak{A} = (A, +, -, f_i)_{i \in I}$ where $(A, +, -)$ is a boolean algebra and every operator f_i is *normal* ($f_i(a_0, \ldots, 0, \ldots, a_{n-1}) = 0$) and *additive* ($f_i(a_0, \ldots, a_i + a'_i, \ldots, a_{n-1}) = f_i(a_0, \ldots, a_i, \ldots, a_{n-1}) + f_i(a_0, \ldots, a'_i, \ldots, a_{n-1})$).

So it seems to be time to say something about the algebras involved. The basic idea is that one starts with algebras consisting of a set of binary relations, together with some natural operations on them. The next step is to abstract away from this class of concrete algebras to the class of all boolean algebras with operators of the appropriate similarity type. One aim of the algebraic theory is then to find necessary and sufficient conditions for the representability of an abstract algebra as a concrete algebra of relations.

Definition 4.1 Let U be a set; consider the following operations on $\mathcal{P}(U \times U)$:

$$\begin{aligned} R \mid S &= \{(u,v) \in U \times U \mid \exists w ((u,w) \in R \ \& \ (w,v) \in S)\} \\ R^{-1} &= \{(u,v) \in U \times U \mid (v,u) \in R\} \\ Id &= \{(u,v) \in U \times U \mid u = v\}. \end{aligned}$$

Any algebra of the form $\mathfrak{A} = (\mathcal{P}(U \times U), \cup, -, \mid, (\cdot)^{-1}, Id)$ is called a full relational set algebra; the class of such algebras is denoted by FRA. The class RRA of representable relation algebras is defined as the class **ISPFRA**, i.e., isomorphic copies of subalgebras of direct products of full relational set algebras.

A relational type algebra is a boolean algebra augmented with the following operators: a binary ;, a unary $\smile$ and a constant 1'.

A full relational set algebra contains *all* binary relation on a given set U. RRA consists of all 'real' relation algebras: algebras in which the elements can be seen as binary relations, and the operations behave in the expected manner. In order to enumerate the equational theory of the class RRA, which happens to be a variety, Tarski proposed the following:

Definition 4.2 A relation algebra $\mathfrak{A} = (A, +, -, ;, \smile, 1')$ is a relational type algebra satisfying the following axioms:
(*RA0*)   Axioms stating that $(A, +, -)$ is a Boolean Algebra
(*RA1*) $(x+y);z = x;z + y;z$
(*RA2*) $(x+y)^\smile = x^\smile + y^\smile$
(*RA3*) $(x;y);z = x;(y;z)$
(*RA4*) $x;1' = x$
(*RA5*) $(x^\smile)^\smile = x$
(*RA6*) $(x;y)^\smile = y^\smile;x^\smile$
(*RA7*) $x^\smile; - (x;y) \leq -y$.
The class of relation algebras is denoted by RA.

For an introduction to the theory of relation algebras we refer to Jónsson 1982, 1991 or Maddux 1991. It soon turned out that the RA-axioms do not exhaustively generate all valid principles governing binary relations, in other words: RA is only an *approximation* of RRA. In fact, RRA is not axiomatizable by a finite (purely equational) axiomatization, as was shown by Monk (cf. Monk 1969). This negative result has been considerably strengthened by various authors, cf. Andréka 1991 for the most recent version. Infinite axiomatizations of RRA are known, cf. for instance Lyndon 1950.

One way to overcome the negative results is to allow derivation systems that are not in equational form, cf. Wadge 1975, Maddux 1983 or Orlowska 1991 where Gentzen-type axiomatizations are given in which variables referring to elements of the base set U occur, or Venema 1991. Another possibility is to widen the class of representable relation algebras, for instance by starting with power set algebras in which the top set is not necessarily a *full* cartesian square, but a *subset* of it, and to *relativize* the operations to this subset:

Definition 4.3 Let W be a subset of the cartesian square $U \times U$, for some set U. The operations $-^W$, $\cup^W$, $|^W$, $(\cdot)^{-1:W}$ and Id^W are defined as the relativizations to W of $-$, $\cup$, $|$, $(\cdot)^{-1}$ and Id respectively, i.e.,

$$-^W R = -R \cap W$$
$$R \mid^W S = (R \mid S) \cap W$$
etc.

Any algebra of the form $\mathfrak{A} = (\mathcal{P}(W), -^W, \cup^W, |^W, (\cdot)^{-1:W}, Id^W)$ with $\mathcal{P}(W)$ and the operations as defined above, is called a relativized relational power set algebra; the class of such algebras is denoted by FRA$_\emptyset$; FRA$_{rs}$ is the subclass of FRA$_\emptyset$ in which the algebras have a reflexive, symmetric relation W as their top element. The class RRA$_\emptyset$ of relativized representable relation algebras is defined as **ISFRA**$_\emptyset$ (isomorphic copies of subalgebras of relativized relational power set algebras), RRA$_{rs}$ as **ISFRA**$_{rs}$.

It is an easy and instructive exercise to prove that RRA $\subseteq$ RRA$_\emptyset$ by showing that any representable relation algebra is (isomorphic to) a subalgebra of a relativized relational power set algebra where the top set W is an *equivalence* relation.

The variety RRA$_\emptyset$ *does* allow a finite axiomatization, cf. Kramer 1991, or Marx 1995 for a simpler system. A nice result by Maddux (cf. Maddux 1982) states that RRA$_{rs}$ is precisely the class WA of *weakly associative* algebras which arises if we replace the associativity axiom $(RA4)$ in the definition of RA by its weaker variant

(*WA*) $\qquad\qquad ((1' \cdot x);1);1 = (1' \cdot x);(1;1).$

Now one of the most important connections with arrow logic lies in the fact that relation type algebras are the *complex algebras* of arrow frames; complex algebras form one of the fundamental structural operations in the duality theory of relational frames and BAOs. To explain what a complex algebra is, consider a relational frame $\mathfrak{F} = (W, R_i)_{i \in I}$. With each $n+1$-ary

relation R_i we associate an n-ary *operation* f_{R_i} on the power set $\mathcal{P}(W)$:
$$f_{R_i}(X_1,\ldots,X_n) = \{x_0 \in W \mid \exists x_1 \ldots x_n (R_i x_0 x_1 \ldots x_n \ \& \bigwedge_{1 \leq i \leq n} x_i \in X_i)\}.$$
The complex algebra $\mathfrak{Cm}\mathfrak{F}$ of $\mathfrak{F}$ is given as
$$\mathfrak{Cm}\mathfrak{F} = (\mathcal{P}(W), \cup, -, f_{R_i})_{i \in I}.$$
For a class K of relational structures, we let **CmK** denote the class of its complex algebras. In the case of arrow logic we find that the full relation algebras are the complex algebras of the square arrow frames, and a similar result applies to the relativized squares:

Proposition 4.4
$$\begin{aligned} \mathsf{FRA} &= \mathbf{CmSQ} \\ \mathsf{FRA}_\emptyset &= \mathbf{CmRSQ} \\ \mathsf{FRA}_{\mathsf{rs}} &= \mathbf{CmLSQ} \end{aligned}$$

Proof. Straightforward; note for instance that
$$\begin{aligned} R \mid S &= \{(u,w) \in U \times U \mid \exists v[(u,v) \in R \ \& \ (v,w) \in S]\} \\ &= \{(u,w) \in U \times U \mid \exists v[C(u,w)(u,v)(v,w) \ \& \\ & \qquad \& \ (u,v) \in R \ \& \ (v,w) \in S]\} \\ &= \{(u,w) \in U \times U \mid \exists rs[C(u,w)rs \ \& \ r \in R \ \& \ s \in S]\} \\ &= f_C(R,S). \end{aligned}$$
□

These observations have important consequences for the axiomatics of arrow logic. First, observe that, modulo a trivial translation τ (with clauses like $\tau(\phi \vee \psi) = \tau\phi + \tau\psi$, $\tau(\phi \circ \psi) = \tau\phi; \tau\psi$, etc.), the *formulas* of arrow logic are the *terms* of the algebraic language for relation type algebras. Now the following proposition is almost immediate:

Proposition 4.5
$$\begin{aligned} \mathsf{SQ} \models \phi &\iff \mathsf{RRA} \models \tau\phi = 1 \\ \mathsf{RSQ} \models \phi &\iff \mathsf{RRA}_\emptyset \models \tau\phi = 1 \\ \mathsf{LSQ} \models \phi &\iff \mathsf{RRA}_{\mathsf{rs}} \models \tau\phi = 1. \end{aligned}$$

These propositions form a basic reason for the applicability of techniques from algebraic logic in arrow logic (and vice versa!): axiomatizing the modal theory of SQ and the equational theory of RRA are two sides of the same coin. However, note that axiomatics is not the only area in which duality theory proves it use; in fact, almost all properties of logics have an algebraic counterpart.

5 Characterization Results

In this section we discuss some basic model theory of arrow logic. First we work on the level of arrow *models*, for which we define arrow bisimulations as the fundamental notion of similarity between two arrow models. In the second part of the section we move to the level of frames, considering the question, how classes of arrow frames and properties of arrow frames can be defined in the language of arrow models.

So let us start with considering the notion of a bisimulation between structures. Bisimulations are the key tools to compare (labeled) transition systems, cf. van Benthem and Bergstra 1993, van Benthem et al. 1993, and thus of fundamental importance to computer science, cf. Park 1981 for a first reference. It is interesting to note that in the latter area, a bisimulation is usually defined as a relation between *states*, whereas in arrow logic, we need to compare *transitions*:

Definition 5.1 Let $\mathfrak{M} = (W, C, R, I, V)$ and $\mathfrak{M}' = (W', C', R', I', V')$ be two arrow models. A relation $Z \subseteq W \times W'$ is an arrow bisimulation if it satisfies, for any propositional variable p:

if aZa', then $a \in V(p)$ iff $a' \in V'(p)$,

the following 'forth'-conditions ($a, b, c \in W$ and $a' \in W'$):

$(ZZ)_C^{\rightarrow}$ $Cabc$ and aZa' only if there are $b', c' \in W'$ such that
                                    bZb', cZc' and $C'a'b'c'$,
$(ZZ)_R^{\rightarrow}$ Rab and aZa' only if there is a $b' \in W'$ s.t. bZb' and $Ra'b'$,
$(ZZ)_I^{\rightarrow}$ Ia and aZa' only if $I'a'$,

and the converse 'back'-conditions $(ZZ)_C^{\leftarrow}$, $(ZZ)_R^{\leftarrow}$ and $(ZZ)_I^{\leftarrow}$ (with the obvious definition).

We write $\mathfrak{M}, a \leftrightarrow \mathfrak{M}', a'$ if there is an arrow bisimulation Z between $\mathfrak{M}$ and $\mathfrak{M}'$ such that aZa'.

In de Rijke 1993, the crucial role of bisimulations in the model theory of modal logic is investigated, the slogan being that bisimulations are to modal logic what partial isomorphisms are to first order logic. For instance, a nice result of De Rijke (in terms of arrow logic) states that two *rooted*[6] arrow models have the same arrow theory iff they have bisimilar ultraproducts. In particular, the existence of a bisimulation between two arrow models is an indication of the fact that the two models are very much alike. The following theorem states that the modal language of arrow logic cannot distinguish two bisimilar arrows:

Theorem 5.2 *Assume that* $\mathfrak{M}, a \leftrightarrow \mathfrak{M}', a'$. *Then for every arrow formula*

---

[6] A rooted arrow model is an arrow model with a distinguished arrow, as in the definition of $\mathfrak{M}, a \leftrightarrow \mathfrak{M}', a'$; the theory of a rooted arrow frame is the set of formulas holding at its root.

ϕ:
$$\mathfrak{M}, a \Vdash \phi \iff \mathfrak{M}', a' \Vdash \phi.$$

This theorem has an easy proof by induction on the complexity of arrow formulas. Now another manifestation of the fundamental importance of bisimulations lies in the fact, that a (restricted) kind of converse of Theorem 5.2 holds as well: a property of arrows that is first-order definable, can be expressed by a modal formula if and only if it is invariant under bisimulations. To be a bit more precise:

Theorem 5.3 *Let $\alpha(x)$ be a formula in the first order language over arrow models*[7]. *Assume that truth of $\alpha(x)$ is invariant under arrow bisimulations, i.e. $\mathfrak{M}_0, a_0 \underline{\leftrightarrow} \mathfrak{M}_1, a_1$ implies that*

$$\mathfrak{M}_0 \models \alpha[x \mapsto a_0] \iff \mathfrak{M}_1 \models \alpha[x \mapsto a_1].$$

Then $\alpha(x)$ is equivalent to an arrow formula, i.e. there is a formula ϕ in the modal language of arrow logic such that for every arrow model $\mathfrak{M}$ and every arrow a in $\mathfrak{M}$

$$\mathfrak{M} \models \alpha[x \mapsto a] \iff \mathfrak{M}, a \Vdash \phi.$$

A proof of Theorem 5.3 (for the ordinary modal case with one diamond $\diamond$) can be found in van Benthem 1985.

Let us now move on to the level of arrow *frames*. As a corollary of Theorem 5.2 we find that the validity of arrow formulas is preserved under certain operations on frames. As an example, we consider zigzag morphisms:

Definition 5.4 *Let $\mathfrak{F}_0 = (W_0, C_0, R_0, I_0)$ and $\mathfrak{F}_1 = (W_1, C_1, R_1, I_1)$ be two arrow frames. A function $z : W_0 \mapsto W_1$ is a zigzag morphism from $\mathfrak{F}_0$ to $\mathfrak{F}_1$ if its graph*[8] *satisfies the back & forth conditions of Definition 5.1. We call $\mathfrak{F}_1$ a zigzagmorphic image of $\mathfrak{F}_0$, if there is a total, surjective zigzag morphism from $\mathfrak{F}_0$ to $\mathfrak{F}_1$.*

Theorem 5.5 *Let $\mathfrak{F}_1$ be a zigzagmorphic image of $\mathfrak{F}_0$. Then for every arrow formula ϕ:*

$$\mathfrak{F}_0 \models \phi \Rightarrow \mathfrak{F}_1 \models \phi.$$

As an example of a zigzag morphism, consider the square frame $\mathfrak{F}$ over the set $\mathbb{Z}$ of integers, and the arrow frame $\mathfrak{Z}$ of the additive group over $\mathbb{Z}$, cf. (4). We invite the reader to check that the map $q : \mathbb{Z} \times \mathbb{Z} \mapsto \mathbb{Z}$ defined by

$$q(y, z) = z - y,$$

---

[7] This is to say that $\alpha(x)$ may use the following fixed predicate symbols: a ternary C, a binary R and a unary I, and further arbitrary unary predicate symbols $P_0, P_1, \ldots$.

[8] The graph of a function $f : W_0 \mapsto W_1$ is the relation $\{(x_0, x_1) \in W_0 \times W_1 \mid x_1 = f(x_0)\}$.

is a zigzag morphism from $\mathfrak{F}$ onto 3.[9]

This example takes us to the second part of the section, where we consider *characterization results* in arrow logic. We say that a set Γ of arrow formulas *defines* or *characterizes* a class K of arrow frames if for any arrow frame $\mathfrak{F}$, validity of Γ in $\mathfrak{F}$ is equivalent to membership of $\mathfrak{F}$ in K. Then it follows from Theorem 5.5 that a class of frames is *not* modally definable if it is not closed under taking zigzagmorphic images. So, the previous example shows that the class of (relativized) squares cannot be characterized in the language of arrow logic. We will come back to this point later on.

Let us first consider some positive results: it turns out that many natural *properties* of arrow frames do allow a modal characterization. We say that a modal formula ϕ *defines* or *characterizes* a property P of arrow frames if it defines the class of frames having this property. Now consider the following list of formulas:[10]

$(A1)$ $\quad \neg\otimes p \qquad\qquad \to \quad \otimes\neg p$
$(A2)$ $\quad \otimes\neg p \qquad\qquad \to \quad \neg\otimes p$
$(A3)$ $\quad \otimes\otimes p \qquad\qquad \to \quad p$
$(A4)$ $\quad \otimes(p \circ q) \qquad\quad \to \quad \otimes q \circ \otimes p$
$(A5)$ $\quad p \circ \neg(\otimes p \circ q) \quad \to \quad \neg q$
$(A6)$ $\quad \iota\delta \qquad\qquad\qquad \to \quad \otimes\iota\delta$
$(A7)$ $\quad \iota\delta \circ p \qquad\qquad \to \quad p$
$(A8)$ $\quad p \qquad\qquad\qquad \to \quad \iota\delta \circ p$
$(A9)$ $\quad p \circ (q \circ r) \qquad\; \leftrightarrow \quad (p \circ q) \circ r$

$A1$ defines the arrow frames $\mathfrak{F}$ in which the reverse relation R is *serial*, i.e. $\mathfrak{F} \models \forall x \exists y Rxy$. For, let R be serial and suppose that $\mathfrak{M}, x \Vdash \neg\otimes p$ for some model $\mathfrak{M}$ based on $\mathfrak{F}$. By assumption, x has an R-successor y, and as p cannot be true at y, we have $y \Vdash \neg p$. So $\mathfrak{M}, x \Vdash \otimes\neg p$. Conversely, suppose that R is not serial; then there is an x in $\mathfrak{F}$ without an R-successor. Consider the valuation V with $V(p) = \emptyset$, then $x \Vdash \neg\otimes p$, but not $x \Vdash \otimes\neg p$.

Likewise, we can show that $A2$ characterizes the arrow frames in which R is *functional*, i.e. $\mathfrak{F} \models \forall xyz((Rxy \wedge Rxz) \to y = z)$, and that among these functional frames, $A3$ defines the ones with *idempotent* functions, i.e. $ffx = x$. Let us call a frame $\mathfrak{F}$ an f-frame if R is serial, functional and idempotent. In the sequel it will be convenient to represent such frames as $\mathfrak{F} = (W, C, f, I)$, i.e., in the signature we replace the relation R by the function f.

Now consider a C-triple (a, b, c) in an f-frame $\mathfrak{F}$:

---

[9] In fact, the Cayley representation of an arbitrary group induces a zigzag morphism from the square over the carrier of the group to the group itself.

[10] Note that all of these formulas are in so-called *Sahlqvist* form; thus it is immediate that they correspond to first-order properties of frames (cf. Sahlqvist 1975).

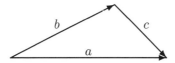

It seems to be a very natural property that given this C-triangle, *all* of the triples of the following set CYC_{abc} should be in C:

$$CYC_{abc} = \{(a,b,c), (fa,fc,fb), (b,a,fc), (fb,c,fa), (c,fb,a), (fc,fa,b)\}.$$

This property is well known from the theory of relation algebras; in Lyndon 1950 triples (a,b,c) satisfying $CYC_{abc} \subseteq C$ are called *cycles*. The property "$CYC_{abc} \subseteq C$ for all arrows a, b and c with $Cabc$" is taken care of by $A4$ and $A5$:

$$\mathfrak{F} \models A4 \iff \mathfrak{F} \models \forall xyz\, (Cfxyz \to Cxfzfy)$$
$$\mathfrak{F} \models A5 \iff \mathfrak{F} \models \forall xyz\, (Cxyz \to Czfyx).$$

To show the equivalence for $A5$ first assume that

$$\mathfrak{F} \models \forall xyz\, (Cxyz \to Czfyx),$$

and that in some model over $\mathfrak{F}$ there is an arrow x at which $p \circ \neg(\otimes p \circ q)$ is true. We will derive a contradiction from the assumption that $x \Vdash q$. Note that there must be y and z such that $Cxyz$, $y \Vdash p$ and $z \Vdash \neg(\otimes p \circ q)$. So $fy \Vdash \otimes p$, whence by $Czfyx$ and our assumption we find $z \Vdash \otimes p \circ q$: the desired contradiction. Conversely, if

$$\mathfrak{F} \not\models \forall xyz\, (Cxyz \to Czfyx),$$

then there must be x, y and z such that $Cxyz$ and not $Czfyx$. Now consider a valuation V with $V(p) = \{y\}$ and $V(q) = \{x\}$; the crucial observation is that now fy is the *only* arrow at which $\otimes p$ is true, and x the *only* one with $x \Vdash q$. So $\otimes p \circ q$ must be *false* at z, whence $x \Vdash p \circ \neg(\otimes p \circ q)$. As we also have $x \Vdash q$, this gives $\mathfrak{F} \not\models A5$.

Of the formulas involving the identity, and in the class of f-frames, $A6$ characterizes the arrow frames in which reverses of identity arrows are identity arrows ($\forall x(Ix \to Ifx)$), $A8$ defines those frames in which every arrow has a starting arrow ($\forall x \exists y(Iy \wedge Cxyx)$), and $A7$ defines the following (perhaps most essential) property of an identity arrow y:

$$\forall xyz\, ((Cxyz \wedge Iy) \to x = z).$$

Now we turn to the *associativity* axiom $A9$, which plays a very interesting role in arrow logic, as we will see in the next section. It has a first order correspondent on the class of arrow frames as well:

$$\mathfrak{F} \models A9 \iff \mathfrak{F} \models \forall xyuv\, (\exists z(Cxyz \wedge Czuv) \leftrightarrow \exists w(Cxwv \wedge Cwyu)),$$

viz. the picture below:

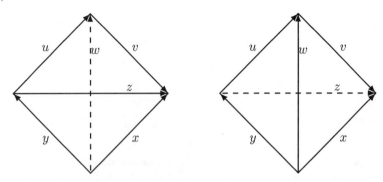

The formulas $A1, \ldots, A9$ have been chosen for a special reason; taken together, they constitute the arrow-counterpart of Tarski's relation algebras:

Proposition 5.6 *Let $\mathfrak{F}$ be an arrow frame. Then*
$$\mathfrak{F} \models A1 \ldots A9 \iff \mathfrak{Cm}\mathfrak{F} \text{ in } \mathsf{RA}.$$

Now we return to the question of what to do with classes of frames which are not modally definable. Even when in modal logic *characterizations* of frame classes do not lead automatically to complete *axiomatizations*, this issue is of importance. We will only treat the cases of the squares and the relativized squares.

The non-definability of RSQ leaves open the possibility that one can find a set of formulas characterizing the class $\mathbf{H_f}\mathsf{RSQ}$ of *zigzagmorphic images* of relativized squares. And indeed, this aim can be achieved, as a result by Marx shows:

Theorem 5.7 *Let $\mathfrak{F}$ be an arrow frame. Then $\mathfrak{F}$ is a zigzagmorphic image of a relativized square iff the following formulas are valid in $\mathfrak{F}$:*

(A^r1)	$\otimes \neg p$	$\to$	$\neg \otimes p$
(A^r2)	$\otimes \otimes p$	$\to$	p
(A^r3)	$p \wedge \iota\delta$	$\to$	$\otimes p$
(A^r4)	$\iota\delta$	$\to$	$\iota\delta \circ \iota\delta$
(A^r5)	$\iota\delta \circ p$	$\to$	p
(A^r5')	$p \circ \iota\delta$	$\to$	p
(A^r6)	$p \circ \neg \otimes p$	$\to$	$\neg \iota\delta$
(A^r7)	$\otimes p \circ \neg(p \circ q)$	$\to$	$\neg q$
(A^r7')	$\neg(p \circ q) \circ \otimes q$	$\to$	$\neg p$
(A^r8')	$((p \wedge \iota\delta) \circ q) \circ r$	$\leftrightarrow$	$(p \wedge \iota\delta) \circ (q \circ r)$
(A^r8'')	$(p \circ (q \wedge \iota\delta)) \circ r$	$\leftrightarrow$	$p \circ ((q \wedge \iota\delta) \circ r)$
(A^r8''')	$(p \circ q) \circ (r \wedge \iota\delta)$	$\leftrightarrow$	$p \circ (q \circ (r \wedge \iota\delta))$

Note that the formulas listed above *all* have first-order correspondents, as each of them is in Sahlqvist form. We refer the reader to chapter 2 for the precise form of these first-order correspondents as well as for the proof of the theorem.

The situation for the squares is rather different, because the modal theory of SQ is not finitely axiomatizable by standard means (cf. the results by Monk and Andréka in the previous section). Therefore, we have to look for a different *kind* of characterization. First, define the operator D as the following abbreviation:

(5) $\qquad D\phi \equiv \neg\iota\delta \circ (\phi \circ \top) \vee (\top \circ \phi) \circ \neg\iota\delta.$

For an arbitrary frame $\mathfrak{F}$, we can define an accessibility relation R_D for D by setting

$$R_D = \{(a,b) \mid \mathfrak{F} \models \exists x_1 x_2 y \left((Cax_1x_2 \wedge \neg Ix_1 \wedge Cx_2by) \vee (Cax_1x_2 \wedge \neg Ix_2 \wedge Cx_1yb) \right)\},$$

i.e. we have in every arrow model $\mathfrak{M}$:

$$\mathfrak{M}, a \Vdash D\phi \text{ iff there is a } b \text{ with } R_D ab \text{ and } \mathfrak{M}, b \Vdash \phi.$$

The reader is invited to check that in a *square* model we have:

$$\mathfrak{M}, a \Vdash D\phi \text{ iff there is a } b \text{ with } a \neq b \text{ and } \mathfrak{M}, b \Vdash \phi.$$

In other words, over the class of squares we have defined the *difference operator*. Now our characterization theorem below states that this property (viz. that the defined difference operator has the inequality relation as its accessibility relation), precisely characterizes the squares among the RA-frames (a proof can be distilled from Venema 1991):

Theorem 5.8 SQ *consists precisely of the arrow frames* $\mathfrak{F}$ *satisfying*
(i) $\quad \mathfrak{F} \models A1 \ldots A9$,
(ii) $\quad \mathfrak{F} \models \forall xy (R_D xy \leftrightarrow x \neq y)$.

6   Completeness and Decidability

The axiomatics and decidability aspects of arrow logics have been investigated intensively; in this section we will try to give a sketchy overview of what is known.

Completeness

To start with axiomatics, let us first define the kinds of derivation systems that have been considered in the literature.

Definition 6.1 A derivation system for arrow logic is a pair (A, R) with A a set of axioms and R a set of derivation rules. A derivation system is called normal if A contains the following axioms:

$(CT)\quad$ all classical tautologies
$(DB)\quad (p \to p') \circ q \to (p \circ q \to p' \circ q)$
$ p \circ (q \to q') \to (p \circ q \to p \circ q')$
$ \otimes(p \to q) \to (\otimes p \to \otimes q)$

and R contains the rules of Modus Ponens, Universal Generalization and Substitution:

(MP) $\phi, \phi \to \psi \ / \ \psi$
(UG) $\phi \ / \ \phi \circ \psi, \psi \circ \phi$
         $\phi \ / \ \otimes \phi$
(SUB) $\phi \ / \ \sigma \phi$

where σ is a map uniformly substituting formulas for propositional variables in formulas.

A normal derivation system is called orthodox if (MP), (UG) and (SUB) are the only derivation rules of the system. For a set Σ of formulas, we let $\Omega(\Sigma)$ denote the orthodox logic generated by Σ, i.e. Ω is the orthodox derivation system having as axioms Σ, (CT) and (DB). A formula is a theorem of the derivation system $\Delta = (A, R)$, notation: $\Delta \vdash \phi$, if ϕ is the last item ϕ_n of a sequence $\phi_0, \ldots, \phi_n$ of formulas such that each ϕ_i is either an axiom or the result of applying a rule to formulas of $\{\phi_0, \ldots, \phi_{i-1}\}$. A derivation system Δ is sound with respect to a class K of frames if every theorem of Δ is valid in K, complete if every K-valid formula is a theorem of Δ.

The notion of a normal arrow logic is the straightforward generalization to the similarity type $\{\circ, \otimes, \iota\delta\}$ of the notion of a normal modal logic in the similarity type with one diamond $\Diamond$. Orthodox modal logics are the direct modal correspondents of standard algebraic equational axiomatizations for varieties of boolean algebras with operators. For instance, if the orthodox modal logic generated by the set of axioms Σ is sound and complete with respect to a class K of frames, then the set $\{\tau \sigma = 1 \mid \sigma \in \Sigma\}$ [11] completely axiomatizes the class $\mathbf{CmK}$ (and, equivalently, the variety generated by it).

A nice aspect of arrow logic is that almost all interesting modal formulas are in Sahlqvist form — for instance, all formulas that we encountered in section 5. Therefore, many easy completeness results are obtainable:

Proposition 6.2 *Let Σ be a subset of the formulas occurring in section 5. Then $\Omega(\Sigma)$ is sound and complete with respect to the class of frames that is characterized by (the first order equivalents of the formulas in) Σ.*

The proof of Proposition 6.2 follows from the *completeness* part of Sahlqvist's theorem, cf. Sahlqvist 1975

For the two-dimensional semantics the situation is more complicated. For instance, the squares are not finitely axiomatizable by an orthodox system (this follows from the non-finite axiomatizability result for Representable Relation Algebras mentioned in section 4). Fortunately, many classes of *relativized squares* allow finite orthodox axiomatizations. For

---

[11] τ is the trivial translation from arrow logic *formulas* to relation algebraic *terms*, cf. the paragraph preceding Proposition 4.5.

instance, the following theorem is an immediate consequence of Proposition 6.2 and the result by Maddux that WA = RRA$_{rs}$ (cf. section 3):

Theorem 6.3 *Let Σ be the set of axioms A1 ... A8 and*
$(A10)$ $((\iota\delta \wedge p) \circ \top) \circ \top \leftrightarrow (\iota\delta \wedge p) \circ (\top \circ \top).$
Then $\Omega(\Sigma)$ is sound and complete with respect to the class LSQ *of locally square arrow frames.*

In a similar way, the next theorem (due to M. Marx, cf. 1995), follows from Proposition 6.2 and Theorem 5.7:

Theorem 6.4 *Let Σ be the set of axioms $A^r1 \ldots A^r8'''$ from Theorem 5.7. Then $\Omega(\Sigma)$ is sound and complete with respect to* RSQ.

If one allows unorthodox derivation rules, full square validity can be axiomatized by a finite system too. The proof of the following result can be found in Venema 1991.

Theorem 6.5 *Let $\mho$ be the derivation system having as its axioms the formulas A1 ... A9 of section 4, and as its rules the orthodox derivation rules and*

(6) $(p \wedge \neg Dp) \to \phi \; / \; \phi, \quad \text{provided } p \text{ does not occur in } \phi,$

where D is the operator defined in (5). Then $\mho$ is sound and complete with respect to SQ.

The use of rather odd looking derivation rules like (6) to axiomatize theories lacking a finite orthodox axiomatization originates with Gabbay 1981, Burgess 1980, and is discussed in detail in Venema 1993.

Decidability

With one exception, we will be rather sketchy concerning decidability and related matters, as most of the known results can be found in other contributions to this volume. The watershed in the hierarchy of arrow logics seems to be the associativity axiom:[12]

$(A9)$ $p \circ (q \circ r) \leftrightarrow (p \circ q) \circ r.$

From results in chapter 3 and Kurucz et al. 1993 we can distill the following theorem:

Theorem 6.6 *Let Λ be any non-trivial[13] arrow logic containing A9. Then it is undecidable whether a given formula is derivable in Λ.*

As the proof of Theorem 6.6 contains some nice arguments of which the central ideas are not too difficult to follow, we will give a proof sketch here:

---

[12] Recent results have shown that undecidability already occurs under weaker conditions than associativity of ∘. We refer the reader to chapter 3.

[13] We prefer to keep this notion informal, referring to chapter 3 for a precise formulation; as an *example* of a sufficient condition for the theorem to hold, one might demand that every theorem of Λ is valid on squares of arbitrary size.

Proof. Assume that Λ is the orthodox arrow logic axiomatized by $(A9)$. Recall that a semigroup is an algebra $\mathfrak{G} = (G, \cdot)$ such that $\cdot$ is an associative operation on G, and that a quasi-equation is an expression of the form

(7) $\qquad s_1 = t_1 \wedge \ldots \wedge s_n = t_n \rightarrow s_0 = t_0,$

where the s_i's and t_i's are terms. We will show that Λ is undecidable by reducing the (undecidable!) quasi-equational theory of semigroups to it[14].

Now let Q be a quasi-equation in the language of semigroups; we will first show that

(8) $\qquad \mathsf{SG} \models Q \iff \mathsf{BA}^\circ_{A9} \models Q,$

where SG is the variety of semigroups and BA°_{A9} is the class of boolean algebras with an associative, normal and additive operator $\circ$. Note that by duality, the *equational* theory of BA°_{A9} is decidable iff Λ is decidable.

To prove (8), the direction $\Rightarrow$ is trivial, as the $\circ$-reduct of any algebra in BA°_{A9} *is* a semigroup. For the other direction, let $\mathfrak{G}$ be a semigroup such that $\mathfrak{G} \not\models Q$. Without loss of generality we may assume that $\mathfrak{G}$ is a monoid, i.e. that $\mathfrak{G}$ has a designated element e such that $\mathfrak{G} \models ex = xe = x$. (For, if $\mathfrak{G}$ is not a monoid, we can *embed* it in a monoid $\mathfrak{G}'$; if $\mathfrak{G} \not\models Q$, then Q will fail to hold in $\mathfrak{G}'$ as well). Now we will embed $\mathfrak{G}$ in a complex algebra $\mathfrak{A} = (\mathcal{P}(W), \circ)$ in BA°_{A9}. $\mathfrak{A}$ is defined by

$$W = \{(u, uv) \mid u, v \in G\},$$
$$X \circ Y = X \mid^W Y,$$

i.e. $\mathfrak{A}$ is the $\{+, -, ; \}$-reduct of a power set algebra of a relativized square over G. Note that $\mid^W$ is identical to the relational composition $\mid$, as W is a transitive relation over G. So $\circ$ is associative, and hence $\mathfrak{A}$ is in BA°_{A9}.

Now to embed $\mathfrak{G}$ in $\mathfrak{A}$, we define a map h reminiscent of the Cayley representation of groups:

$$h(u) = \{(v, vu) \mid v \in G\}.$$

To show that h is a homomorphism, first assume $(x, y) \in h(uu')$. Then $y = x(uu') = (xu)u'$, so by $(x, xu) \in h(u)$ and $(xu, (xu)u') \in h(u')$ we obtain $(x, y) \in h(u) \mid h(u')$. Thus we find $h(uu') \subseteq h(u) \mid h(u')$. For the other direction, assume that there is a z with $(x, z) \in h(u)$ and $(z, y) \in h(u')$. By definition of h, $z = xu$ and $y = zu'$, so $y = (xu)u' = x(uu')$, implying that $(x, y) \in h(uu')$. Therefore, $h(u) \mid h(u') \subseteq h(uu')$. So to show that h is an embedding it suffices to prove injectivity. Assume $h(u) = h(u')$; then $(e, u) \in h(u')$, and as u' is the *only* element $x \in G$ such that $(e, x) \in h(u')$, this gives $u = u'$. *This proves (8)*.

---

[14]Note that this is a very strong simplification of the proof method used in the cited papers. The 'core' of the undecidability of arrow logic with associativity is not (just) the undecidability of the quasi-equational theory of semigroups itself, but (also) the fact that various classes of (semi-)groups have *recursively inseparable* quasi-equational theories.

The second part of the proof consists in showing that over BA°_{A9}, the validity problem of quasi-equations is reducible to that of equations, by defining a (recursive!) translation E from quasi-equations to equations such that
(9) $$\mathsf{BA}^{\circ}_{A9} \models Q \iff \mathsf{BA}^{\circ}_{A9} \models E(Q).$$
It would go too far to define the translation which works for BA°_{A9} itself; to capture the essential idea, let us look at classes $\mathsf{K} \subseteq \mathsf{BA}^{\circ}_{A9}$ for which there is a term $c(x)$ satisfying, for all $\mathfrak{A}$ in K[15]:
$$\mathfrak{A} \models \begin{cases} c(a) = 0 & \text{if } a = 0, \\ c(a) = 1 & \text{if } a \neq 0, \end{cases}$$
In other words, we are in a *discriminator* variety of boolean algebras with operators. Now for such a class K, we can define the equation $E(Q)$ as
$$c(s_0 \oplus t_0) \leq c(s_1 \oplus t_1) + \ldots + c(s_n \oplus t_n)$$
where Q is as in (7) and $\oplus$ is the the symmetric difference operation given by $x \oplus y = x \cdot -y + -x \cdot y$.

We leave it to the reader to prove (9) by showing that for all $\mathfrak{A}$ in K, $\mathfrak{A} \models Q$ iff $\mathfrak{A} \models E(Q)$; the crucial observation is that in any $\mathfrak{A}$ in K, $c(s \oplus t)$ is either 0 or 1, according to whether $s \neq t$ or $s = t$. □

Many logics not containing $A9$ are decidable, cf. chapter 2, as are many theories in simpler languages, cf. the literature mentioned in section 7. Two quite interesting examples of decidable logics are formed by the Dynamic Arrow Logic of van Benthem 1994 and the modal theory of relativized squares, cf. chapter 2. The most common method to prove that a logic Λ is decidable is to show that it has the finite model property (fmp), i.e. to prove that every non-theorem of the logic is falsifiable in a model based on a *finite* frame on which Λ is valid (and to show that it is decidable whether this condition is met by the frame). For instance, this method is used to prove the two results mentioned above.

It is interesting to note that for pair arrow logics, one can define an even stronger property that we will call the *finite base property* (fbp), which applies to a logic Λ if each of its non-theorems is falsifiable in a two-dimensional Λ-frame based on a finite *base set*[16]. This property is strictly stronger than the fmp, as a relatively simple example involving a dense linear ordering will show.

Interpolation

Let us finish this section by noting that with respect to other logical properties, the landscape of arrow logic still seems to be terra incognita. An

---
[15]As an example of a term satisfying this condition, take the term $c(x) = 1; x; 1$, then (9) holds for every **FRA**.
[16]Recently, it was shown by H. Andréka that non-square arrow logics have this fbp.

exception concerns interpolation properties; in Sain 1990 and Venema 1991 negative results can be found for logics *with* the associativity axiom. For many classes of relativized frames, positive results are proved by Sain, Németi and Marx; an overview of known results is given in Marx 1995.

7 The Family

In this section we briefly discuss, or just mention a number of systems that are closely related to the basic arrow logic, and/or fragments or extensions of it.

Dynamic Arrow Logic

In many of the application areas of arrow logic, e.g. linguistics, computer science or the algebraic theory of binary relations, *iteration* plays an essential role; therefore the Kleene star would be the first candidate to add to the language of arrow logic. Van Benthem introduces the system of dynamic arrow logic, as a proposal for the 'computational core calculus' of dynamic logics (cf. van Benthem 1994). Here the focus is on abstract arrow semantics; the truth definition of $*$ is given by

$\mathfrak{M}, a \Vdash *\phi \iff \mathfrak{M}, a \Vdash \phi$ or a can be C-decomposed into some finite sequence of arrows, each satisfying ϕ in $\mathfrak{M}$,

where C-decomposition means that there is at least one way of successive composition of the arrows in the sequence so as to obtain a. As axioms van Benthem proposes, besides $A1, \ldots, A8$ of section 5, the following:

$$\phi \rightarrow *\phi$$
$$*\phi \circ *\phi \rightarrow *\phi$$

and the induction principle can be found in the rule

$$\phi \rightarrow \alpha, \ \alpha \circ \alpha \rightarrow \alpha \ / \ *\phi \rightarrow \alpha.$$

Both a completeness result and a decidability result are proved by van Benthem.

Arrow Logic, Bulgarian style

In chapter 7 systems are studied that are closely related to the arrow logics discussed in this introduction, and that are also called arrow logics. Here the universe of a model consists of arrows as well, but the signature of these arrow models is different, and so is the similarity type of the modal language. The basic system has frames of the form $\mathfrak{F} = (W, R_{ij})_{i,j \in \{l,r\}}$ with every R_{ij} a binary relation on the set W of arrows. In the intended frames, W is the set of edges in a multigraph (cf. section 2) and the accessibility relations are defined by

$$R_{ij}ab \text{ iff } i(a) = j(b),$$

for instance $R_{lr}ab$ holds if the left point of a is the right point of b. The modal language has a diamond $\Diamond_{ij}$ for every relation R_{ij}. Note that in the full square semantics, these operators are definable in the language of ordinary arrow logic, for instance $\Diamond_{lr}\phi$ as $\otimes\phi \circ \top$. Vakarelov introduces an interesting construction called *copying*, and proves several completeness and decidability results (cf. also Venema 1992 for a completeness result for the full square semantics). In Arsov and Marx 1994 the connection between the two approaches to arrow logic are investigated; one of the main results of the paper is a completeness result with respect to the class of multigraphs, for a language combining the operators of the two approaches.

Modal Transition Logics

Arrow logic occupies an interesting position in the landscape of what we will call modal transition logics. With a modal transition logic we understand (informally) a modal formalism (in the sense of abstract modal logic, cf. the discussion in the first section) with an intended semantics in which at least part of the well-formed expressions are interpreted as sets of transitions. To classify such systems, the main criteria seem to be (i) the relation of states versus transitions in the intended semantics, and (ii) the set of modal operators governing transitions, states and their connection. The extremity of arrow logic in this landscape lies in the fact that its models consists of transitions only. This goes too for Dynamic Implication Logic (Blackburn and Venema 1993) which considers a relatively poor fragment of arrow logic, and for Action Logic (Pratt 19900, which deals with the connectives $*$, $\vee$, $\circ$ and the residuals / and \ of $\circ$ (to be discussed below).

Note that the 'classical' dynamic logic, Propositional Dynamic Logic (PDL, cf. Harel 1984) is a modal transition logic as well: although the formulas of PDL are evaluated at states, the programs are evaluated at transitions. (It is interesting to note that the decidability of PDL depends on the impossibility to express the equivalence of two programs, cf. Németi 1981) The system of Dynamic Modal Logic (DML, cf. de Rijke to appear) is similar to PDL in that it has an algebra of diamonds as well; the main difference being that the algebra of diamonds is not dynamic but relational.

Finally, there are hybrid systems: languages with two sorts of formulas, referring to states and transitions respectively, and a rich set of operators including modalities that relate the two sorts. We refer the reader to van Benthem 1994 for an abstract approach, chapter 5 for results on concrete one- and two-dimensional interpretation of sorted transition systems, and de Rijke 1993 for a very rich language, and completeness results for the full square case.

Residuals and Conjugates

In section 3 we already mentioned the dynamic interpretation of the Lambek calculus due to van Benthem (cf. van Benthem 1991a); in this semantics, the connectives of categorial logic (\, / and ∘) are interpreted in two-dimensional models (with a transitive and irreflexive universe $W \subseteq U \times U$) as follows:

$\mathfrak{M}, (x,y) \Vdash \phi \circ \psi$ if   there are $(x,z), (z,y) \in W$ with $\mathfrak{M}, (x,z) \Vdash \phi$ and $\mathfrak{M}, (z,y) \Vdash \psi$,

$\mathfrak{M}, (x,y) \Vdash \phi/\psi$ if   for all $(y,z) \in W$: $\mathfrak{M}, (y,z) \Vdash \psi$ implies $\mathfrak{M}, (x,z) \Vdash \phi$,

$\mathfrak{M}, (x,y) \Vdash \phi\backslash\psi$ if   for all $(z,x) \in W$: $\mathfrak{M}, (z,x) \Vdash \phi$ implies $\mathfrak{M}, (z,y) \Vdash \psi$.

In other words, the slashes are the *residuals* of ∘ — the following derivation rules form the basis of the Lambek calculus:

$$\phi \to \chi/\psi \iff \phi \circ \psi \to \chi \iff \psi \to \phi\backslash\chi.$$

Note that the residuals can be defined in the language of arrow logic, e.g. / by $\phi/\psi \equiv \neg(\neg\phi \circ \otimes\psi)$.

Van Benthem observed that with the interpretation defined above, the Lambek calculus is sound with respect to the class of relativized squares where the universe is a transitive binary relation. In Andréka and Mikulás 1994 completeness is proved, via an algebraic representation theorem (cf. also Kurtonina to appear for a derivation system related to the Gentzen calculus for RRA discussed in Wadge 1975).

In chapter 6, Mikulás studies the composition operator ∘ together with its *conjugates*, i.e. connectives $\circ_1$ and $\circ_2$ with the following truth definition on arrow models:

$\mathfrak{M}, b \Vdash \phi \circ_1 \psi$ if   there are a, c with $Cabc$, $\mathfrak{M}, a \Vdash \phi$ and $\mathfrak{M}, c \Vdash \psi$,

$\mathfrak{M}, c \Vdash \phi \circ_2 \psi$ if   there are a, b with $Cabc$, $\mathfrak{M}, a \Vdash \phi$ and $\mathfrak{M}, b \Vdash \psi$.

Multi-Dimensional Modal Logic

Let us call an abstract modal system multi-dimensional if it has an intended semantics in which the 'possible worlds' are tuples over some base set, cf. Venema 1991 for an overview. Then arrow logic is an example of a two-dimensional logic, as part of its intended semantics is formed by the (relativized) squares. Therefore it is interesting to see which phenomena of arrow logic are shared by other multi-dimensional modal logics. We give a few examples: to start with, in Venema 1995 a formalism of cylindric modal logic is developed which can be seen as the modal version of a restricted kind of first order logic. The main result of that paper, an axiomatization of the class of cubes (higher-dimensional analogues of squares) is closely related to Theorem 6.5. Our second example is the paper Németi 1995; (cf. also chapter 10) the author shows how the undecidability of first-order logic vanishes if we generalize the class of models in the same way that the relativized squares form a generalization of the class of full squares. In Andréka et al. 1995 the model theory of this 'restricted first-order logic' is developed. Finally, in chapter 7 a higher-dimensional version of the Bulgarian-style arrow logic is studied.

8 Questions

We finish this introduction with indicating some lines for further research, and listing a few technical open problems. There are still a lot of areas to be explored in the landscape of arrow logic.

Recall that in the introduction, we mentioned a list of possible entities that arrows might represent. One direction of research could be to extend this list with other application areas for arrow logic, and to develop the arrow-logical theory of various applications; for instance, what can be said about the arrow logic of various kinds of *categories*? Note that one need not only think here of questions concerning axiomatics or decidability; it would also be very interesting to see some (more) general results on other logical properties like interpolation, Beth definability, functional completeness, etc. In particular, we mentioned in section 1, that notions like *partiality* or *constraints* on the valuations might be needed for applications of arrow logic in for instance dynamic semantics. These are areas which have not yet been exploited.

These question can also be approached from a more theoretical perspective: the general question would be to study the *lattice of arrow logics*. In particular, it would be nice to have more results along the lines of Theorem 6.6 — it is interesting to know exactly where in the lattice of arrow logics the right combinations of complexity and expressiveness are situated.

Another area in which quite general questions can be asked, is given by the connections between arrow logics and *substructural logics*. In section 7 we mentioned the connections between arrow logic and the Lambek Calculus. Now the Lambek Calculus is only one particular spot in the landscape of *substructural logics* (cf. Došen and Schroeder-Heisler to appear), i.e. Gentzen-style calculi where the ordinary structural rules are absent. It would be interesting to establish more general connections between the hierarchy of substructural logics and formalisms of arrow logics.

Mentioning Gentzen-style derivation systems takes us to an undeveloped area of arrow logic — its *proof theory*. It would be nice to have complete sequent calculi for logics which allow finite Hilbert-style axiomatizations (like the relativized squares or the locally square arrow frames), in particular calculi with nice properties like cut-elimination and decidability. For some known derivation systems in the Gentzen-style, the reader is referred to Wadge 1975, Maddux 1983 or Orlowska 1991; of these, the second paper reveals some nice connections between various varieties of relation (type) algebras and the number of variables needed in a proof.

A huge research field opens up if we shift some of the parameters underlying the *similarity type* of arrow logic. For instance, we may study languages for arrow frames having modal operators different from ∘, ⊗ and $\iota\delta$. We already encountered the difference operator D, the residuals / and \, and the conjugates $\circ_1$ and $\circ_2$, but there is of course an infinite hierarchy of modal languages that can be interpreted in arrow frames. Another parameter shift would be to study other (almost) basic arrow relations besides C, R and I. For instance, the *parallel* composition operator of processes in process algebra might be generalized into a ternary relation of arrows, for which a binary operator could be added to the language.

An even further generalization of the arrow logic ideology would be the following. Recall from section 3 that over the class of square models, arrow logic is equivalent to the three variable fragment of the predicate calculus of binary relations. The general arrow semantics can then be seen as an abstraction of the usual semantics for first order logic: dyadic predicates are interpreted as sets of arrows which may but *need not* be identified with pairs. This abstraction from concrete pairs to abstract arrows can just as well be made in the higher-dimensional case: n-ary relations may be interpreted as sets of n-dimensional arrows instead of as sets of n-tuples over some domain. Some first exercises in this area have been made[17], but as yet, the questions outnumber the answers by far.

Finally, there are some *technical* nuts to crack:

1. *Axiomatize the square logic of* $\{\circ, \iota\delta, \wedge, \bot\}$!
   For various fragments of the language of arrow logics, axiomatizations have been found, or non-finite axiomatizability results obtained (cf. chapter 2). For the fragment listed above, the question is still open whether there is a finite axiomatization of the set of formulas that are valid in the class **SQ**. This problem is interesting because the fragment plays an important role in *situation theory*, cf. Moss and Seligman 1994.

2. *Does the Lambek Calculus have the finite base property?*
   Recall from section 6 that the fbp of a logic states that any non-theorem of the logic can be falsified in a two-dimensional model based on a finite base set. For the associative Lambek Calculus the problem is the following: given two formulas ϕ and ψ, built up using only /, \ and ∘ and such that $\phi \to \psi$ is not a theorem of the Lambek Calculus, is there a finite two-dimensional model (with a transitive universe)

---

[17] Cf. the part on multi-dimensional modal logics in the previous section; note however that most of these authors concentrate on non-standard, but still multi-dimensional models.

in which $\phi \to \psi$ can be falsified? For more information, cf. Andréka and Mikulás 1994.

3. *Is the class* RF *of representable arrow frames elementary?*
We call an arrow frame $\mathfrak{F}$ representable if its complex algebra $\mathfrak{Cm}\mathfrak{F}$ is representable, i.e. if $\mathfrak{Cm}\mathfrak{F}$ is in **SP**(FRA). Note that if an arrow frame $\mathfrak{F}$ is *not* representable, then this fact is witnessed by a formula ϕ from the modal theory of squares, which is not valid in $\mathfrak{F}$. So the class of non-representable arrow frames is closed under taking ultrapowers. Therefore, to solve this open problem, it suffices to answer the question whether RF itself is closed under taking ultrapowers. A positive solution would be given by finding a (necessarily infinite) orthodox axiomatization of the class SQ using only Sahlqvist axioms. For more information, cf. Venema 1991.

Acknowledgments. I would like to thank the following people for reading and commenting on earlier versions of the manuscript: Hajnal Andréka, Patrick Blackburn, Jan van Eijck, Maarten Marx, Istvan Németi, Szabolcs Mikulás, Ildikó Sain and Jerry Seligman. Special thanks are due to Johan van Benthem.

References

Andréka, H. 1991. Complexity of the equations valid in algebras of relations. Thesis for D.Sc. (a post–habilitation degree) with Math. Inst. Hungar. Ac. Sci. Budapest. A slightly updated version will appear in *Annals of Pure and Applied Logic*.

Andréka, H., and Sz. Mikulás. 1994. Lambek calculus and its relational semantics: completeness and incompleteness. *Journal of Logic, Language and Information* 3(1):1–38.

Andréka, H., J. D. Monk, and I. Németi (ed.). 1991. *Algebraic Logic (Proc. Conf. Budapest 1988)*. Amsterdam. Colloq. Math. Soc. J. Bolyai, North-Holland, Amsterdam.

Andréka, H., J. van Benthem, and I. Németi. 1995. Back and Forth between Modal Logic and Classical Logic. *Bulletin of the Interest Group in Pure and Applied Logics* 3(3):685–720.

Arsov, A., and M. Marx. 1994. Basic Arrow Logic with Relation Algebraic Operators. In *Proceedings of the 9th Amsterdam Colloquium*, ed. P. Dekker and M. Stokhof, 93–112. Universiteit van Amsterdam. Institute for Logic, Language and Computation.

van Benthem, J. 1985. *Modal Logic and Classical Logic*. Bibliopolis.

van Benthem, J. 1991a. *Language in Action (Categories, Lambdas and Dynamic Logic)*. Studies in Logic, Vol. 130. North-Holland, Amsterdam.

van Benthem, J. 1991b. Logic and the flow of information. In *Proceedings of the 9th International Congress of Logic, Methodology and Philosophy of Science*. North Holland, Amsterdam.

van Benthem, J. 1994. A Note on Dynamic Arrow Logics. In van Eijck and Visser 1994, 15–29.

van Benthem, J., and J. Bergstra. 1993. Logic of Transition Systems. Technical Report CT–93–03. Universiteit van Amsterdam: Institute for Logic, Language and Computation.

van Benthem, J., J. van Eijck, and V. Stebletsova. 1993. Modal logic, transition systems and processes. Computer Science/Department of Software Technology Report CS–R9321. Amsterdam: Centrum voor Wiskunde en Informatica.

Bergstra, J., and J.W. Klop. 1984. Process algebra for synchronous communication. *Information and Control* 60:109–137.

Blackburn, P., and Y. Venema. 1993. Dynamic squares. Logic Group Preprint Series 92. Department of Philosophy, Utrecht University. to appear in *Journal of Philosophical Logic*.

Brink, C. 1993. Power Structures. *Algebra Universalis* 30:177–216.

Burgess, J.P. 1980. Decidability for branching time. *Studia Logica* 39:203–218.

Došen, K., and P. Schroeder-Heisler (ed.). to appear. *Substructural Logics*. Oxford University Press.

van Eijck, J. 1993. Presuppositions and dynamic logic. In *Papers from the 2nd CSLI Workshop on Logic, Language and Computation*, ed. M. Kanazawa, C. Piñon, and H. de Swart. CSLI Publications. to appear.

van Eijck, J., and A. Visser (ed.). 1994. *Logic and Information Flow*. MIT Press, Cambridge (Mass.).

Freyd, P., and A. Scedrov. 1990. *Categories, Allegories*. North-Holland, Amsterdam.

Gabbay, D.M. 1981. An irreflexivity lemma with applications to axiomatizations of conditions on linear frames. In *Aspects of Philosophical Logic*, ed. U. Mönnich. 67–89. D. Reidel Publishing Company.

Gärdenfors, P. 1988. *Knowledge in Flux; Modeling the Dynamics of Epistemic States*. Cambridge (Mass.): Bradford Books/MIT Press.

Goldblatt, R.I. 1989. Varieties of Complex Algebras. *Annals of Pure and Applied Logic* 38:173–241.

Groenendijk, J., and M. Stokhof. 1991. Dynamic Predicate Logic. *Linguistics and Philosophy* 14:39–100.

Harel, D. 1984. Dynamic Logic. In *Handbook of Philosophical Logic*, ed. D.M. Gabbay and F. Guenther. 497–604. Dordrecht: Reidel.

Jónsson, B. 1982. Varieties of relation algebras. *Algebra Universalis* 15:273–298.

Jónsson, B. 1991. The theory of binary relations. In Andréka et al. 1991, 245–292.

Jónsson, B., and A. Tarski. 1951. Boolean algebras with operators. Parts I and II. *American Journal of Mathematics* 73:891–939. and 74:127–162, 1952.

Kramer, R. 1991. Relativized relation algebras. In Andréka et al. 1991, 293–349.

Kurtonina, N. to appear. The Lambek calculus: relational semantics and the method of labeling. *Studia Logica*.

Kurucz, Á, I. Németi, I. Sain, and A. Simon. 1993. Undecidable varieties of semilattice-ordered semigroups, of boolean algebras with operators, and logics extending Lambek Calculus. *Bulletin of the IGPL* 1(1):91–98.

Lyndon, R. 1950. The representation of relational algebras. *Annals of Mathematics* 51:73–101.

Maddux, R. 1991. Introductory course on relation algebras, finite dimensional cylindric algebras, and their connections. In Andréka et al. 1991, 361–392.

Maddux, R. D. 1982. Some varieties containing relation algebras. *Transactions of the American Mathematical Society* 272:501–526.

Maddux, R. D. 1983. A sequent calculus for relation algebras. *Annals of Pure and Applied Logic* 25:73–101.

Marx, M. 1995. *Algebraic Relativization and Arrow Logic*. Doctoral dissertation, Institute for Logic, Language and Computation, University of Amsterdam. ILLC Dissertation Series 1995–3.

Milner, R. 1980. *A Calculus of Communicating Systems*. Berlin: Springer Verlag.

Monk, J. D. 1969. Nonfinitizability of classes of representable cylindric algebras. *Journal of Symbolic Logic* 34:331–343.

Moss, L., and J. Seligman. 1994. Classification domains and information links: a brief survey. In van Eijck and Visser 1994.

Németi, I. 1981. Dynamic algebras of programs. In *Fundamentals of Computation Theory '81 (Proceedings Conference Szeged 1981)*, Lecture Notes in Computer Science, Vol. 117, 281–290. Springer-Verlag, Berlin.

Németi, I. 1991. Algebraizations of quantifier logics: an overview. *Studia Logica* L(3/4):485–569.

Németi, I. 1995. Decidability of weakened versions of first–order logic. In *Logic Colloquium '92*, ed. L. Csirmaz, D. Gabbay, and M. de Rijke, 177–242. Studies in Logic, Language and Information, No. 1. Stanford. CSLI Publications.

Orlowska, E. 1991. Relational interpretation of modal logics. In Andréka et al. 1991, 443–471.

Park, D. 1981. Concurrency and automata on infinite sequences. In *Proceedings 5th GI Conference*, 167–183. New York. Springer.

Pratt, V. 1990. Action Logic and Pure Induction. In *Logics in AI*, ed. J. van Eijck, Lecture Notes in Artificial Intelligence, Vol. 478, 97–120. Springer–Verlag.

de Rijke, M. 1993. *Extending Modal Logic*. Doctoral dissertation, Institute for Logic, Language and Computation, Universiteit van Amsterdam. ILLC Dissertation Series 1993–4.

de Rijke, M. to appear. A system of dynamic modal logic. *Journal of Philosophical Logic*.

Sahlqvist, H. 1975. Completeness and Correspondence in the First and Second Order Semantics for Modal Logic. In *Proc. of the Third Scandinavian Logic Symposium Uppsala 1973*, ed. S. Kanger. Amsterdam. North–Holland.

Sain, I. 1990. Beth's and Craig's properties via epimorphisms and amalgamation in algebraic logic. In *Algebraic logic and universal algebra in computer science*, ed. C. H. Bergman, R. D. Maddux, and D. L. Pigozzi. Lecture Notes in Computer Science, Vol. 425, 209–226. Springer-Verlag, Berlin.

Sambin, G., and V. Vaccaro. 1989. A Topological Proof of Sahlqvist's Theorem. *Journal of Symbolic Logic* 54:992–999.

Tarski, A. 1941. On the calculus of relations. *Journal of Symbolic Logic* 6:73–89.

Veltman, F. to appear. Defaults in update semantics. *Journal of Philosophical Logic*.

Venema, Y. 1991. *Many–Dimensional Modal Logic*. Doctoral dissertation, Institute for Logic, Language and Computation, Universiteit van Amsterdam.

Venema, Y. 1992. A note on the tense logic of dominoes. *Journal of Philosophic Logic* 21:173–182.

Venema, Y. 1993. Derivation rules as anti-axioms in modal logic. *Journal of Symbolic Logic* 58:1003–1034.

Venema, Y. 1995. Cylindric Modal Logic. *Journal of Symbolic Logic* 60(2):591–623.

Visser, A. 1994. Actions under presuppositions. In van Eijck and Visser 1994, 196–233.

Wadge, W. 1975. A complete natural deduction system for the relation calculus. Theory of Computation Report 5. University of Warwick.

2
Investigations in Arrow Logic
MAARTEN MARX, SZABOLCS MIKULÁS, ISTVÁN NÉMETI
AND ILDIKÓ SAIN

ABSTRACT. We give an extensive overview of arrow logics of *pair-frames* with respect to the following four properties:
- finite axiomatizability with an orthodox derivation system,
- decidability,
- Craig interpolation, and
- Beth definability.

Since the universe of a pair-frame is a binary relation, we can consider classes of such frames satisfying conditions like reflexivity, symmetry and transitivity. Our main result is that an arrow logic of pair-frames has any of the above properties if and only if it contains frames with non-transitive universes. Transitivity of the universe is closely connected to *associativity* of composition, because composition is associative on a pair-frame if its universe is a transitive relation. So the conclusion of this paper can also be stated as follows: we can give arrow logic the intuitive pair-frame semantics and have all the four above mentioned positive properties if we are willing to give up associativity of composition.

Arrow logic comes in two versions: the "abstract arrows" version and the concrete "ordered pairs" version. Our results point in the direction that the abstract arrows version seems to have no advantages (like decidability etc.) over the ordered pairs version.

For motivation and an introduction to arrow logic we refer to chapter 1. We use the definitions and notation developed there.

This chapter is organized as follows. In the first section we sketch a hierarchy of arrow logics, and state our main results. The proofs can be found in

---

The research of the Dutch author supported by NWO grant SIR 12-1114. The research of the Hungarian participants was supported by Hungarian National Foundation for Science, Research Grants No. F17452 and No. T16448

Arrow Logic and Multi-Modal Logic
M. Marx, L. Pólos and M. Masuch, eds.
Copyright © 1996, CSLI Publications.

the next three sections. Section 2 is about completeness, section 3 about decidability, and section 4 about interpolation and definability.

Section 5 discusses a research direction which is complementary to the one taken in this paper. Besides investigating arrow logics with a semantics which differs from the square semantics we can also change the similarity type. Several important connectives (like the residuals of composition or the difference operator) are term-definable on the square semantics, but in general not on the relativized square semantics. We review for several connectives what is known when we "put them back" to arrow logic with the new semantics.

We close the chapter with a short overview of the existing literature on the subjects treated here, and give a list of suggestions for further reading.

1 An Arrow Logical Hierarchy of Distinguished Classes of Pair-Frames

1.1 The Hierarchy

We start with the definition of a *pair-frame*. If x is a pair $\langle u, v \rangle$, then we use x_0 and x_1 to denote u and v, respectively. Recall from chapter 1 the notion of an arrow frame and related notation.

Definition 1.1 (Pair-frames) (i) An arrow frame $\mathcal{F} = \langle F, C, R, I \rangle$ is called a *pair-frame* if $F \subseteq U \times U$ for some set U and the relations C, R, I agree with C_F, R_F, I_F, respectively. These are defined as follows (we use the subscript F in the relations to indicate that they are relations on pair-frames which are uniquely determined by the universe F). For any $x, y, z \in F$,

$$C_F xyz \stackrel{\text{def}}{\iff} x_0 = y_0, y_1 = z_0 \ \& \ z_1 = x_1$$
$$R_F xy \stackrel{\text{def}}{\iff} x_0 = y_1 \ \& \ x_1 = y_0$$
$$I_F x \stackrel{\text{def}}{\iff} x_0 = x_1.$$

(ii) For a pair-frame $\mathcal{F} = \langle F, \ldots \rangle$ with $F \subseteq U \times U$, F is called the *domain* of $\mathcal{F}$. If $V \subseteq U \times U$, then $\mathcal{F}_{pair}(V)$ denotes the pair-frame with domain V. The *base* of this pair-frame, denoted by $\mathsf{Base}(V)$, is defined as $\{u \in U : (\exists x \in V)(u = x_0 \text{ or } u = x_1)\}$.

(iii) KP denotes the class of all pair-frames.

We use the following convention: if $\mathcal{F}$ is an arrow frame then most often we denote its domain by F. But if we want to be more specific, then we denote the domain of a pair-frame by V and that of an arbitrary arrow frame by W.

A great advantage of pair-frames is that we can draw pictures which immediately explain the meanings of the formulas. While in Kripke frames one usually draws the elements of the domain (the "worlds") as points, and indicates the (accessibility) relations by arrows, in pair frames the "worlds" are pairs $\langle u, v \rangle$ drawn as arrows going from u to v, and the accessibility relations need not be drawn, since they are implicit in the arrows. One can say that in pair-frames the accessibility relations are coded inside the worlds. In figure 1 we establish our convention for drawing accessibility relations.

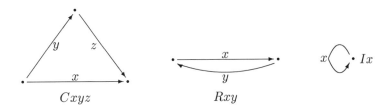

FIGURE 1 (Accessibility) Relations in pair-frames.

If K is a class of arrow frames we use AL(K) to denote the arrow-logic of that class (see definitions 1.2 and 1.3 in chapter 1).

In chapter 1 three classes of pair-frames are discussed: the *squares*, where the universes of the frames are full Cartesian products (i.e. $U \times U$ for some base set U), the *local squares*, where the universes are reflexive and symmetric relations and the *relativized squares*, where the universe can be any binary relation. Here we introduce a more refined classification of pair-frames as follows.

Definition 1.2 Let $H \subseteq \{R, S, T\}$ where R stands for 'reflexive', S for 'symmetric' and T for 'transitive'. Let $V \subseteq U \times U$ for some set U. We say that V *is an H relation* if V has the properties mentioned in H. For example, if $H = \{R, S\}$ then V is an H-relation if it is reflexive and symmetric. Define:

$$\mathsf{KP}_H \stackrel{\text{def}}{=} \{\mathcal{F} \in \mathsf{KP} : F \text{ is an } H \text{ relation}\};$$
$$\mathsf{KP}_{SQ} \stackrel{\text{def}}{=} \{\mathcal{F} \in \mathsf{KP} : F = U \times U \text{ for some set } U\}.$$

The classes just defined can be ordered as given in figure 2, where $X \to Y$ denotes $X \supseteq Y$. It is easy to see that all the inclusions are strict except the one labeled with $=$. This does not imply that the arrow logics of these classes are different too. As the next proposition shows the squares validate the same formulas as the frames whose universes are equivalence relations.

Definition 1.3 For two classes K_1 and K_2 of arrow frames, we say that K_1 and K_2 are equivalent if any arrow logical formula ϕ is K_1-valid if and only if it is K_2-valid. We denote this equivalence by $K_1 \equiv_{AL} K_2$.

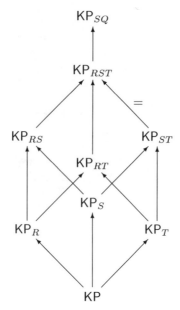

FIGURE 2 A hierarchy of distinguished classes of pair-frames

Proposition 1.4 $\mathsf{KP}_{ST} = \mathsf{KP}_{RST} \equiv_{\mathrm{AL}} \mathsf{KP}_{SQ}$. *These are the only equivalences occuring in Figure 2.*

Proof. Transitivity and symmetry together imply reflexivity. The equivalence follows from the fact that every frame from KP_{RST} is a disjoint union of frames from KP_{SQ}, and arrow-logical truth is preserved under disjoint unions.

The differences between the classes can be indicated by the following formulas; the reader is invited to complete the proof.

For reflexivity: $\iota\delta \circ p \leftrightarrow p$, ($\iota\delta$ denotes identity, $\circ$ is composition)
For symmetry: $\otimes \otimes p \leftrightarrow p$ ($\otimes$ denotes converse), and
For transitivity: $(p \circ q) \circ r \leftrightarrow p \circ (q \circ r)$. □

1.2 The Properties

We now turn to the logical properties we will study for arrow logics. Let K be a class of arrow frames and AL(K) the arrow logic of that class. We define the following five properties.

1. AL(K) *is finitely axiomatizable* if there is an *orthodox* derivation system $\Omega(\Sigma)$ which is sound and complete with respect to K (cf. 6.1 in chapter 1) and where Σ is a *finite* set of axioms.
2. AL(K) *is decidable* if the set $\{\phi : \mathsf{K} \models \phi\}$ is a decidable set.

3. AL(K) *has the strong Craig interpolation property* if for any ϕ, ψ such that K $\models \phi \to \psi$, there exists a θ such that K $\models \phi \to \theta$ and K $\models \theta \to \psi$, with the restriction that all the propositional variables which occur in θ occur in both ϕ and ψ.
4. Define the *global consequence relation* $\Gamma \models_K \phi$ for a set $\Gamma \cup \{\phi\}$ of arrow logical formulas as follows: $\Gamma \models_K \phi$ if every arrow model over a frame from K which validates all formulas in Γ also validates ϕ.
   AL(K) *has the weak Craig interpolation property* if for any ϕ, ψ such that $\phi \models_K \psi$, there exists a θ such that $\phi \models_K \theta$ and $\theta \models_K \psi$, with the same restriction as in 3.
5. Let Fml(P) be the set of arrow logical formulas generated from an infinite set P of variables. Let p and p' be two new propositional variables not in P. Let $\Sigma(p)$ be a set of formulas in the language Fml($P \cup \{p\}$) and let $\Sigma(p')$ be the corresponding set in Fml($P \cup \{p'\}$), formed by replacing p everywhere by p'. We say that $\Sigma(p)$ defines p *implicitly* iff $\Sigma(p) \cup \Sigma(p') \models_K p \leftrightarrow p'$. The set $\Sigma(p)$ is said to define p *explicitly* iff there exists a formula $\theta \in$ Fml(P) such that $\Sigma(p) \models_K \theta \leftrightarrow p$.
   AL(K) *has the Beth definability property* if every implicit definition is equivalent to an explicit definition.

Now we are ready for stating the promised theorem which we summarize in table 1. In the last collumn we indicate the sections where we deal with these results.

	$H \subseteq \{R, S\}$ AL(KP$_H$)	$T \in H$ AL(KP$_H$)	section
• finitely axiomatizable	yes	no	2
• decidable	yes	no	3
• strong Craig interpolation	yes	no	4
• weak Craig interpolation	yes	no	4
• Beth definability	yes	no	4

TABLE 1 Results about the hierarchy of classes of pair-frames (cf. figure 2).

Theorem 1.5 *Let $H \subseteq \{R, S, T\}$. For each of the properties* finite axiomatizability, decidability, weak and strong interpolation *and* definability, *the arrow logic* AL(KP$_H$) *has the property if and only if $T \notin H$.*

On the Proof. The theorem consists of two parts: a negative and a positive one. The negative results are not treated here. The negative results about finite axiomatizability can be found in Andréka 1991, Andréka 1994. The negative results about decidability are treated in chapter 3. The failure of the Beth and Craig properties are due to Németi-Sain (cf. Andréka 1994, Andréka and Németi 1994).

The positive side of the theorem will be proved in the next three sections. These sections are all set up in the same way: we give the crucial lemma, show how the relevant part of the theorem follows from it and continue with a proof of the lemma. We start with some preliminaries that we will use later on.

1.3 Correspondence and Arrow Frames

In the next proposition we give a list of axioms plus the properties of arrow frames which they characterize[1]. As mentioned in chapter 1 all axioms are in Sahlqvist form. The first twelve are *valid* on all pair-frames. Condition (C_{13}) is valid on *symmetric* pair-frames (i.e., pair-frames whose domains are symmetric relations) and conditions (C_{14}) and (C_{15}) on *reflexive* ones.

Their meaning is most easily grasped by checking these validities using the proposed way of drawing pair-frames. We briefly go through the list. The meaning of (C_1) and (C_{10}) is easy: every arrow has at most one converse and the converse relation is symmetric (i.e., if we look at the relation R_V as a partial function f it says that if fx is defined then so is ffx and it equals x). (C_{13}) says that every arrow has a converse. Conditions (C_2) and (C_3) state that an identity arrow is its own converse, and can be decomposed into itself. The meaning of (C_4) and (C_5) becomes clear by the following pictures:

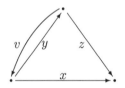 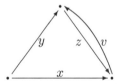

Condition (C_6) states that if an identity arrow is decomposed into y and z, then y and z are converses. Conditions $(C_7) - (C_9)$ express that arrows which "meet" share the same identity arrow at this meeting-point. If an arrow is a pair $\langle u, v \rangle$, then we call $\langle u, u \rangle$ its domain, and $\langle v, v \rangle$ its range. Conditions (C_{14}) and (C_{15}) state that for every arrow its domain and range are defined. The meaning of conditions 11 and 12 is obvious.

Proposition 1.6 *For $1 \leq i \leq 15$ every formula (A_i) given below characterizes the frame condition (C_i) of arrow frames $\mathcal{F} = \langle W, C, R, I \rangle$. The characterization of 2, and 7–9 hold only when 11 and 12 are assumed. The*

---
[1] In the sense of chapter 1, defined three paragraphs below theorem 5.5.

characterization of 1 holds only when 10 is assumed.

(C_1)	$\forall xy(Rxy \Rightarrow Ryx)$	(A_1)	$p \wedge \otimes \top \to \otimes \otimes p$
(C_2)	$\forall x(Ix \Rightarrow Cxxx)$	(A_2)	$\iota\delta \to \iota\delta \circ \iota\delta$
(C_3)	$\forall x(Ix \Rightarrow Rxx)$	(A_3)	$p \wedge \iota\delta \to \otimes p$
(C_4)	$\forall xyzv(Cxyz \& Ryv \Rightarrow Czvx)$	(A_4)	$(\otimes p) \circ \neg (p \circ q) \wedge q \to \bot$
(C_5)	$\forall xyzv(Cxyz \& Rzv \Rightarrow Cyxv)$	(A_5)	$\neg(p \circ q) \circ \otimes q \wedge p \to \bot$
(C_6)	$\forall xyz(Cxyz \& Ix \Rightarrow Rzy)$	(A_6)	$p \circ \neg \otimes p \to \neg \iota\delta$
(C_7)	$\forall xyzv(Cxyz \& Cxvx \& Iv)$	(A_7)	$[(p \wedge \iota\delta) \circ q] \text{or} \leftrightarrow$
	$\Leftrightarrow Cxyz \& Cyvy \& Iv)$		$(p \wedge \iota\delta) \circ [q \circ r]$
(C_8)	$\forall xyzv(Cxyz \& Cyyv \& Iv)$	(A_8)	$[p \circ (q \wedge \iota\delta)] \text{or} \leftrightarrow$
	$\Leftrightarrow Cxyz \& Czvz \& Iv)$		$p \circ [(q \wedge \iota\delta) \circ r]$
(C_9)	$\forall xyzv(Cxyz \& Czzv \& Iv)$	(A_9)	$[p \circ q] \circ (r \wedge \iota\delta) \leftrightarrow$
	$\Leftrightarrow Cxyz \& Cxxv \& Iv)$		$p \circ [q \circ (r \wedge \iota\delta)]$
(C_{10})	$\forall xyz(Rxy \& Rxz \Rightarrow y = z)$	(A_{10})	$\otimes p \wedge \otimes q \to \otimes(p \wedge q)$
(C_{11})	$\forall xyz(Cxyz \& Iy \Rightarrow x = z)$	(A_{11})	$\iota\delta \circ p \to p$
(C_{12})	$\forall xyz(Cxyz \& Iz \Rightarrow x = y)$	(A_{12})	$p \circ \iota\delta \to p$
(C_{13})	$\forall x \exists y(Rxy)$	(A_{13})	$\otimes \top$
(C_{14})	$\forall x \exists yz(Iy \& Cxyz)$	(A_{14})	$\iota\delta \circ \top$
(C_{15})	$\forall x \exists yz(Iy \& Cxzy)$	(A_{15})	$\top \circ \iota\delta$

Proof. Immediate since they are all Sahlqvist formulas. □

Partial Functions for Converse, Domain and Range. By conditions (C_7), (C_9), (C_{10})-(C_{12}) we have three *partial* functions living in our frames (cf. proposition 1.7 below). If one assumes $(C_{13}) - (C_{15})$ as well, they will be *total*. It's useful to make them explicit, so define

$$fx = y \stackrel{\text{def}}{\iff} Rxy \quad \text{otherwise } f \text{ is undefined;}$$
$$x_l = y \stackrel{\text{def}}{\iff} Cxyx \& Iy \quad \text{otherwise } (\cdot)_l \text{ is undefined;}$$
$$x_r = y \stackrel{\text{def}}{\iff} Cxxy \& Iy \quad \text{otherwise } (\cdot)_r \text{ is undefined.}$$

So, if they are defined, fx gives the converse arrow of x, and the functions x_l and x_r give the domain and range of x, respectively.

It is convenient to have explicit symbols in our language corresponding to the two defined functions; so define $\mathsf{t}\phi = (\iota\delta \wedge \phi) \circ \top$ and $\mathsf{h}\phi = \top \circ (\iota\delta \wedge \phi)$. The intuition is that $\mathsf{t}\phi$ is true at an arrow, if ϕ is true at the domain (the "head") of the arrow, and similarly for $\mathsf{h}\phi$ and "tail". Their meanings are given as follows. This is easy to see by writing out the definitions.

$$M, x \Vdash \otimes\phi \iff M, fx \Vdash \phi,$$
$$M, x \Vdash \mathsf{t}\phi \iff M, x_l \Vdash \phi,$$
$$M, x \Vdash \mathsf{h}\phi \iff M, x_r \Vdash \phi.$$

Using the functions defined above we can formulate the following useful consequences of the frame conditions given above (see proposition 1.7 below)[2].

(T_0) f, $_l$ and $_r$ are partial functions and f is idempotent.
(T_1) $Ix \Rightarrow x = f(x) = x_l = x_r$,
       (in particular: if Ix then all three functions are defined on x).
(T_2) $Rxy \Rightarrow x_l = y_r$ & $x_r = y_l$.
(T_3) $Cxyz \Rightarrow x_l = y_l$ & $y_r = z_l$ & $z_r = x_r$.
(T_4) f, $_l$ and $_r$ are total.

Note that from (T_1) it follows that $x_l = (x_l)_l = (x_l)_r$ and $x_r = (x_r)_r = (x_r)_l$, since by definition it holds that Ix_l and Ix_r.

Proposition 1.7
(i) (C_1)-$(C_{12}) \models (T_0)$-(T_3);
(ii) (C_1)-$(C_{15}) \models (T_0)$-(T_4).

Proof. (T_0): functionality follows from $(C_7), (C_9), (C_{10}) - (C_{12})$. Idempotency of f follows from (C_1).
For (T_1):

$$Ix \stackrel{(C_2) \& (C_3)}{\Rightarrow} Cxxx \& Rxx \stackrel{\text{def}}{\iff} x = fx = x_l = x_r.$$

(T_3): this is just a short way of summarizing (C_7)-(C_9).
(T_2) : Suppose Rxy and suppose x_l is defined. Then Rxy & Cxx_lx. Then by (C_5) we get Cx_lxy, hence by (T_3) it follows that $(x_l)_r = y_r$. Finally by (T_1) we get $(x_l)_r = x_l = y_r$. Now suppose x_l is not defined: assume that y_r is defined and derive a contradiction in the same way as above using (C_4). Prove the other one in a similar way.
(T_4): That these three functions are total follows from (C_{13})-(C_{15}). □

2 Completeness

We want to find finite orthodox derivation systems for the four classes of pair-frames KP_H with $H \subseteq \{R, S\}$. We use the following strategy. We define four classes of *arrow frames* (cf. definition 1.1 in chapter 1) which are definable by a finite set of *Sahlqvist* formulas. Then, by proposition 6.2 in chapter 1, these classes are finitely axiomatizable by orthodox derivation systems. Subsequently we show that these four classes consist of zigzagmorphic images of the above mentioned four classes of pair-frames KP_H. (For a definition of zigzagmorphisms, see 5.4 in chapter 1.) Hence these classes of arrow frames are equivalent to the classes of pair-frames, and we found an axiomatization of the logic of the latter.

---

[2] If we use two partial functions f and g, then $fx = gy$ means that either both fx and gy are undefined, or they are both defined and $fx = gy$. With idempotency of a partial function f, we mean that if fx is defined, then $ffx = x$.

2.1 The Lemma and Applications

Of course conditions (C_1)–(C_{15}) are not chosen arbitrarily. They can be used to define classes of arrow frames which have the same arrow logical theory as several classes of pair frames.

Definition 2.1

$$\begin{aligned}
\mathsf{KA} &\stackrel{\text{def}}{=} \{\mathcal{F} = \langle W, C, R, I \rangle : \mathcal{F} \models (C_1) - (C_{12})\} \\
\mathsf{KA}_S &\stackrel{\text{def}}{=} \{\mathcal{F} \in \mathsf{KA} : \mathcal{F} \models (C_{13})\} \\
\mathsf{KA}_R &\stackrel{\text{def}}{=} \{\mathcal{F} \in \mathsf{KA} : \mathcal{F} \models (C_{14}), (C_{15})\} \\
\mathsf{KA}_{RS} &\stackrel{\text{def}}{=} \mathsf{KA}_R \cap \mathsf{KA}_S.
\end{aligned}$$

Theorem 2.2 *For all $H \subseteq \{R, S\}$, the classes of arrow frames KA_H are finitely axiomatizable by an orthodox derivation system.*

Proof. By proposition 1.6 and proposition 6.2 in chapter 1. □

The next lemma is the crucial one for our completeness result of the pair-frames. We will prove it in section 2.4. Recall the notion of a zigzag morphism from chapter1.

Lemma 2.3 *(i) Every arrow frame $\mathcal{F} \in \mathsf{KA}$ (i.e., satisfying conditions $(C_1) - (C_{12})$) is a zigzagmorphic image of a pair-frame $\mathcal{G} \in \mathsf{KP}$.*
(ii) Let $X \subseteq \{(C_{13}), (C_{14}), (C_{15})\}$. If in addition $\mathcal{F} \models X$, then also $\mathcal{G} \models X$.

The lemma leads to the following result.

Corollary 2.4 $(\forall H \subseteq \{R, S\}) \, \mathsf{KA}_H \equiv_{\text{AL}} \mathsf{KP}_H$.

Proof. By the validity of the conditions $(C_1) - (C_{15})$ on (reflexive or symmetric) pair-frames, $\mathsf{KP}_H \subseteq \mathsf{KA}_H$, for all $H \subseteq \{R, S\}$. On the other hand, lemma 2.3 implies that for $H \subseteq \{R, S\}$, if $\mathcal{F} \in \mathsf{KA}_H$, then $\mathcal{F}$ is a zigzagmorphic image of a frame $\mathcal{G} \in \mathsf{KP}_H$. It follows that the two classes are equivalent. □

Application to Theorem 1.5. This corollary together with theorem 2.2 imply the positive completeness part of theorem 1.5.

In the next two sections we also use this equivalence heavily. We prove the decidability, interpolation and definability results for the classes of arrow frames KA_H. These proofs are relatively straightforward. By the above stated equivalence the results hold for the arrow logics of equivalent classes of pair-frames as well.

The remaining part of this section is devoted to the proof of the last lemma. To get familiar with the step-by-step construction we will use, we start with a simple case. After that we introduce the notion of a mosaic (a small arrow frame), and show that every mosaic is a zigzagmorphic image of a pair-frame. We use these mosaics as building blocks to construct a pair-frame satisfying the conditions of the lemma. About this proof-strategy one can find more in remark 2.7; readers familiar with these constructions may skip the next subsection.

2.2 Warm-up: Booleans with Composition

To warm up we will prove a representation theorem for the reduct which only contains the composition operator and the Booleans. Rather surprisingly it is not needed to make any assumptions on the frames, in order to represent them as pair-frames. After that we prove the above lemma. Define:

$$\mathsf{K}^C \stackrel{def}{=} \{\mathcal{F} = \langle W, C \rangle : W \text{ is a set and } C \subseteq W \times W \times W\},$$
$$\mathsf{K}^C_{set} \stackrel{def}{=} \{\mathcal{F} = \langle V, C_V \rangle : V \subseteq U \times U \text{ for some set } U\}.$$

The next lemma is stated for finite frames only because we wanted to make the example as simple as possible. In the proof of lemma 2.3 we will represent frames of any cardinality. We say that a function from one arrow frame to another has the zigzag property if it satisfies the "back"-condition of the arrow-bisimulation (cf. definition 5.1 in chapter 1).

Lemma 2.5 *Every finite $\mathcal{F} \in \mathsf{K}^C$ is a zigzagmorphic image of a $\mathcal{G} \in \mathsf{K}^C_{set}$.*

Proof. Let $\mathcal{F} = \langle W, C \rangle \in \mathsf{K}^C$ be finite. Step by step we will construct a set of pairs V and a function $l : V \longrightarrow W$. It is convenient to think of l as a labelling function which labels each element of V with an element of W. In each step $n+1$ we will add pairs to V_n such that l is a homomorphism and for all pairs in V_n the function l has the zigzag property. The function won't have the zigzag property for the pairs added in step $n+1$, but we repair them in the next step, so after ω steps our zigzagmorphism is complete.

Construction. Let U be a set of size ω.

Step 0. In this step we ensure that l is surjective. Let $V_0 \subset U \times U$ such that (1) $|W| = |V_0|$ and (2) V_0 is irreflexive and completely disconnected, that is $(\forall s, r \in V_0)(\forall i, j \leq 1) : s_i = r_j \iff s = r \;\&\; i = j$. Let l_0 be any bijection between V_0 and W.

Step $n+1$. Let X_n be the (finite) set of pairs which were added in the previous step. Do for each $x \in X_n$ and $y, z \in W$ such that $Cl_n(x)yz$ the following:

Take an element $v \in U$ which is *not used before* and add $\langle x_0, v \rangle$ and $\langle v, x_1 \rangle$ to V_n and set $l_{n+1}(\langle x_0, v \rangle) = y$ and $l_{n+1}(\langle v, x_1 \rangle) = z$. (see the picture below). Define V_{n+1} as the result of all these additions to V_n, and l_{n+1} as the result of these additions to l_n.

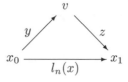

Step ω. Define $V \stackrel{\text{def}}{=} \bigcup_{n<\omega} V_n$ and $l \stackrel{\text{def}}{=} \bigcup_{n<\omega} l_n$.
End of construction

Claim 1 l is a zigzagmorphism from $\mathcal{G} = \langle V, C_V \rangle$ onto $\mathcal{F} = \langle W, C \rangle$.

Proof of Claim. It is surjective by step 0. The zigzag property is immediate by the construction. To show that l is a homomorphism we show by induction that it is a homomorphism after every step. This is clear for V_0, since $(\forall xyz \in V_0) : \neg C_V xyz$. Suppose it holds for step n and suppose $C_V xyz$, where at least one of $\{x, y, z\}$ were added in the $n+$ first step. Since we took a new point from U for every repair we made, it follows from our construction that $(\exists y', z' \in W) : Cl_n(x)y'z'$ & $l_{n+1}(y) = y'$ & $l_{n+1}(z) = z'$. But then $Cl_{n+1}(x)l_{n+1}(y)l_{n+1}(z)$. Hence l is a homomorphism after $n+1$ steps. Clearly l is a homomorphism after the limit step as well.   ◄

We finished the proof of the lemma.   □

Remark 2.6 The construction used above can be seen as a generalization of the *unraveling construction* from standard modal logic to binary modalities (cf. Sahlqvist 1975, de Rijke 1993).

The construction does not depend on the finiteness of $\mathcal{F}$. A similar construction can be used to represent abstract frames of any cardinality. The only difference will be that in general we have to make infinitely many repairs in the inductive step. In the next section we show what changes we have to make in the construction in order to represent infinite frames as well.

Remark 2.7 To get the idea how we will prove a similar representation theorem for arrow frames the following might be useful. Think of the frame which is to be represented as being built up by little frames $\langle \{x, y, z\}, Cxyz \rangle$ (where x, y, z need not be different). Later we will call these little frames *mosaics*. Clearly each such frame is a zigzagmorphic image of a triangle $\{\langle u, v \rangle, \langle u, w \rangle, \langle w, v \rangle\}$. If we had to make a repair in the above construction we added the representation of a mosaic to the partially constructed graph *using a fresh point*. The intuitive idea is that in the construction we play a kind of domino game where the tiles ("repaired mosaics") may want one or more tiles being laid next to them. If we play this game infinitely long we can fulfill the desires of each tile, thereby creating the zigzagmorphic pre-image of the frame which was to be represented. The function will be a homomorphism precisely because we always took fresh points.

2.3 Representation by Mosaics

In remark 2.7 we announced that a slight modification of the last proof would do for proving the same statement about the full language. The crucial point in that proof was that we used fresh points every time we made a repair. Clearly the situation is now more complex, but the complexity is in a sense very local. To see this the concept of a mosaic is useful.

In the next definition we expand the similarity type of our arrow frames with the two partial functions $_l$ and $_r$. When a mosaic belongs to the class KA these functions are definable (see proposition 1.7).

Definition 2.8 (Mosaics) Let $\mathcal{F} = \langle W, C, f, (\cdot)_l, (\cdot)_r, I \rangle$ be an expanded arrow frame where $f, (\cdot)_l$ and $(\cdot)_r$ are partial functions.

(i). We call $\mathcal{F}'$ an $\langle x, y, z \rangle$ mosaic iff $\{x, y, z\} \subseteq W$ (where x, y, z need not be different), $Cxyz$ holds and there is no proper subset of W which contains $\{x, y, z\}$ and which is closed under the functions $f, (\cdot)_l$ and $(\cdot)_r$. We call the elements x, y, z the *generators* of the $\langle x, y, z \rangle$ mosaic.

(ii). We say that a $\langle x, y, z \rangle$ mosaic $\mathcal{F}'$ is *repairable* if there exists a pair-frame $\langle \langle u, v \rangle, \langle u, w \rangle, \langle w, v \rangle \rangle$ mosaic $\mathcal{G}_{pair}(V)$ with base $\{u, v, w\}$ (u, v, w need not be different) and a surjective function $l : V \longrightarrow W$ such that

- $l\langle u, v \rangle = x$, $l\langle u, w \rangle = y$ and $l\langle w, v \rangle = z$,
- l is a homomorphism for C_V,
- $(\forall s \in V) : I_V s \iff Il(s)$ and
- l commutes with $f, (\cdot)_l$ and $(\cdot)_r$ in the following strong sense: $lf_V s = fls$ means that if one side is defined, then also the other, or both sides are undefined, and similarly for $_l$ and $_r$.

We call the tuple $\langle \mathcal{G}_{pair}(V), l, (u, v, w) \rangle$ a *repair* of $\mathcal{F}$.

It is easy to see that in the class of locally square pair-frames KP_{RS} every mosaic is one of the three *square* pair-frames in the figure below. The mosaic at the right is generated by one identity arrow, in the middle mosaic, one of the three generators is an identity arrow, and in the largest, none of the generators is an identity arrow.

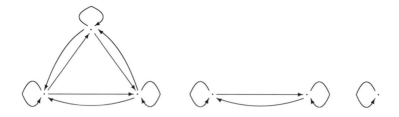

Mosaics will be used in the step-by-step construction to repair a situation in which there is a pair s in the partially constructed graph, and $Cl_n(s)yz$ & $\neg Iy$ & $\neg Iz$ holds in the frame to be represented. Then we will add a repair of the $\langle l_n(s), y, z \rangle$ mosaic to the partially constructed graph. The following facts will be useful later on.

Proposition 2.9 Let $\mathcal{F}$ be an $\langle x, y, z \rangle$ mosaic and $\langle \mathcal{G}_{pair}(V), l, (u, v, w) \rangle$ be a repair of $\mathcal{F}$. Then the following holds:
(i) For every point $x \in F$ and for every pair $s \in V$ such that $l(s) = x$ and for every function $f, (\cdot)_l, (\cdot)_r$ it holds that the function is defined on x if and only if it is defined on s.
(ii) If we view the partial functions as relations, then:

- l is a surjective homomorphism from $\mathcal{G}_{pair}(V)$ onto $\mathcal{F}$ and
- l has the zigzag property for $I, f, (\cdot)_l$ and $(\cdot)_r$

Proof. Immediate by the definitions. □

Repairing Mosaics. The next proposition contains the heart of the proof of lemma 2.3. We only need the axioms in the proof of this proposition. For glueing mosaics together no additional axioms are needed, just as in the "composition only" case. Recall that KA is the class of arrow frames which satisfy conditions $(C_1) - (C_{12})$ from proposition 1.6.

Proposition 2.10 (i) Every $\langle x, y, z \rangle$ mosaic $\mathcal{F} \in$ KA *is repairable.*
(ii) *Every* $\langle x, y, z \rangle$ *mosaic* $\mathcal{F} \in$ KA *is -up to isomorphism- uniquely repairable. Hence given a base* $\{u, v, w\}$ *we can speak about the repair of* $\mathcal{F}$.
(iii) *For all* $H \subseteq \{R, S\}$, *it holds that if the* $\langle x, y, z \rangle$ *mosaic* $\mathcal{F}$ *belongs to* KA_H, *then its repair belongs to* KP_H.

The proof consists of checking several cases. We continue here with the main line of the completeness argument and defer the proof of this proposition to the appendix.

Mosaics of an Arrow Frame. Mosaics are very small arrow frames which tend to live in bigger frames. For an arrow frame $\mathcal{F} \in$ KA we define the notion of an $\langle x, y, z \rangle$ mosaic of $\mathcal{F}$ as follows. An $\langle x, y, z \rangle$ mosaic $\mathcal{G}$ is an $\langle x, y, z \rangle$ mosaic of $\mathcal{F}$ if $\mathcal{G} \subseteq \mathcal{F}$ and the relations in $\mathcal{G}$ are the restrictions of the relations in $\mathcal{F}$ to G.

Proposition 2.11 (i) *If* $\mathcal{F} \in$ KA, *then any* $\langle x, y, z \rangle$ *mosaic of* $\mathcal{F}$ *is in* KA. *Also, for* $H \subseteq \{R, S\}$, *if* $\mathcal{F} \in \mathsf{KA}_H$, *then any* $\langle x, y, z \rangle$ *mosaic of* $\mathcal{F}$ *is in* KA_H.
(ii) *For* $\mathcal{F} \in$ KA *and* $\{x, y, z\} \subseteq F$ *such that* $Cxyz$ *there exists a unique* $\langle x, y, z \rangle$ *mosaic of* $\mathcal{F}$.

Proof. Left to the reader. □

2.4 Proof of the Main Lemma

Now we are ready for the proof of lemma 2.3, where we will use the mosaics to imitate the simple construction we used for the "composition only" reduct. We first give a sketch of the proof, after that we define the construction formally.

Proof-idea. Let $\mathcal{F} = \langle W, C, R, I \rangle \in \mathsf{KA}$ be arbitrary. Let M be the set of all $\langle x, y, z \rangle$ mosaics of $\mathcal{F}$. We want to copy the step by step procedure given in the proof of lemma 2.5. There our mosaics were much simpler and all of them could be repaired in the same way. Just a little reflection is needed to see that in the present proof, if we repair whole mosaics at once, we only need to repair situations where we have $Cl(x)yz$ & $\neg Iy$ & $\neg Iz$. Since we know that we can repair each mosaic, we know that we can make any repair necessary. So again by the construction we can ensure that the function l has the zigzag property. The more difficult question is whether the function l is also a homomorphism: this will follow from the fact that we only need to repair mosaics.

Proof. (Lemma 2.3) (i). Let $\mathcal{F} = \langle W, C, R, I \rangle \in \mathsf{KA}$ be arbitrary. Without loss of generality, we may assume that $\mathcal{F}$ is point-generated[3], say by point d. There are two cases: either d is part of a mosaic, or it is not. In the latter case, d is not an identity arrow, and the functions $(\cdot)_r$ and $(\cdot)_l$ are not defined on d or fd (use the conditions from proposition 1.7). So the frame consists only of d or of $\{d, fd\}$, and there are no C-relations. These frames are zigzagmorphic images of the pair-frames $\mathcal{F}_{pair}(\{\langle u, v \rangle\})$ and $\mathcal{F}_{pair}(\{\langle u, v \rangle, \langle v, u \rangle\})$ ($u \neq v$), respectively.

So, let d be part of a mosaic of $\mathcal{F}$. Let M be the set of all $\langle x, y, z \rangle$ mosaics of $\mathcal{F}$. By fact 2.11 every mosaic in M belongs to KA so, by Prop.2.10, all of them are repairable. We now define the construction one has to use, interspersing it with the argument why it works.

Construction

Let U be an infinite set such that $|U| \geq |W|$. We will use the elements of U to construct the pairs of the representation. The condition ensures that U is large enough for this purpose.

Step 0. Let $\mathcal{M}$ be an $\mathcal{F}$-mosaic containing d. Let $\langle \mathcal{H}_{pair}(V^*), l^*, (u, v, w) \rangle$, with $u, v, w \in U$, be the repair of $\mathcal{M}$. We set $V_0 = V^*$ and $l_0 = l^*$.

Claim 1 l_0 is a homomorphism and satisfies the zigzag property for I, R and the functions $(\cdot)_l$ and $(\cdot)_r$.

Proof of Claim. Immediate by proposition 2.10. ◀

Step n+1. Let X_n be the set of pairs which were added in the previous step. For each $s \in X_n$ a function[4] create

$$g_s : \{yz : Cl_n(s)yz \ \& \ \neg Iy \ \& \ \neg Iz\} \longrightarrow (U \setminus \mathsf{Base}(V_n))$$

such that

- all g_s are injective,

---

[3] Because every frame is a zigzagmorphic image of the disjoint union of all its point generated subframes.

[4] If $f : X \longrightarrow Y$ then we use $f^*(X)$ or — if the domain is clear by the context — f^* to denote the range of f.

- the ranges of the g_s are pairwise disjoint, and
- $|U \setminus \bigcup_{s \in X_n}(g_s^*)| = |U|$.

Such a set of functions clearly exists. They guarantee that we use a brand new element from U for every mosaic we add, and that the set $U \setminus \text{Base}(V_{n+1})$ remains large enough to continue the construction using new elements.

For every $s \in X_n$ and for every $y, z \in W$ such that $Cl_n(s)yz, \neg Iy, \neg Iz$, we repair the $\langle l_n(s), y, z \rangle$ mosaic of $\mathcal{F}$ by the $\langle\langle s_0, s_1\rangle, \langle s_0, g_s(yz)\rangle, \langle g_s(yz), s_1\rangle\rangle$ mosaic, and add that representation to V_n, in this way creating V_{n+1}. Define l_{n+1} as the extension of l_n where the new pairs are mapped as given by the repairs of the mosaics. This can be described formally as follows: define REP_n as the set

$$\{\langle \mathcal{F}_{pair}(V), l, (s_0, s_1, g_s(yz))\rangle : (\exists s \in X_n)(\exists yz \in W)(Cl_n(s)yz \;\&\; \neg Iy \;\&\; \neg Iz)$$

and $\langle \mathcal{F}_{pair}(V), l, (s_0, s_1, g_s(yz))\rangle$ is the repair of the $\langle l_n(s), y, z\rangle$ mosaic$\}$. Then define V_{n+1} and l_{n+1} as follows:

$$V_{n+1} = V_n \cup \bigcup \{V : \langle \mathcal{F}_{pair}(V), l, (s_0, s_1, g_s(yz))\rangle \in REP_n\}$$
$$l_{n+1} = l_n \cup \bigcup \{l : \langle \mathcal{F}_{pair}(V), l, (s_0, s_1, g_s(yz))\rangle \in REP_n\}.$$

For every repair $\langle \mathcal{F}_{pair}(V), l, (s_0, s_1, g_s(yz))\rangle \in REP_n$, and for every $r \in V \cap V_n$, it holds that $l(r) = l_n(r)$: thus l_{n+1} is well-defined.

Claim 2 Let $n < \omega$ be arbitrary. Then (i)–(iii) below hold.

(i) l_n is a surjective homomorphism;

(ii) l_n satisfies the zigzag property for I and R;

(iii) Let $n > 0$. l_n satisfies the zigzag property for C for all elements of V_{n-1}, i.e., $(\forall s \in V_{n-1})(\forall yz \in W) : (Cl_{n-1}(s)yz \Rightarrow (\exists y', z' \in V_n) : C_V sy'z' \;\&\; l_n(y') = y \;\&\; l_n(z') = z)$.

Proof of Claim. The proof is by induction on n. By the previous claim, (i), (ii) and (iii) restricted to the substitution functions hold for l_0 and V_0. Assume they hold for l_n and V_n.

(i). Suppose a pair s was added in the n+1-th step. Since s is part of a mosaic, homomorphism is guaranteed for I_V and R_V. For C_V we are in precisely the same simple situation as before. Since we always used fresh points, s will only stand in C_V relations with elements of the mosaic it is a part of, and they are guaranteed by the mosaic proposition.

(ii). That l_{n+1} has the zigzag property for I and R is, given that l_n has it, immediate because we represented whole mosaics.

(iii). Suppose s was added in step n and $Cl_n(s)yz$. If $\neg Iy$ and $\neg Iz$, we added the needed pre-images in step n+1. In the three other cases, the needed pre-images are either s itself or $\langle s_0, s_0\rangle$ or $\langle s_1, s_1\rangle$, and since s is part of a mosaic, these were — by induction hypothesis — already in V_n. ◀

Step ω. Define
$$V \stackrel{\text{def}}{=} \bigcup_{n<\omega} V_n \text{ and } l \stackrel{\text{def}}{=} \bigcup_{n<\omega} l_n.$$
End of construction

Claim 3 l is a zigzagmorphism from the pair-frame $\mathcal{F}_{pair}(V)$ onto the arrow frame $\mathcal{F}$.

Proof of Claim. Since l_n is a homomorphism for every n, so is l. It is zigzag because each added point is repaired in the next step. The function is surjective, because the frame $\mathcal{F}$ is generated by d, and we created a pre-image of d in step 0. ◂

With the last claim we finished the proof of part (i) of the lemma.

(ii). If $\mathcal{F}$ satisfies one or more of the conditions $(C_{13}), (C_{14}), (C_{15})$, that means that the corresponding function is always defined, hence (propositions 2.10 and 2.11) in every repair of a mosaic the corresponding function is always defined. □

With the proof of the crucial lemma we finish this section on completeness.

3 Decidability

We will prove decidability using a technique which is well-known from modal logic: *filtration*. Since we know that for $H \subseteq \{R, S\}$, the class of arrow frames KA_H is equivalent to the class of pair-frames KP_H, for which we want to prove decidability, it is sufficient to show that the arrow logics $\mathsf{AL}(\mathsf{KA}_H)$ are decidable.

Preliminaries. The notions in this paragraph are well-known from modal logic. For proofs see e.g., Hughes and Creswell 1984.

Definition 3.1 (Filtration) Let $\mathsf{M} = \langle \mathcal{F}, \mathsf{v} \rangle$ be a model, $\mathcal{F} = \langle W, C, R, I \rangle$, and Σ a set of arrow-logical formulas which is closed under taking subformulas and under the Boolean connectives. Define an equivalence relation $\equiv_\Sigma \subseteq W \times W$ as follows:
$$(\forall w, v \in W) : w \equiv_\Sigma v \stackrel{\text{def}}{\iff} (\forall \phi \in \Sigma)\, [\mathsf{M}, w \Vdash \phi \iff \mathsf{M}, v \Vdash \phi].$$
Let $\overline{w}$ denote the equivalence class $w/\equiv_\Sigma$, and equate equivalence classes $\overline{w}$ with the sets of formulas $\{\phi \in \Sigma : \mathsf{M}, w \Vdash \phi\}$.

We call a model $\mathsf{M}^* = \langle \mathcal{F}^*, \mathsf{v}^* \rangle$ a *filtration* of M *through* Σ if conditions (i)–(iv) below hold.

(i) $W^* \stackrel{\text{def}}{=} \{\overline{w} : w \in W\}$;
(ii) $\mathcal{F}^* = \langle W^*, C^*, R^*, I^* \rangle$ is an arrow frame;
(iii) $\mathsf{v}^*(p) \stackrel{\text{def}}{=} \{\overline{w} \in W^* : p \in \overline{w}\}$, for all *variables* $p \in \Sigma$;
(iv) **min** and **max**, given below, hold.

min $Cxyz \Rightarrow C^*\overline{x}\,\overline{y}\,\overline{z}$
$Rxy \Rightarrow R^*\overline{x}\,\overline{y}$
$Ix \iff I^*\overline{x}$
max $(\forall \phi \circ \psi \in \Sigma) : ((C^*\overline{x}\,\overline{y}\,\overline{z}\ \&\ \phi \in \overline{y}\ \&\ \psi \in \overline{z}) \Rightarrow \phi \circ \psi \in \overline{x})$
$(\forall \otimes \phi \in \Sigma) : ((R^*\overline{x}\,\overline{y}\ \&\ \phi \in \overline{y}) \Rightarrow \otimes \phi \in \overline{x}).$

We call a filtration *minimal* if all relations are defined minimally, as in

$$C^*\overline{x}\,\overline{y}\,\overline{z} \stackrel{\text{def}}{\iff} (\exists x' \equiv_\Sigma x, y' \equiv_\Sigma y, z' \equiv_\Sigma z) : Cx'y'z'.$$

As their names indicate, **min** provides a lower-bound, and **max** an upper-bound for the relations. They are designed to make the following lemma true. The proof is by a straightforward induction on the complexity of ϕ.

Lemma 3.2 (Truth lemma) *Let* $M^* = \langle \mathcal{F}^*, v^* \rangle$ *be a filtration of* $M = \langle \mathcal{F}, v \rangle$ *through* Σ. *Then* (**T**) *below holds.*

$$(\forall \phi \in \Sigma) : M \models \phi \iff M^* \models \phi. \qquad (\mathbf{T})$$

Definition 3.3 (Allows Filtrations) Let K be a class of arrow frames. We say that K *allows filtrations* if for any finite set of formulas X and any model $M = \langle \mathcal{F}, v \rangle$ where $\mathcal{F} \in K$, there exists a set $\Sigma \supseteq X$ and a filtration $\langle \mathcal{F}^*, v^* \rangle$ of M through Σ such that $\mathcal{F}^*$ *is finite and it belongs to* K.

We will call the set $\Sigma \supseteq X$ as it appears in the above definition the *closure set* of X. It will, by definition, be closed under taking sub-formulas and under the Boolean connectives.

Lemma 3.4 (Filtration Lemma) *Let* K *be a class of arrow frames. If* K *is basic elementary (i.e., definable by a single FO sentence) and allows filtrations, then the arrow logical theory* $\{\phi : K \models \phi\}$ *of* K *is decidable.*

The Lemma and Applications

Lemma 3.5 *The frame classes* KA, KA_R, KA_S *and* KA_{RS} *all allow filtrations.*

The positive decidability part of theorem 1.5 now follows from this lemma, the filtration lemma and the fact that for $H \subseteq \{R, S\}$, $KA_H \equiv_{AL} KP_H$.

Proof. (Lemma 3.5) We will be able to prove the lemma if we close the closure set under taking t, h and $\otimes$. The next claim states that we can do that without losing finiteness.

Claim 1 *Let* X *be a finite set of formulas closed under sub-formulas and let* $CL(X)$ *be the smallest set containing* $X \cup \{\iota\delta\}$ *which is closed under the Boolean connectives,* t, h *and* $\otimes$. *Then for any arrow logic stronger than* AL(KA) *the set* $CL(X)$ *is logically finite.*

Proof of Claim. Because the classes KA and KP are logically equivalent we can prove the claim by reasoning in KP. The same claim for the class

KP follows in a straightforward way (see e.g. Marx 1995 lemma 3.1.11) if we can show that every $I, (\cdot)_l, ((\cdot)_r, f$ point-generated subframe of KP is finite, and that there are — up to isomorphism — only finitely many of them. But that is immediate: let $\mathcal{F}_{pair}(V)$ be $I, (\cdot)_l, ((\cdot)_r, f$-generated from $\{\langle 0, 1\rangle\}$. Then V is a subset of $\{\langle 0, 1\rangle, \langle 1, 0\rangle, \langle 0, 0\rangle, \langle 1, 1\rangle\}$. ◀

We first prove the lemma for the class KA. Let $\mathcal{F} = \langle W, C, R, I\rangle \in$ KA and M = $\langle \mathcal{F}, \mathsf{v}\rangle$ be a model and let X be a finite set of formulas. Let the closure set $CL(X)$ be the smallest set containing $X \cup \{\iota\delta\}$ which is closed under sub-formulas, h, t, ⊗ and the Booleans. Let $\langle \mathcal{F}^*, \mathsf{v}^*\rangle$, where $\mathcal{F}^* = \langle W^*, C^*, R^*, I^*\rangle$, be the *minimal* filtration of M through $CL(X)$. The next claim implies that KA allows filtrations.

Claim 2 (i) W^* is finite and (ii) $\mathcal{F}^* \in$ KA.

Proof of Claim. (i). Claim 1 implies that the closure set is finite modulo KA-equivalence. So there are only finitely many equivalence classes, and W^* is finite.

(ii). To show that $\mathcal{F}^* \in$ KA, we have to prove that $\mathcal{F}^* \models (C_1) - (C_{12})$ since these are the conditions defining that class. In order to do that efficiently we need another claim. From now on we suppress the subscript $_{CL(X)}$ in $\equiv_{CL(X)}$.

Claim 3 The following statements hold for the above given $\mathcal{F}$ and $\equiv$.

(1) $\quad\quad\quad\quad Ix \ \& \ x \equiv x' \ \Rightarrow \ Ix'$

(2) $\quad Rxy \ \& \ Rx'z \ \& \ x \equiv x' \ \Rightarrow \ y \equiv z$

(3) $\quad\quad\quad Rxy \ \& \ x \equiv x' \ \Rightarrow \ (\exists z) : Rx'z$

(4) $\quad Cxyx \ \& \ Iy \ \& \ x \equiv x' \ \Rightarrow \ (\exists z) : Cx'zx' \ \& \ Iz \ \& \ y \equiv z$

(5) $\quad Cxxy \ \& \ Iy \ \& \ x \equiv x' \ \Rightarrow \ (\exists z) : Cx'x'z \ \& \ Iz \ \& \ y \equiv z$.

Proof of Claim. (1). Immediate since $\iota\delta \in CL(X)$.

(2). Suppose the antecedent. We compute:
$\phi \in \overline{y} \iff$ (since $CL(X)$ is closed under ⊗) $\otimes\phi \in \overline{x} \iff \otimes\phi \in \overline{x'} \iff \phi \in \overline{z}$.

(3). Suppose the antecedent. Then $\otimes\top \in \overline{x}$, hence $\otimes\top \in \overline{x'}$, so $\exists z : Rx'z$.

(4). Suppose the antecedent. Then $\mathsf{t}\top \stackrel{def}{=} \iota\delta \circ \top \in \overline{x}$, hence also in $\overline{x'}$, so there exists a z such that $Cx'zx' \ \& \ Iz$. To show that $y \equiv z$, suppose that $\phi \in \overline{y}$. We compute:
$\phi \in \overline{y} \iff$ (since $CL(X)$ is closed under taking t) $\mathsf{t}\phi \in \overline{x} \iff \mathsf{t}\phi \in \overline{x'} \iff \phi \in \overline{z}$.

(5) is similar to (4); now use closure under h. ◀

Now we are ready to prove that $\mathcal{F}^*$ validates the conditions (C_1)-(C_{12}). Conditions (C_1)-(C_3) are immediate because of the minimal filtration. For (C_4) suppose $C^*\overline{x}, \overline{y}, \overline{z} \ \& \ R^*\overline{y}, \overline{v}$, then (by the *minimal* definition of C^*

and R^*) there are ($x' \equiv x, z' \equiv z, v' \equiv v$ & $y' \equiv y'' \equiv y$) such that $Cx'y'z'$ & $Ry''v'$. By (2) and (3) we find $Ry'v''$ & $v'' \equiv v'$ for some v''. Then, by (C_4), $Cz'v''x'$, so, by definition of C^*, $C^*\overline{z}, \overline{v}, \overline{x}$. Condition ($C_5$) can be shown similarly. For (C_6) use (1). Conditions (C_7) − (C_9) are all similar; we show the ⇒ side of (C_8) as an example. Suppose the antecedent, then there are ($x' \equiv x, y''' \equiv y'' \equiv y' \equiv y, z' \equiv z, v'' \equiv v' \equiv v$) such that $Cx'y'z'$ & $Cy''y'''v'$ & Iv''. By (1) also Iv', so by (C_{12}) we have $y''' = y''$ and by (5) and (1), ($\exists v''' \equiv v'$) : $Cy'y'v''$ & Iv'''. By (C_8) we obtain $Cz'v'''z'$ and, by definition of C^*, $C^*\overline{z}, \overline{v}, \overline{z}$. Condition ($C_{10}$) is immediate by (2) and conditions (C_{11}) and (C_{12}) by (1). ◂

We have finished the proof for KA. The three other classes KA_R, KA_S and KA_{RS} are obtained from KA by adding one or more of the conditions (C_{13})−(C_{15}). If we filtrate a frame satisfying one or more of these conditions, then the filtration will also satisfy them, because we took a minimal filtration. So these three classes also allow filtrations. □

Remark 3.6 The filtration proof is a bit complicated since we were dealing with partial functions. If the functions $f, (\cdot)_l, (\cdot)_r$ are total, as in the case of KA_{RS}, we can streamline the proof considerably by defining these functions in the filtration and showing that, for $i \in \{l, r\}$, statements (6) and (7) below hold.

(6) $\qquad (\forall x, y \in W) : x \equiv y \Rightarrow x_i \equiv y_i$

(7) $\qquad (\forall x, y \in W) : (\overline{x})_{i*} = \overline{x_i}$ and $f^*\overline{x} = \overline{fx}$.

Having shown decidability, we now turn to interpolation and definability.

4 Interpolation and Definability

We prove the interpolation and definability results using a general lemma. We repeat the lemma here and give a bit of intuition about the main construction in the lemma.

Preliminaries. In the next definition we use the notions of substructure and (sub)direct product in their first order model-theoretic sense (cf. Chang and Keisler 1973).

Definition 4.1 Let $\mathcal{F}$ and $\mathcal{G}$ be two frames of the same type. Any substructure of the direct product $\mathcal{F} \times \mathcal{G}$ where projections are surjective *zigzag-morphisms*, is called a *zigzag product* of $\mathcal{F}$ and $\mathcal{G}$.

A useful way of thinking about a zigzag product is as a *subdirect product* where the projections also have the zigzag property. For $\mathcal{F}, \mathcal{G}$ frames and $Z \subseteq F \times G$ we denote the substructure of the binary product $\mathcal{F} \times \mathcal{G}$ with universe Z by $(\mathcal{F} \times \mathcal{G}) \restriction_Z$.

The notion of a zigzag product is closely connected to the notion of *bisimulation* (cf. chapter 1, definition 5.1). A bisimulation $Z \subseteq F \times G$

between two frames $\mathcal{F}$ and $\mathcal{G}$ is called a *zigzag connection* if the domain of Z equals F and its range equals G. The next proposition states the connection between bisimulations and zigzag products. Its easy proof is left to the reader.

Proposition 4.2 *Let $\mathcal{F}$ and $\mathcal{G}$ be arrow frames and $Z \subseteq F \times G$.*

Z is a zigzag connection if and only if $(\mathcal{F} \times \mathcal{G}) \upharpoonright_Z$ is a zigzag product.

We are ready to formulate the lemma from which our results follow almost immediately.

Lemma 4.3 *Let K be a class of frames which is axiomatizable by a set of Sahlqvist formulas. If K is closed under zigzag products then the logic of K has the properties mentioned in (i)–(iii) below.*

(i) *strong Craig interpolation property;*
(ii) *weak Craig interpolation property;*
(iii) *Beth's definability property.*

Proof. Cf. Marx 1995. □

The Lemma and Applications

Lemma 4.4 *Let K_Δ be any class of arrow frames which is defined by a set $\Delta \subseteq \{(C_1) - (C_{15})\}$. Then K_Δ is closed under zigzag products.*

Proof. Conditions $(C_1) - (C_{12})$ are universal Horn sentences, hence they are preserved under subdirect products, so a fortiori under zigzag products. The three existential sentences $(C_{13}) - (C_{15})$ are not preserved under subdirect products, but it is easy to see that they are preserved under zigzag products. □

Application to Theorem 1.5. The last two lemmas imply that the arrow logics of the classes of arrow frames KA_H for $H \subseteq \{R, S\}$ have the two interpolation and the definability properties. But then — by the equivalence — the same holds for the classes of pair-frames KP_H.

5 Expanding the Language of Arrow Logic

On the preceding pages we have studied just a few "natural" connectives on arrows. In this section we will introduce a few more. All the connectives we will study in this section are term-definable in the language $\{\wedge, \neg, \circ, \otimes, \iota\delta\}$ on *square pair-frames*, but they aren't anymore on arbitrary pair-frames. All these connectives are introduced in chapter 1.

What we will show here is that we can add these connectives to arrow logics of non-transitive pair-frames *without losing the positive properties* we have studied above. The last statement holds for all the properties we have studied — finite axiomatizability, decidability, interpolation and definability — except for the combination difference operator and interpolation/definability.

In chapter 5, the language is expanded with the Kleene star, a connective which is not even definable on the square pair-frames. There again axiomatizability and decidability are preserved.

Define the following connectives on pair-frames: the universal modality $\Diamond$, the difference operator D, the $\langle ij \rangle$ modalities from Vakarelov's work on arrow logic (cf., chapter 7), and the residuals of composition $\backslash$ and $/$ ("slashes"). Let $M = \langle \mathcal{F}, v \rangle$ be a model where $\mathcal{F}$ is a pair-frame.

$$M, x \Vdash \Diamond \phi \stackrel{\text{def}}{\iff} (\exists y) : M, y \Vdash \phi$$
$$M, x \Vdash D\phi \stackrel{\text{def}}{\iff} (\exists y) : x \neq y \ \& \ M, y \Vdash \phi$$
$$M, x \Vdash \langle ij \rangle \phi \stackrel{\text{def}}{\iff} (\exists y) : x_i = y_j \ \& \ M, y \Vdash \phi \ (i,j \in \{0,1\})$$
$$M, x \Vdash \phi \backslash \psi \stackrel{\text{def}}{\iff} (\forall yz) : (x_0 = y_1, y_0 = z_0, z_1 = x_1 \ \& \ M, y \Vdash \phi)$$
$$\Rightarrow M, z \Vdash \psi$$
$$M, x \Vdash \phi/\psi \stackrel{\text{def}}{\iff} (\forall yz) : (x_0 = y_0, x_1 = z_0, z_1 = y_1 \ \& \ M, z \Vdash \psi)$$
$$\Rightarrow M, y \Vdash \phi.$$

Here is the meaning of the new connectives in pictures:

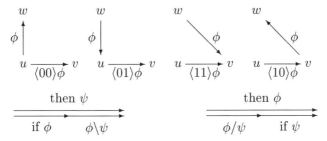

Proposition 5.1 (i) *In the arrow logic of squares* $\mathsf{AL}(\mathsf{KP}_{SQ})$ *the above given connectives are term-definable in the way presented below. None are definable in the arrow logic of relativized squares* $\mathsf{AL}(\mathsf{KP})$. *(Since composition is associative in* $\mathsf{AL}(\mathsf{KP}_{SQ})$ *we leave out the brackets.)*

$$\Diamond \phi \leftrightarrow \top \circ \phi \circ \top \qquad\qquad \langle 00 \rangle \phi \leftrightarrow \phi \circ \top$$
$$D\phi \leftrightarrow (\top \circ \phi \circ \neg \iota \delta) \vee (\neg \iota \delta \circ \phi \circ \top) \qquad \langle 01 \rangle \phi \leftrightarrow \otimes \phi \circ \top$$
$$\phi \backslash \psi \leftrightarrow \neg (\otimes \phi \circ \neg \psi) \qquad\qquad \langle 11 \rangle \phi \leftrightarrow \top \circ \phi$$
$$\phi / \psi \leftrightarrow \neg (\neg \phi \circ \otimes \psi) \qquad\qquad \langle 10 \rangle \phi \leftrightarrow \top \circ \otimes \phi$$

Proof. Left to the reader. □

The slashes are definable once we have a symmetric domain, the $\langle ij \rangle$ modalities once we have symmetric and reflexive domains (for definitions see Marx 1995). It is not hard to show that the universal modality and the difference operator are not definable on the classes KP_H when $H \subseteq \{R, S\}$.

The following results are known to us about expansions with these operators. We state the results without proof. The proofs can be found in the given references.

1. The expansion of AL(KP$_H$) for $H \subseteq \{R, S\}$ with the four $\langle ij \rangle$ modalities is finitely axiomatizable, decidable and has both Craig and Beth properties (cf. Marx 1995).
2. The same holds for expansions with the universal modality (cf. Marx 1995).
3. The expansion of AL(KP$_{RS}$) with the difference operator is finitely axiomatizable (cf. Marx et al. 1995), and decidable (cf. Mikulás 1995). Both interpolation properties and the definability property fail (cf. Marx 1995).
4. In chapter 6 an arrow logic is studied where instead of converse the slashes are primitive connectives. It is easy to see that converse is definable on reflexive pair-frames (e.g. $\otimes \phi \leftrightarrow \neg(\iota\delta\backslash\neg\phi)$). In Marx 1995 a finite axiomatization is provided for the class of all pair-frames with respect to this language. From the axiomatization it follows that the logic has both interpolation properties and the definability property. We conjecture that this logic is decidable as well.

6 Main Literature and Further Reading

In this section we suggest some literature to the related subject of *Generalized Relation Algebra* (in particular Relativized Relation Algebra) and provide some historic notes on the theorems discussed in this paper. For literature on arrow logic we refer to chapter 1.

As explained in chapter 1, the arrow logic of pair-frames is very closely connected to the field of (Relativized) Representable Relation Algebras in particular, and to Algebras of Relations in general. The study of Algebras of Relations goes back to de Morgan, Schröder and Peirce in the 19th century. Works on Algebras of Relations which are also related to the subject matter of the present paper include Andréka 1991, Andréka et al. 1996, Andréka and Németi 1996, Givant 1991, Jipsen 1992, Jipsen et al. 1996, Jónsson 1991, Jónsson and Tshinakis 1991, Maddux 1991b, Marx 1995, 1993, Mikulás 1995, Németi 1991, Pratt 1992, Tarski and Givant 1987.

A couple of results which are captured in our main theorem 1.5 were previously proved in an algebraic form. In the present paper we discuss these results in their equivalent logical formulation. Monk 1964 showed that the class KP$_{RST}$ of pair-frames, where the universe is an equivalence relation, is not finitely axiomatizable by any orthodox inference system. Andréka 1991 strengthened this result, e.g., by showing that adding unorthodox rules does not help as far as the new rules remain strongly sound.

She also proved that adding new logical connectives (like e.g. generalized quantifiers) does not help (assuming these connectives satisfy certain conditions). Tarski-Givant 1987 show that the arrow logical theory of this class is highly undecidable. Further results in this line are in Andréka et al. 1996. For the positive side: Maddux 1982 showed that the class KP_{RS} of locally square pair-frames is finitely axiomatizable and Németi 1987 showed that it is decidable. Kramer 1991 showed that the class of all pair-frames is finitely axiomatizable. This result is provided with a simple proof and is generalized in Marx 1995.

Venema 1991 contains a finite *unorthodox* derivation system for the class KP_{RST} using the, so called, irreflexivity rule. Similar unorthodox completeness methods are in Simon 1991 and in Mikulás 1995 (see the result based on Andréka et al. 1996).

Results concerning Craig interpolation and Beth-type definability properties for Arrow Logics and for algebras of relations are in Madarász 1996, Marx 1995, Sain 1993.

Acknowledgements. Hajnal Andréka, András Simon and Yde Venema helped substantially with ideas, discussions and comments. Special thanks are due to Johan van Benthem.

Appendix

Now follows the proof of proposition 2.10.

Proof. (i). Let $\mathcal{F} \in \mathsf{KA}$ be an $\langle x, y, z \rangle$ mosaic. In the proof we will use the KA theorems (T_0)–(T_3) from proposition 1.7. Recall that these theorems should be read as if the functions are partial (e.g., $x_l = y_l$ means that either they are both undefined or that they are both defined and $x_l = y_l$).

The first insight we get by looking at the possible I "valuation" of the generators x, y and z. The third column in the table below expresses the fact that if two of the three generators are identity arrows then so is the third one. So we only have the cases with three, one or zero identity generators. This follows from conditions (C_{11}) and (C_{12}). The results in the last column follow from $(T_0) - (T_3)$ and (C_6) and will become obvious when we look at the cases separately below.

	x	y	z	result	size of domain F
1	I	I	I	$x = y = z$	$|F| = 1$
2	I	I		impossible	
3	I		I	impossible	
4	I			$x \neq y, x \neq z$	$2 \leq |F| \leq 4$
5		I	I	impossible	
6		I		$x = z \neq y$	$2 \leq |F| \leq 4$
7			I	$x = y \neq z$	$2 \leq |F| \leq 4$
8					$1 \leq |F| \leq 9$

We will look at the remaining cases one by one.

Case 1. In case 1, by (T_1), the mosaic consists of just one element, and clearly that is isomorphic to the pair-frame defined by $\{\langle u, u\rangle\}$.

Case 4. In case 4 we find, by (C_6) and (C_1), that $f(y) = z$ & $f(z) = y$. By (T_3) and (T_1) we have $x = x_l = x_r = fx = y_l = z_r$. If y_r is not defined then the domain of the mosaic equals $\{x, y, z\}$ and we repair it by the pair-frame in figure 8. An x attached to an arrow $\langle u, v\rangle$ means that x is represented by $\langle u, v\rangle$ (in other words $l\langle u, v\rangle = x$).

It follows from the argument given above that the $\langle\langle u, u\rangle, \langle u, w\rangle, \langle w, u\rangle\rangle$ mosaic and the function l as given in figure 8 form a repair of this mosaic. In the sequel we ask the reader to check this using the provided pictures.

(8)

We continue with case 4. If y_r is defined, then (C_4) (using that $Cyyy_r \wedge Ryz$) implies that $Cy_r zy$. By (T_1) again $y_r = (y_r)_r = (y_r)_l = f(y_r)$. Again by (T_3), $z_l = y_r$ so we add $\langle w, w\rangle$ to the mosaic and set $l\langle w, w\rangle = y_r$ (see figure 9).

This picture really covers two cases: the one where $y \neq z$ and the one where $y = z$ (which implies that $x = y_r$). For the argument given above this distinction does not matter, so we covered both cases. In the sequel we will not mention these cases because we will always give arguments which cover the cases when some of the points in the mosaic happen to be equal.

In a case 4 mosaic the functions f, l and $_r$ cannot generate more points so we are done.

(9)

Case 6 and 7. If the functions are all defined, case 6 and 7 are very similar to case 4. We treat case 6 only. If x_r and $f(x)$ are not defined, we represent x by $\langle u, v\rangle$ ($u \neq v$) and y by $\langle u, u\rangle$ and we are done (see figure 10).

(10)

If x_r is defined we add $\langle v, v\rangle$ as well and set $l\langle v, v\rangle = x_r$. If $f(x)$ is defined then (C_5) implies $Cyxf(x)$ and (if x_r is defined as well) $Cx_r f(x)x$, arriving in the two situations of case 4. We treated all possible case 6 mosaics. Case 7 mosaics are handled similarly.

Case 8. Case 8 finally will be repaired with a mosaic consisting of at least 3 non reflexive pairs. There are many possibilities here, depending whether or not the functions $f, (\cdot)_l, (\cdot)_r$ are defined. The heart of the representation will be a triangle $\{\langle u,v\rangle, \langle u,w\rangle, \langle w,v\rangle\}$, where $l\langle u,v\rangle = x$, $l\langle u,w\rangle = y$ and $l\langle w,v\rangle = z$ and u, v, w are all different. Depending on the presence of other arrows in the mosaic we have to add more pairs. First suppose x_l is defined, then by (T_3), $x_l = y_l$ and we can represent it by $\langle u,u\rangle$. Similarly for $y_r = z_l$ and $z_r = x_r$. If $f(y)$ is defined we need to represent that by $\langle w,u\rangle$ and we get $C_V \langle w,v\rangle, \langle w,u\rangle, \langle u,v\rangle$, but by (C_4) also $Czf(y)x$ (see figure 11).

(11)
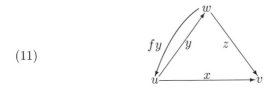

Use (C_5) in the similar situation where $f(z)$ is defined and (C_4) and (C_5) when two or more of $f(x), f(y)$ and $f(z)$ are defined. To see that the function l behaves correct on parts like $\{\langle u,u\rangle, \langle u,w\rangle, \langle w,u\rangle, \langle w,w\rangle\}$ reason as in case 4. If all functions were defined the representation would look like figure 12.

(12)
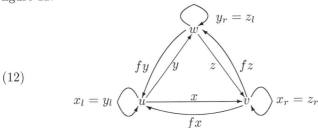

We covered all case 8 mosaics. So we covered all possible mosaics, by which we finish the proof of part (i) of the proposition. Parts (ii) and (iii) are immediate by the proof of part (i) and fact 2.9(i). □

References

Andréka, H. 1991. Complexity of the equations valid in algebras of relations. Thesis for D.Sc. (a post-habilitation degree) with Math. Inst. Hungar. Ac. Sci. Budapest. A slightly updated version will appear in *Annals of Pure and Applied Logic*.

Andréka, H., J. D. Monk, and I. Németi (ed.). 1991. *Algebraic Logic (Proc. Conf. Budapest 1988)*. Amsterdam. Colloq. Math. Soc. J. Bolyai, North-Holland, Amsterdam.

Andréka, H., and I. Németi. 1994. Craig Interpolation does not imply amalgamation after all. Preprint. Math. Inst. Hungar. Acad. Sci., Budapest.

Andréka, H., I. Németi, and I. Sain. 1994. On Interpolation, Amalgamation, Universal Algebra and Boolean Algebras with Operators. manuscript. Math. Inst. Hungar. Acad. Sci., Budapest, september.

Andréka, H., S. Givant, Sz. Mikulás, I. Németi and A. Simon 1996. Rectangular density implies representability. Manuscript.

Andréka, H., S. Givant and I. Németi 1996. Decision problems for equational theories of relation algebras. To appear in the Memoirs of the AMS.

Andréka H. and I. Németi 1996. Axiomatization of identity-free equations valid in relation algebras. Submitted to Algebra Universalis.

Chang, C.C., and H.J. Keisler. 1973. *Model Theory*. Amsterdam: North-Holland.

Givant, S. 1991. Tarski's development of logic and mathematics based on the calculus of relations. In Andréka et al. Andréka et al. 1991, 189–215.

Hughes, G., and M. Creswell. 1984. *A Companion to Modal Logic*. Methuen.

Jipsen P. 1992. Computer aided investigations of relation algebras. PhD Dissertation, Vanderbilt University, Nashville TN (USA).

Jipsen, P., B. Jónsson and J. Rafter 1996. Adjoining units to residuated Boolean algebras. To appear in Algebra Universalis.

Jónsson, B. 1991. The theory of binary relations. In Andréka et al. Andréka et al. 1991, 245–292.

Jónsson, B. and C. Tshinakis, 1991. Relation algebras as residuated Boolean algebras Preprint, Dept. of Math., Vanderbilt Univ., Nashville, TN (USA).

Kramer, R. 1991. Relativized relation algebras. In Andréka et al. Andréka et al. 1991, 293–349.

Madarász, J., 1996. Interpolation and definability in arrow logics and in residuated lattices. Mathematical Institute Budapest, Manuscript.

Maddux, R. 1982. Some varieties containing relation algebras. *Transactions of the American Mathematical Society* 272:501–526.

Maddux, R. 1991a. Introductory course on relation algebras, finite dimensional cylindric algebras, and their connections. In Andréka et al. Andréka et al. 1991, 361–392.

Maddux, R. 1991b. The Origin of Relation Algebras in the Development and Axiomatization of the Calculus of Relations. *Studia Logica* L(3/4):421–456.

Marx, M., 1993. Axiomatizing and deciding relativized relation algebras. Report 93-87, CCSOM, University of Amsterdam. Submitted.

Marx, M. 1995. *Algebraic Relativization and Arrow Logic*. Doctoral dissertation, Institute for Logic, Language and Computation, University of Amsterdam. ILLC Dissertation Series 1995-3.

Marx, M., Sz. Mikulás, and I. Németi. 1995. Taming Logic. *Journal of Logic, Language and Information* 4:207–226.

Mikulás, Sz. 1995. *Taming Logics*. Doctoral dissertation, Institute for Logic, Language and Computation, University of Amsterdam. ILLC Dissertation Series 95-12.

Monk, J. D. 1964. On representable relation algebras. *Michigan Mathematical Journal* 11:207–210.

Németi, I. 1987. Decidability of relation algebras with weakened associativity. *Proc. Amer. Math. Soc.* 100, 2:340–344.

Németi, I. 1991. Algebraizations of quantifier logics: an overview (version 11.2). Preprint. Math. Inst. Hungar. Acad. Sci., Budapest. A short version without proofs appeared in *Studia Logica*, 50(3/4):485–569, 1991.

Pratt V. R., 1992. Origins of the Calculus of Binary Relations Proc. IEEE Symp. on Logic in Computer Science, 248-254, Santa Cruz, CA (USA).

de Rijke, M. 1993. *Extending Modal Logic.* Doctoral dissertation, Institute for Logic, Language and Computation, Universiteit van Amsterdam. ILLC Dissertation Series 1993-4.

Sahlqvist, H. 1975. Completeness and Correspondence in the First and Second Order Semantics for Modal Logic. In *Proc. of the Third Scandinavian Logic Symposium Uppsala 1973*, ed. S. Kanger. Amsterdam. North-Holland.

Sain, I., 1993. Beth's and Craig's properties via epimorphisms and amalgamation in algebraic logic. In: Algebraic logic and universal algebra in computer science (Proc. Conf. Ames USA, 1988, Bergman, C.H. Maddux, R.D. and Pigozzi, D.L. eds.), Springer Lecture Notes in Computer Science, Vol. 425, 209–226.

Simon, A., 1991. Finite schema completeness for typeless logic and representable cylindric algebras. In: H. Andréka, J.D. Monk, and I. Németi, eds, Algebraic Logic, North-Holland, Amsterdam, 665–670.

Tarski, A., and S. Givant. 1987. *A Formalization of Set Theory without Variables.* AMS Colloquium publications, Providence, Rhode Island.

Venema, Y. 1991. *Many-Dimensional Modal Logic.* Doctoral dissertation, Institute for Logic, Language and Computation, Universiteit van Amsterdam.

3
Causes and Remedies for Undecidability in Arrow Logics and in Multi-Modal Logics

HAJNAL ANDRÉKA, ÁGNES KURUCZ, ISTVÁN NÉMETI, ILDIKÓ SAIN AND ANDRÁS SIMON

> ABSTRACT. This paper discusses decidability questions concerning the lattice of arrow logics. We try to identify those parts (sublattices, intervals, etc.) of this lattice which consist of undecidable logics. Our results also apply to a wide variety of multi-modal logics, related to the "dynamic trend" (e.g. Pratt's action logics, modal logics embedding Lambek Calculus). We show that associativity of modal operator ∘ almost always causes undecidability, even if ∘ is commutative as well. An other, quite innocent looking postulate — the Euclidean axiom — can also cause undecidability. On the other hand, a weaker form of associativity leaves even strong arrow logics decidable, even if one adds new logical connectives like difference operator D, universal modality E, stratified modalities and Kleene star. We can also obtain associative but decidable logics by omitting axiom (K) for ∘ as well as by weakening the Boolean fragment.

1 Introduction and Preliminaries

Arrow logics (van Benthem 1994, Venema 1991) are among the many logics which were found useful in various (sometimes almost unrelated) areas (e.g. in linguistics, artificial intelligence, computer science). Other examples of such logics are dynamic logic (both computer science and linguistic versions), Pratt's action logics, the family of logics Pratt calls two dimen-

---

Research supported by the Hungarian National Foundation for Scientific Research grants Nos. T7255, T16448, F17452. Research of the second author is also supported by a grant from the Logic Graduate School, Eötvös Loránd University Budapest.

Arrow Logic and Multi-Modal Logic
M. Marx, L. Pólos and M. Masuch, eds.
Copyright © 1996, CSLI Publications.

sional (2D) logics, resource-sensitive and substructural logics, fragments of linear logic, extensions of the Lambek Calculus, cf. e.g. van Benthem 1991, Pratt 1994, Roorda 1991, Roorda 1992. Some authors say that many of these logics, at least partially, originate from the historical paper of McCarthy and Hayes 1969, while e.g. Pratt 1994 traces many of their features back to relevance logic.

One common feature in all these logics is that — beside the usual Boolean connectives — a new binary connective, say ∘, is available. So if φ, ψ are formulas then $(\varphi \circ \psi)$ is a formula too. A further common feature is that connective ∘ behaves like a ◇-type modality, i.e. the 'binary version' of axiom (K) of modal logic holds for ∘. This motivates our first definition.

Definition 1.1 A logic $\mathcal{L}$ is called a *logic with a binary modality* iff (i–iii) below hold.

(i) The language (alphabet) of $\mathcal{L}$ includes the usual Boolean connectives and a binary connective ∘. (There may be further connectives in the language.) Formulas of $\mathcal{L}$ ($F_\mathcal{L}$) are built up from an infinite set $\{p_0, p_1, \ldots\}$ of propositional variables with the help of the connectives in the usual way.

(ii) $\mathcal{L}$ has (PT) and (K°) below among its axioms.

(PT)   all propositional tautologies;
(K°) $((p_0 \vee p_1) \circ p_2) \leftrightarrow ((p_0 \circ p_2) \vee (p_1 \circ p_2))$
       $(p_0 \circ (p_1 \vee p_2)) \leftrightarrow ((p_0 \circ p_1) \vee (p_0 \circ p_2))$.

(iii) The set of validities (theorems) of $\mathcal{L}$ is closed under substitution, modus ponens and the following 'replacement of equivalents':

$$\frac{\varphi \leftrightarrow \psi}{(\varphi \circ \chi) \leftrightarrow (\psi \circ \chi)} \qquad \frac{\varphi \leftrightarrow \psi}{(\chi \circ \varphi) \leftrightarrow (\chi \circ \psi)}.$$

A logic with a binary modality is called *normal* iff the set of validities of $\mathcal{L}$ is also closed under the necessitation (modal generalization) rules:

$$\frac{\varphi}{\neg(\neg\varphi \circ \neg\psi)} \qquad \frac{\psi}{\neg(\neg\varphi \circ \neg\psi)}.$$

The ∘-*fragment* of a logic $\mathcal{L}$ with a binary modality is the set of those validities of $\mathcal{L}$ which include only ∘ and Boolean connectives.

A logic $\mathcal{L}$ is *minimal* among logics having some properties if the following hold: (i) the language of $\mathcal{L}$ is the 'smallest possible' (e.g. in case of logics with a binary modality, the language consists of the Booleans and ∘ only); (ii) for any logic $\mathcal{L}'$ having the particular properties, and for any $T \cup \{\varphi\} \subseteq F_\mathcal{L}$, $T \vdash_\mathcal{L} \varphi$ implies $T \vdash_{\mathcal{L}'} \varphi$.

We note that, by a classical theorem of Jónsson and Tarski 1951, the minimal normal logic with a binary modality is strongly sound and complete[1] w.r.t. the following class F of Kripke-frames:

$$\mathsf{F} \stackrel{\text{def}}{=} \{\langle W, C \rangle : W \text{ is a set}, C \subseteq W \times W \times W\}.$$

A model corresponding to such a frame is triplet $\mathfrak{M} = \langle W, C, V \rangle$, where $\langle W, C \rangle \in \mathsf{F}$ and V is a function mapping the propositional variables to subsets of W. The truth definitions are standard for the Booleans, and also for ∘:

$\mathfrak{M}, w \Vdash \varphi \circ \psi$ iff $(\exists u, v \in W)(C(w, u, v)$ and $\mathfrak{M}, u \Vdash \varphi$ and $\mathfrak{M}, v \Vdash \psi)$.

In the following definition we define some special kinds of frames.

Definition 1.2 A *square frame with base* U is a frame $\langle W, C \rangle$, where

$W = U \times U$ for some set U and
$C = \{\langle \langle a, c \rangle, \langle a, b \rangle, \langle b, c \rangle \rangle : a, b, c \in U\}$.

That is, square frames have pairs as worlds and the accessibility relation C act as composition between them.

Recall that a *semigroup* is a structure $\mathfrak{S} = \langle S, ; \rangle$, where ; is an associative binary operation. By the *semigroup frame* over some semigroup $\langle S, ; \rangle$ we mean the frame $\langle S, C \rangle$, where

$C = \{\langle a ; b, a, b \rangle : a, b \in S\}$

(see also Došen 1992). A semigroup frame over a group is called a *group frame*.

In the above context, arrow logics are examples for normal logics with a binary modality (cf. chapter 1).

Definition 1.3 $\mathcal{L}$ is an *arrow logic* iff (i–iv) below hold.

(i) The language of $\mathcal{L}$ consists of the Boolean connectives, a binary connective ∘, a unary connective ⊗ and a constant $\iota\delta$.
(ii) $\mathcal{L}$ is a normal logic with a binary modality ∘.
(iii) $\mathcal{L}$ also has (K$^\otimes$) below among its axioms.
   (K$^\otimes$) $\otimes(p_0 \vee p_1) \leftrightarrow (\otimes p_0 \vee \otimes p_1)$.
(iv) The set of validities of $\mathcal{L}$ is also closed under the rule

$$\frac{\varphi}{\neg \otimes \neg \varphi}.$$

A *square arrow frame with base* U is a frame $\langle W, C, F, I \rangle$, where

$\langle W, C \rangle$ is a square frame with base U,
$F = \{\langle \langle a, b \rangle, \langle b, a \rangle \rangle : a, b \in U\}$, and
$I = \{\langle a, a \rangle : a \in U\}$.   □

The *group arrow frame* over some group $\langle G, ; ,^{-1}, e \rangle$ is the frame $\langle G, C, F, I \rangle$, where

---
[1] In the sense of Andréka et al. 1994a

$$C = \{\langle a\,;b,a,b\rangle \;:\; a,b \in G\},$$
$$F = \{\langle a^{-1},a\rangle \;:\; a \in G\},$$
$$I = \{e\}.$$

A set Γ of formulas is called *decidable* iff there is an algorithm deciding whether a formula of $F_{\mathcal{L}}$ belongs to Γ or not. A *logic* $\mathcal{L}$ is *decidable* iff the set of validities (theorems) of $\mathcal{L}$ is decidable. Two sets of formulas are *recursively inseparable* iff there is no decidable set of formulas which contains one of them and is disjoint from the other.

We will look at the lattice of logics with a binary modality (and as a special case, at the lattice of arrow logics) and will study how postulates on connective ∘ influence the logic's behaviour with respect to decidability.

The structure of the paper is the following. In section 2 we demonstrate an old technique which can be used for proving undecidability of logics with an associative binary modality, in case the algebraic counterpart of the logic in question is a so-called *discriminator variety*[2]. One of the main aims of this paper is to elaborate methods which work for logics whose algebraic counterparts are not discriminator varieties. In section 3 we prove by these methods that "non-trivial" logics with an associative binary modality are all undecidable (Theorems 3.1–3.4). Our sharpest results are stated in an algebraic form as Theorem 3.6. In section 4 we list some possible applications of the results to other logics of the above mentioned "landscape", e.g. to Lambek Calculus, to Pratt's action logics, to intuitionistic modal logic, etc. Section 5 discusses some non-associative logics which are also "hereditarily" undecidable in many directions. As contrast, in section 6 we summarize the results about decidable arrow logics. Namely, a weaker form of associativity leaves even strong arrow logics decidable, even if we add new logical connectives, like difference operator D, universal modality E, stratified or graded modalities and Kleene star (Theorems 6.1–6.2). Also, associative arrow logics with a weaker Boolean fragment as well as associative arrow logics without axiom (K∘) are also decidable (Theorems 6.3–6.4).

All the logical results are proved in an algebraic setting, with the help of the correspondence between logics and their algebraic counterparts, see e.g. Andréka et al. 1994a, Blok and Pigozzi 1989, chapters 3.4, 5.6 of Henkin et al. 1985 about this connection in general. In case of logics having a Boolean fragment (thus in case of logics with a binary modality) this correspondence can be described as follows. To any such logic $\mathcal{L}$, one can associate a similarity type $t_{\mathcal{L}}$ of usual first-order logic (having only function symbols) by considering any n-ary connective of $\mathcal{L}$ as an n-ary function symbol of

---

[2] For discriminator varieties as well as for universal algebra basics, see e.g. Burris and Sankappanavar 1981.

$t_\mathcal{L}$. In this way a formula of $\mathcal{L}$ containing propositional variables, say, p_5, p_{19} can be considered as a term of type $t_\mathcal{L}$ having (algebraic) variables p_5, p_{19}. For any $T \subseteq F_\mathcal{L}$, let

$$\mathfrak{A}_\mathcal{L}^T \stackrel{\text{def}}{=} \langle A_\mathcal{L}^T, f \rangle_{f \in t_\mathcal{L}}$$

be the *formula-algebra* (or Lindenbaum-Tarski algebra) of T in $\mathcal{L}$, that is,

$$A_\mathcal{L}^T \stackrel{\text{def}}{=} \{(\varphi)_T : \varphi \in F_\mathcal{L}\},$$

where $(\varphi)_T$ is the congruence class $\{\psi \in F_\mathcal{L} : T \vdash_\mathcal{L} (\varphi \leftrightarrow \psi)\}$. Then the *algebraic counterpart* $\mathsf{Alg}(\mathcal{L})$ of $\mathcal{L}$ is the following class:

$$\mathsf{Alg}(\mathcal{L}) \stackrel{\text{def}}{=} \{\mathfrak{A}_\mathcal{L}^T : T \subseteq F_\mathcal{L}\}.$$

Now for each $\varphi \in F_\mathcal{L}$,

φ is a validity (theorem) of $\mathcal{L}$ $\iff$ $\mathsf{Alg}(\mathcal{L}) \models (\varphi = 1)$.

As a consequence we obtain the following fact:

Fact 1.4 *A logic $\mathcal{L}$ is decidable iff its algebraic counterpart $\mathsf{Alg}(\mathcal{L})$ has a decidable equational theory, that is, the set*

$$Eq(\mathsf{Alg}(\mathcal{L})) \stackrel{\text{def}}{=} \{(\tau = \sigma) : \tau, \sigma \text{ are terms and } \mathsf{Alg}(\mathcal{L}) \models (\tau = \sigma)\}$$

is decidable.

In case of logics with a binary modality, if every extra-Boolean connective satisfies the corresponding (K)-axiom, then the algebraic counterpart of the logic in question is a Boolean algebra with operators.

Definition 1.5 An algebra $\mathfrak{A} = \langle A, +, \cdot, -, 0, 1, f_i \rangle_{i \in I}$ is a *Boolean algebra with operators* if $\langle A, +, \cdot, -, 0, 1 \rangle$ is a Boolean algebra and every function f_i ($i \in I$) distributes over $+$ in each of its arguments:

$$f_i(a_0, \ldots, a_j + b_j, \ldots, a_{n-1}) =$$
$$= f_i(a_0, \ldots, a_j, \ldots, a_{n-1}) + f_i(a_0, \ldots, b_j, \ldots, a_{n-1}).$$

Such a function is called an *(extra-Boolean) operator*. We note that monotonicity of the operators follows from their distributivity.

BAO denotes the class of all Boolean algebras with operators.

As an example, the algebraic counterpart of an arrow logic is always a subclass of BAO. For more about the logic-algebra correspondence in case of arrow logics, see chapter 1.

Some of the results of the present paper were announced in Kurucz et al. 1995, which also gives an easily understandable demonstration of some proof-techniques.

2 A Simple Example for Undecidability in Arrow Logic

There were results and methods available showing that many logics with an associative binary connective, having *discriminator varieties* as algebraic counterparts are undecidable. Tarski already around 1950 had such results, see e.g. Pixley 1971, Andréka et al. to appear for such methods and references. In this section we demonstrate these techniques with a simple arrow logic example.

Let AL_{AB} denote the minimal arrow logic having the following axioms:

(A) $(p_0 \circ p_1) \circ p_2 \leftrightarrow p_0 \circ (p_1 \circ p_2)$
(B) $\top \circ \neg (\top \circ p_0) \to \neg p_0$
      $\neg (p_0 \circ \top) \circ \top \to \neg p_0$,

where $\top$ abbreviates the formula "$p_0 \to p_0$". We note that one can define unary modalities $\Diamond_0 \varphi \stackrel{\text{def}}{=} \top \circ \varphi$, $\Diamond_1 \varphi \stackrel{\text{def}}{=} \varphi \circ \top$ as well as their usual duals $\Box_0$, $\Box_1$, and then (B) is nothing but the Brouwerian axiom $p_0 \to \Box_i \Diamond_i p_0$ ($i < 2$). We also note that the first axiom of (B) is a consequence of axiom ($A5$) of chapter 1.

Theorem 2.1 AL_{AB} *is undecidable.*

Proof. We will prove that already the $\circ$-fragment of AL_{AB} is undecidable. Let K denote the algebraic counterpart of this $\circ$-fragment that is,

$$\begin{aligned}
\mathsf{K} \; = \; & \{ \mathfrak{A} \in \mathsf{BAO} \,:\, \mathfrak{A} = \langle A, +, \cdot, -, 0, 1, ; \rangle \text{ and} \\
& \mathfrak{A} \models x\,;0 = 0\,;x = 0 \\
& \mathfrak{A} \models (x\,;y)\,;z = x\,;(y\,;z) \\
& \mathfrak{A} \models 1\,; -(1\,;x) \leq -x \\
& \mathfrak{A} \models -(x\,;1)\,;1 \leq -x \;\}.
\end{aligned}$$

We will prove that the equational theory of K is undecidable (cf. Fact 1.4).

First, by a well-known result of Post (cf. e.g. Davis 1977), the halting problem of Turing machines is equivalent to the word problem of semigroups. That is, the unsolvability of the halting problem implies that the quasiequational theory of semigroups is undecidable.

Second, we will show that one can translate any quasiequation q of the language of semigroups to an equation $e(q)$ such that

(1)   "q holds in all semigroups" iff $\mathsf{K} \models e(q)$.

We will show this by proving that K is a discriminator variety.

Third, since the translation e above (turning quasiequations to equations) turns out to be recursive, (1) proves that the equational theory of K is undecidable.

In the proof we will use the following general definition and lemma which we quote from Sain 1978.[3]

---

[3] As far as we know, the present Lemma 2.3 was first stated and proved by I.Sain at the

Definition 2.2 Let $\mathfrak{A} \in$ BAO. A set $I \subseteq A$ is called an *ideal* of $\mathfrak{A}$ iff there is a congruence R of $\mathfrak{A}$ such that $I = 0^{\mathfrak{A}}/R$. That is, I is an ideal iff it is the congruence class of the smallest element $0^{\mathfrak{A}}$ of $\mathfrak{A}$ according to some congruence relation R of $\mathfrak{A}$.

Lemma 2.3 *Let* $\mathfrak{A} \in$ BAO *and* $I \subseteq A$. *Then* I *is an ideal of* $\mathfrak{A}$ *iff* I *is a Boolean ideal and the following condition holds:*

$(\forall \ n\text{-}ary \ operator \ f \ of \ \mathfrak{A})(\forall p_0, \ldots, p_{n-1} \in A)(\forall i < n)(\forall a \in I)$

$[f(p_0, \ldots, p_{i-1}, a, p_{i+1}, \ldots, p_{n-1}) - f(p_0, \ldots, p_{i-1}, 0, p_{i+1}, \ldots, p_{n-1})] \in I.$

This condition says that I *is closed under the unary term* "$f(\bar{p}, x, \bar{q}) - f(\bar{p}, 0, \bar{q})$", *where* $\bar{p}, \bar{q}$ *is an arbitrary choice of parameters.*

Proof of Lemma 2.3. First, assume that I is an ideal. Then $(\forall a \in I) \ aR0$, thus $f(\bar{p}, a, \bar{q})Rf(\bar{p}, 0, \bar{q})$ for any choice of parameters $\bar{p}, \bar{q}$, which obviously implies that $(f(\bar{p}, a, \bar{q}) - f(\bar{p}, 0, \bar{q}))R0$ that is, $(f(\bar{p}, a, \bar{q}) - f(\bar{p}, 0, \bar{q})) \in I$.

For the other direction, let us observe that for any extra-Boolean operator f, for any $a, b, \bar{p}, \bar{q}$,

$$f(\bar{p}, a, \bar{q}) - f(\bar{p}, b, \bar{q}) = [f(\bar{p}, a - b, \bar{q}) + f(\bar{p}, a \cdot b, \bar{q})] - f(\bar{p}, b, \bar{q}) =$$
$$= [f(\bar{p}, a - b, \bar{q}) - f(\bar{p}, b, \bar{q})] + [f(\bar{p}, a \cdot b, \bar{q}) - f(\bar{p}, b, \bar{q})] =$$
(2) $\quad = f(\bar{p}, a - b, \bar{q}) - f(\bar{p}, b, \bar{q}) \leq f(\bar{p}, a - b, \bar{q}) - f(\bar{p}, 0, \bar{q}).$

always holds. Let

$$aRb \overset{\text{def}}{\iff} a \oplus b \in I,$$

where $\oplus$ denotes symmetric difference. We will prove that R is a congruence of $\mathfrak{A}$ which will prove, by $I = 0/R$, that I is an ideal. R is a Boolean congruence, since I is a Boolean ideal by assumption. Now assume aRb. Then, since $a - b \in I$ and I is downward closed, the condition of the lemma and (2) imply that $(f(\bar{p}, a, \bar{q}) - f(\bar{p}, b, \bar{q})) \in I$, for any operator f and parameters $\bar{p}, \bar{q}$. Similarly, $(f(\bar{p}, b, \bar{q}) - f(\bar{p}, a, \bar{q})) \in I$ also holds.

Now let f be an n-ary operator and assume $a_0 R b_0, \ldots, a_{n-1} R b_{n-1}$. Then

$$f(a_0, a_1, \ldots, a_{n-1}) - f(b_0, b_1, \ldots, b_{n-1}) =$$
$$= [f(a_0, a_1, \ldots, a_{n-1}) - f(b_0, a_1, \ldots, a_{n-1})] +$$
$$+ [f(b_0, a_1, \ldots, a_{n-1}) - f(b_0, b_1, \ldots, a_{n-1})] + \cdots$$
$$\cdots + [f(b_0, b_1, \ldots, a_{n-1}) - f(b_0, b_1, \ldots, b_{n-1})].$$

Now, as we showed above, each member of this 'sum' belongs to I, therefore

---

Algebra Seminar in Budapest in 1978, then in the manuscript Sain 1978, from which it was quoted (but in less general form) in Németi 1980. Note that Lemma 2.3 does *not* assume that our BAO's would be normal (i.e. $f(\bar{p}, 0, \bar{q}) \neq 0$ is allowed). The lemma was reproved in a simpler but less general looking form in Sain 1982.

$(f(a_0, a_1, \ldots, a_{n-1}) - f(b_0, b_1, \ldots, b_{n-1})) \in I$. $(f(\bar{b}) - f(\bar{a})) \in I$ can be proved similarly, proving $(f(\bar{a}) \oplus f(\bar{b})) \in I$. □

Define the following term c of the language of K:

$$c(x) \stackrel{\text{def}}{=} x + (1\,;x) + (x\,;1) + (1\,;x\,;1).$$

Let Sir K denote the class of all subdirectly irreducible members of K.

Claim 2.3.1 *For any* $\mathfrak{A} \in Sir$ K,

$$c^{\mathfrak{A}}[a] = \begin{cases} 0^{\mathfrak{A}}, & \text{if } a = 0^{\mathfrak{A}} \\ 1^{\mathfrak{A}}, & \text{otherwise} \end{cases}$$

that is, K *is a discriminator variety.*

Proof. Clearly, K $\models c(0) = 0$ holds. Now for any $\mathfrak{A} \in $ K, $a \in A$, let

$$Rl_a A \stackrel{\text{def}}{=} \{b \in A : b \leq a\}.$$

We claim that for any $\mathfrak{A} \in $ K, $a \in A$, both $Rl_{c^{\mathfrak{A}}[a]}$ and $Rl_{-c^{\mathfrak{A}}[a]}$ are ideals of $\mathfrak{A}$. Indeed, since ; is normal and monotonous, by Lemma 2.3, it is enough to prove that equations

$$1\,;c(x) \leq c(x)$$
$$c(x)\,;1 \leq c(x)$$
$$1\,;-c(x) \leq -c(x)$$
$$-c(x)\,;1 \leq -c(x)$$

hold in $\mathfrak{A}$. The first two equations are trivial by associativity, monotonicity and by $1\,;1 \leq 1$. For the third one: $1\,;c(x) \leq c(x)$ implies (by taking contraposition and by the monotonicity of ;)

$$1\,;-c(x) \leq 1\,;-(1\,;c(x)) \leq -c(x) \quad \text{(by (B))}.$$

The proof of $-c(x)\,;1 \leq -c(x)$ proceeds similarly, using $c(x)\,;1 \leq c(x)$ and (B) again.

Now let $\mathfrak{A} \in Sir$ K, $0^{\mathfrak{A}} \neq a \in A$ and assume $c^{\mathfrak{A}}[a] \neq 1^{\mathfrak{A}}$. Then, by K $\models x \leq c(x)$, $c^{\mathfrak{A}}[a] = 0^{\mathfrak{A}}$ cannot hold. Therefore both $Rl_{c^{\mathfrak{A}}[a]}A$ and $Rl_{-c^{\mathfrak{A}}[a]}A$ are non-$\{0\}$ ideals. But $Rl_{c^{\mathfrak{A}}[a]}A \cap Rl_{-c^{\mathfrak{A}}[a]}A = \{0^{\mathfrak{A}}\}$ which contradicts the fact that in every subdirectly irreducible algebra there is a smallest non-$\{0\}$ ideal, proving Claim 2.3.1. □

Now we can use the techniques concerning discriminator varieties described e.g. in Németi 1991. Let $Queq$ denote the set of all quasiequations[4] of the language of semigroups (using ; for the semigroup operation). For any $q \in Queq$ of form $[(\tau_1 = \sigma_1)\& \ldots \&(\tau_n = \sigma_n)] \Rightarrow (\tau_0 = \sigma_0)$, let $e(q)$ be

---

[4]A *quasiequation* is a formula of form $(e_1 \& \cdots \& e_n) \Rightarrow e_0$, where $e_0, e_1, \ldots, e_n$ are equations.

the following equation in the language of K

$$e(q) \stackrel{def}{=} \tau_0 \oplus \sigma_0 \leq c(\tau_1 \oplus \sigma_1) + \ldots + c(\tau_n \oplus \sigma_n)$$

(here $\oplus$ denotes symmetric difference, i.e., $a \oplus b \stackrel{def}{=} (a \cdot -b) + (b \cdot -a)$). Then

(3) $\quad\quad\quad$ K $\models e(q) \to q \quad\quad$ (by the definitions) $\quad$ and
(4) $\quad\quad\quad$ Sir K $\models q \to e(q) \quad\quad$ (by Claim 2.3.1)

SG denotes the class of all semigroups.

Claim 2.3.2 *For any $q \in Queq$,*

$$\text{SG} \models q \quad \text{iff} \quad \text{K} \models e(q).$$

Proof. Assume that $\mathfrak{S} \not\models q$ for some semigroup $\mathfrak{S}$. This semigroup $\mathfrak{S} = \langle S, ; \rangle$ can be embedded into the ;-reduct of some member of K in the following way. Let $e \notin S$ be a new element and let us define a new semigroup $\mathfrak{S}^+ = \langle S^+, ; \rangle$ by $S^+ \stackrel{def}{=} S \cup \{e\}$ and by postulating $a ; e = e ; a = a$ for all $a \in S^+$. Let $C_S \stackrel{def}{=} \{\{\langle b, b; a\rangle : b \in S^+\} : a \in S^+\}$, and let $\mathfrak{C}_S \stackrel{def}{=} \langle C_S, | \rangle$, where | is the usual composition of relations. Then $\mathfrak{C}_S$ is a semigroup which embeds $\mathfrak{S}$ (the well-known *Cayley representation* of $\mathfrak{S}$). Now, let $\mathfrak{B}$ denote the Boolean algebra of all subsets of $S^+ \times S^+$ expanded with relation composition | as ;. Then $\mathfrak{B} \in$ K and $\mathfrak{C}_S$ (and thus $\mathfrak{S}$) can be embedded into the ;-reduct of $\mathfrak{B}$. Therefore SG $\not\models q$ implies K $\not\models q$, thus by (3) above, K $\not\models e(q)$ follows.

For the other direction, observe that the ;-reduct of each algebra in K is a semigroup. Therefore SG $\models q$ implies Sir K $\models q$. Now, by (4) above, Sir K $\models e(q)$, which implies K $\models e(q)$. □

3 Nontrivial Arrow Logics with Associativity are All Undecidable

In this section we elaborate techniques for proving undecidability of associative logics, whose algebraic counterparts are far from being discriminator varieties. When mapping a lattice of logics for undecidability, it is not very useful to prove that a single logic is undecidable. Therefore we will prove theorems of the following pattern. "If logic $\mathcal{L}$ is strong enough to prove A but is not too strong in that it does not prove B, then $\mathcal{L}$ is undecidable." Why do we need the condition $\not\vdash_\mathcal{L} B$? Well, because e.g. every inconsistent logic is decidable. But even if $\not\vdash_\mathcal{L} False$, we still may have $\vdash_\mathcal{L}$ "∧ is the same as ∘", i.e. $\vdash_\mathcal{L} (\varphi \wedge \psi) \leftrightarrow (\varphi \circ \psi)$ for all φ, ψ. This again would make $\mathcal{L}$ decidable, since this would reduce $\mathcal{L}$ to classical propositional logic.

Theorems 3.1–3.4 below state that many parts of the lattice of logics with a binary associative modality consist of undecidable logics. The strongest versions of these statements are given in an algebraic setting as

Theorem 3.6 below. After proving this algebraic theorem, we show how the logical results follow from it.

A *language model* is a model on the semigroup frame over some free semigroup of finitely many generators (cf. van Benthem 1991, Buszkowski 1986).

Theorem 3.1 *Any "non-trivial" logic $\mathcal{L}$ with a binary associative modality is undecidable. For $\mathcal{L}$ being "non-trivial", any one of conditions (i–iv) below is sufficient.*

  (i) *The $\circ$-fragment of $\mathcal{L}$ is valid in arbitrarily large finite square frames (i.e., for every $n \in \omega$ there is some set U with $|U| \geq n$ such that the $\circ$-fragment of $\mathcal{L}$ is valid in the square frame with base U).*

  (ii) *The $\circ$-fragment of $\mathcal{L}$ is valid in all language models over a two element alphabet.*

  (iii) *The $\circ$-fragment of $\mathcal{L}$ is valid in all language models in which only finitely many languages are definable.*

  (iv) *There are infinitely many nontrivial finite groups $\mathfrak{G}_n$ ($n \in \omega$) such that the $\circ$-fragment of $\mathcal{L}$ is valid in the group frame over $P_{n \in \omega} \mathfrak{G}_n$ (i.e., over the direct product of groups $\mathfrak{G}_n$ ($n \in \omega$)).*

Corollary 3.2   (i) *The minimal logic with an associative binary modality is undecidable.*

  (ii) *If the $\circ$-fragment of a logic $\mathcal{L}$ with an associative binary modality is valid in all square frames then $\mathcal{L}$ is undecidable.*

  (iii) *If the $\circ$-fragment of a logic $\mathcal{L}$ with an associative binary modality is valid in all language models then $\mathcal{L}$ is undecidable.*

  (iv) *The minimal logic having an associative and commutative[5] binary modality is undecidable.*

Proof of Corollary 3.2. (i) and (ii) obviously follow from Theorem 3.1(i), and (iii) from both Theorem 3.1(ii) and (iii). (iv) follows from Theorem 3.1(iv), since the $\circ$-fragment of such an $\mathcal{L}$ is valid e.g. in the group frame over $(\mathbf{Z}_2)^\omega$ (where $\mathbf{Z}_2$ is the two element commutative group). □

Corollary 3.3   (i) *The minimal associative arrow logic is undecidable.*

  (ii) *The minimal associative and commutative arrow logic is undecidable.*

  (iii) *The arrow logic defined by axioms (A1)–(A5) and (A9) of chapter 1 (cf. also van Benthem 1994) is undecidable.*

  (iv) *Let $\mathcal{L}$ be an extension of the minimal associative arrow logic with further axioms (containing perhaps new connectives). If the $\circ$-fragment of $\mathcal{L}$ is valid in arbitrarily large finite square arrow frames then $\mathcal{L}$ is undecidable.*

---

[5] That is, $\mathcal{L}$ has axiom $p_0 \circ p_1 \leftrightarrow p_1 \circ p_0$.

The following theorem is a joint result with S. Givant.

Theorem 3.4 *Let $\mathcal{L}$ be any extension of the arrow logic defined by axioms (A3), (A5) and (A9) of chapter 1. If the $\langle \circ, \otimes \rangle$-fragment[6] of $\mathcal{L}$ is valid in some infinite group arrow frame, then $\mathcal{L}$ is undecidable.*

To prove Theorems 3.1 and 3.4 above it is enough to prove that the algebraic counterparts of the logics in question all have undecidable equational theories. Theorem 3.6 below states sharper results in even more directions. Before formulating these statements, we define some of the most important classes of algebras occurring in them.

Definition 3.5 An algebra $\mathfrak{A} = \langle A, +, 1, * \rangle$ is called a *weak-implicative*[7] *semilattice* iff

(1) $\langle A, +, 1 \rangle$ is a $+$-semilattice (i.e., $x \leq y$ iff $x + y = y$) with greatest element 1
(2) (i) $(x * y) * (x * y) \leq x * y$
     (ii) $y * y \leq x \implies x * x \leq y * y$
     (iii) $y \leq (x * y) + x$.

$\mathfrak{A} = \langle A, +, 1, *, ; \rangle$ is a *weak-implicative semilattice-ordered semigroup* iff

(3) $\langle A, +, 1, * \rangle$ is a weak-implicative semilattice
(4) $\langle A, ; \rangle$ is a semigroup (i.e., ; is associative)
(5) ; distributes over $+$.

$\mathfrak{A} = \langle A, +, \cdot, ; \rangle$ is a *distributive lattice-ordered semigroup (DL-semigroup)* iff

(6) $\langle A, +, \cdot \rangle$ is a distributive lattice
(7) $\langle A, ; \rangle$ is a semigroup
(8) ; distributes over $+$.

A DL-semigroup is called *bounded* iff it has both smallest and greatest elements.

$\mathfrak{A} = \langle A, +, \cdot, 1, *, ; \rangle$ is a *weak-implicative DL-semigroup* iff

(9) $\langle A, +, 1, * \rangle$ is a weak-implicative semilattice
(10) $\langle A, +, \cdot, ; \rangle$ is a DL-semigroup.

A weak-implicative semilattice-ordered semigroup or a weak-implicative DL-semigroup $\mathfrak{A}$ is called *normal* iff there is a smallest element $0^{\mathfrak{A}}$ of $\mathfrak{A}$ such that $\mathfrak{A} \models 0^{\mathfrak{A}} ; x = x ; 0^{\mathfrak{A}} = x * x = 0^{\mathfrak{A}}$.

$\mathfrak{A} = \langle A, +, \cdot, 1, *, ;, \smile \rangle$ is a *weak-implicative DL-semigroup with inverse* iff

---
[6] I.e., those theorems of $\mathcal{L}$ which contain only Boolean connectives, $\circ$ and $\otimes$
[7] Cf. subsection 4.1 below for motivation of the name "weak implication"

(11) $\langle A, +, \cdot, 1, *, ;\rangle$ is a weak-implicative DL-semigroup
(12) $\check{}$ distributes over $+$
(13) $\mathfrak{A} \models x \cdot (y\,;z) \leq (y \cdot (x\,;z^{\check{}}))\,;z$
   $\mathfrak{A} \models x \cdot (y\,;z) \leq y\,;(z \cdot (y^{\check{}}\,;x))$.

Let $\mathfrak{A}$ be a weak-implicative DL-semigroup with inverse. $\mathfrak{A}$ is called *normal* iff there is a smallest element $0^{\mathfrak{A}}$ of $\mathfrak{A}$ such that $\mathfrak{A} \models 0^{\mathfrak{A}\check{}} = 0^{\mathfrak{A}}\,;x = x\,;0^{\mathfrak{A}} = x*x = 0^{\mathfrak{A}}$.

$\mathfrak{A} = \langle A, +, \cdot, 1, *, ;, \triangleright, \triangleleft\rangle$ is a *weak-implicative DL-semigroup with conjugates* iff

(14) $\langle A, +, \cdot, 1, *, ;\rangle$ is a weak-implicative DL-semigroup
(15) $\triangleright$ and $\triangleleft$ distribute over $+$
(16) $\mathfrak{A} \models x \cdot (y\,;z) \leq (y \cdot (x \triangleleft z))\,;z$
   $\mathfrak{A} \models x \cdot (y\,;z) \leq y\,;(z \cdot (y \triangleright x))$.

Let $\mathfrak{A}$ be a weak-implicative DL-semigroup with conjugates. $\mathfrak{A}$ is called *normal* iff there is a smallest element $0^{\mathfrak{A}}$ of $\mathfrak{A}$ such that $\mathfrak{A} \models 0^{\mathfrak{A}}\,;x = x\,;0^{\mathfrak{A}} = 0^{\mathfrak{A}} \triangleright x = x \triangleright 0^{\mathfrak{A}} = 0^{\mathfrak{A}} \triangleleft x = x \triangleleft 0^{\mathfrak{A}} = x*x = 0^{\mathfrak{A}}$.

Statements (III)–(IV) of Theorem 3.6 below are joint results with S. Givant.

Theorem 3.6

(I) Let K be a class of weak-implicative semilattice-ordered semigroups.

  (i) If any finite semigroup is embeddable into the ;-reduct of some normal member of K then the equational theory $Eq(K)$ of K is undecidable.

  (ii) Let S be any class of finite semigroups such that the quasiequational theory of S and the set of quasiequations falsifiable in SG^8 are recursively inseparable[9]. If any member of S is embeddable into the ;-reduct of some normal member of K then $Eq(K)$ is undecidable.

(II) Let K be a class of weak-implicative DL-semigroups such that the following condition holds for K. For any $n \in \omega$, there are non-trivial finite groups $\mathfrak{G}_0, \ldots, \mathfrak{G}_{n-1}$ such that the ;-reduct of $\mathfrak{G} \stackrel{\text{def}}{=} P_{i<n}\mathfrak{G}_i$ is embeddable into the ;-reduct of some normal $\mathfrak{A}_n \in K$ as an antichain (i.e., $(\forall x \neq y \in G)\, (x \neq 0^{\mathfrak{A}_n}$ and $x \cdot^{\mathfrak{A}_n} y = 0^{\mathfrak{A}_n}))$. Then $Eq(K)$ is undecidable.

---

[8]I.e., the set $\{q \in Queq : q$ fails in some semigroup$\}$
[9]As it is proved in Gurevich and Lewis 1984 (see Theorem 3.17 below), the class FSG of all finite semigroups is an example for such a class S. Thus (I)(i) is a consequence of (I)(ii). For other examples for S (quoted also from Gurevich and Lewis 1984), see Theorem 4.3 below.

(III) Let K *be a class of weak-implicative DL-semigroups with inverse. Assume the followings hold for* K.
- K $\models x^{\smile\smile} \leq x$
- *There is some normal* $\mathfrak{A} \in$ K *containing an infinite antichain G such that G is closed under* ; *and* $^{\smile}$, *and G contains a* ; *-neutral element*[10].

*Then Eq(*K*) is undecidable.*

(IV) Let K *be a class of weak-implicative DL-semigroups with conjugates. Assume the followings hold for* K.
- K $\models (x \triangleright y) ; z = x \triangleright (y ; z)$
  K $\models (x \triangleleft y) ; z = x ; (y \triangleright z)$
- *There is some normal* $\mathfrak{A} \in$ K *containing an infinite antichain G such that G is closed under* ;, $\triangleleft$ *and* $\triangleright$, *and G contains a* ; *-neutral element.*[11]

*Then Eq(*K*) is undecidable.*

In the proof of Theorem 3.6(II)–(IV) we use (among others) the so-called *coordinate-frame method*, originating from von Neumann 1960. This method was used by Lipshitz 1974 to prove that the quasiequational theory of modular lattices is undecidable. Urquhart 1995 generalized the method and proved that any class of DL-semigroups that contains the DL-semigroup of all subspaces of some infinite dimensional vector space has an undecidable quasiequational theory. Andréka et al. to appear used further generalizations of the method for proving that certain classes of relation algebras have undecidable quasiequational (and, being discriminator varieties, equational) theories. The particular coordinate-frame method used in the proof of Theorem 3.6(II) is a common modification of the ones in Urquhart 1995 and in Andréka et al. to appear. The "coordinate-frame part" of the proof of (III) is exactly the same as that of Andréka et al. to appear, therefore we will only give a brief summary of it, in order to make the present paper self-contained. The proof of (IV) will mimic that of (III). We note that, in the absence of *id*, $^{\smile}$ is not term-definable from $\triangleright$ or $\triangleleft$.

Proof of Theorem 3.6. The proofs are based on the fact that quasiequations of the language of semigroups can be "coded" by equations of the language

---

[10] I.e., an element e with $e ; g = g ; e = g$, for all $g \in G$

[11] This condition (and the similar one in (III)) can be replaced with the following weaker one: "$(\forall n \in \omega)$ ($\exists$ normal $\mathfrak{A}_n \in$ K) ($\exists G_n \subseteq A_n$) with $\mid G_n \mid \geq n$ and with the properties above". However, for our purposes these two conditions are equivalent, since all the properties in question are first-order expressible, thus the ultraproduct

$$\mathfrak{A} \stackrel{\text{def}}{=} P_{n \in \omega} \mathfrak{A}_n / U$$

(over any non-principal ultrafilter U) contains an appropriate infinite antichain, and $Eq(K) = Eq(K \cup \{\mathfrak{A}\})$.

of the classes K in question. To each algebra in K we will "associate" a semigroup in such a way that

- the universe of the semigroup is equationally definable;
- the semigroup-operation is term definable;
- every finite semigroup[12] is embeddable into the semigroup "associated" to some algebra in K.

In case of (I), it will be easy to find these "associated" semigroups. In fact, the ;-reducts of the algebras in K will do. In the other cases the definitions will be more involved, they will be given with the help of coordinate-frames.

Definition 3.7 A *4-frame in a DL-semigroup* $\mathfrak{A}$ is a set

$$\mathcal{F} \stackrel{\text{def}}{=} \{a_0, \ldots, a_3\} \cup \{c_{ij} : i \neq j < 4\} \subseteq A$$

satisfying the following equations (F1)–(F4) (called *frame axioms*) in $\mathfrak{A}$:

(F1) $(a_i\,;a_j\,;a_j\,;a_k) \cdot (a_i\,;a_\ell\,;a_\ell\,;a_k) \leq a_i\,;a_k$
$(i,j,k,\ell < 4,\ |\{i,j,k,\ell\}| = 4)$
(F2) $a_i\,;a_j = a_j\,;a_i \quad (i,j < 4)$
(F3) $(c_{ij}\,;c_{jk}) \cdot (a_i\,;a_k) = c_{ik} \quad (i,j,k < 4,\ |\{i,j,k\}| = 3)$
(F4) $a_i\,;a_j \leq a_k\,;a_i\,;a_i\,;a_j$ and
$a_i\,;a_j \leq a_i\,;a_j\,;a_j\,;a_k \quad (i,j,k < 4,\ |\{i,j,k\}| = 3)$.

$a \in A$ is *semi-modular in* $\mathfrak{A}$ iff

$$\mathfrak{A} \models (\forall b,c)(c \leq a \Rightarrow a \cdot (b\,;c) \leq (a \cdot b)\,;c\ \&\ a \cdot (c\,;b) \leq c\,;(a \cdot b)).$$

A 4-frame $\mathcal{F}$ in a DL-semigroup $\mathfrak{A}$ is called *semi-modular* iff for all $i,j,k < 4$, $|\{i,j,k\}| = 3$, $a_i\,;a_j\,;a_j\,;a_k$ is semi-modular in $\mathfrak{A}$.

A *4-frame in a DL-semigroup* $\mathfrak{A}$ *with inverse* (cf. Andréka et al. to appear, Defintion 2.2) is a set

$$\mathcal{F} \stackrel{\text{def}}{=} \{a_0, \ldots, a_3\} \cup \{c_{ij} : i \neq j < 4\} \subseteq A$$

satisfying the following equations in $\mathfrak{A}$:

(F1˘) $(a_i\,;a_j\,;a_j\,;a_k) \cdot (a_i\,;a_\ell\,;(a_k\,;a_\ell)^\smile) \leq a_i\,;a_k$
$(i,j,k,\ell < 4,\ |\{i,j,k,\ell\}| = 4)$
(F2˘) $(a_i\,;a_j\,;a_j\,;a_k) \cdot ((a_\ell\,;a_i)^\smile\,;a_\ell\,;a_k) \leq a_i\,;a_k$
$(i,j,k,\ell < 4,\ |\{i,j,k,\ell\}| = 4)$
(F3) as above.

A *4-frame in a DL-semigroup* $\mathfrak{A}$ *with conjugates* is a set

$$\mathcal{F} \stackrel{\text{def}}{=} \{a_0, \ldots, a_3\} \cup \{c_{ij} : i \neq j < 4\} \subseteq A$$

satisfying the following equations in $\mathfrak{A}$:

---

[12]In the proof of (I)(ii) here we require that every member of S

(F1$^{\bowtie}$) $(a_i\,;a_j\,;a_j\,;a_k)\cdot((a_i\,;a_\ell)\triangleleft(a_k\,;a_\ell))\leq a_i\,;a_k$
  $(i,j,k,\ell<4,\,\{i,j,k,\ell\}=4)$
(F2$^{\bowtie}$) $(a_i\,;a_j\,;a_j\,;a_k)\cdot((a_\ell\,;a_i)\triangleright(a_\ell\,;a_k))\leq a_i\,;a_k$
  $(i,j,k,\ell<4,\,\{i,j,k,\ell\}=4)$
(F3) as above.

$$L_{ij}^{\mathcal{F}}\stackrel{\text{def}}{=}\{x\in A\,:\,x\leq a_i\,;a_j\}\text{ for }i\neq j<4,$$

$$x\otimes^{\mathcal{F}}y\stackrel{\text{def}}{=}(x\,;y)\cdot(a_i\,;a_k)\text{ for }i,j,k<4,\,|\{i,j,k\}|=3,$$
$$x\in L_{ij}^{\mathcal{F}},\,y\in L_{jk}^{\mathcal{F}},$$

$$x\odot^{\mathcal{F}}y\stackrel{\text{def}}{=}(x\otimes c_{12})\otimes(c_{20}\otimes y)\text{ for }x,y\in L_{01}^{\mathcal{F}}.$$

We will omit the superscript $\mathcal{F}$, if it is clear from context.

Lemma 3.8 (Item (ii) is Lemma 2.4 in Andréka et al. to appear) *If*

 (i) either $\mathcal{F}$ is a semi-modular 4-frame in some DL-semigroup
 (ii) or $\mathcal{F}$ is a 4-frame in some DL-semigroup with inverse
 (iii) or $\mathcal{F}$ is a 4-frame in some DL-semigroup with conjugates

then $\langle L_{01}^{\mathcal{F}},\odot^{\mathcal{F}}\rangle$ is a semigroup.

Proof. First, if $x\in L_{ij}$, $y\in L_{jk}$ then $x\otimes y\in L_{ik}$ holds by the definitions, and $c_{ij}\in L_{ij}$ by (F3). Thus, if $x,y\in L_{01}$ then $x\odot y\in L_{01}$ too.

Second, we prove that if $x\in L_{ij}$, $y\in L_{jk}$, $z\in L_{k\ell}$ ($\{i,j,k,\ell\}=4$), then
(5) $\qquad(x\otimes y)\otimes z=(x\,;y\,;z)\cdot(a_i\,;a_\ell)=x\otimes(y\otimes z).$
In case (i),

$(x\,;y\,;z)\cdot(a_i\,;a_\ell)=$
$=[((x\,;y)\cdot(a_i\,;a_j\,;a_j\,;a_k))\,;z]\cdot(a_i\,;a_\ell)\stackrel{(F4)}{=}$
$=[((x\,;y)\cdot(a_i\,;a_j\,;a_j\,;a_k))\,;z]\cdot(a_i\,;a_\ell\,;a_\ell\,;a_k)\cdot(a_i\,;a_\ell)\leq$
  [since $z\leq a_k\,;a_\ell\stackrel{(F2)}{=}a_\ell\,;a_k\stackrel{(F4)}{\leq}a_i\,;a_\ell\,;a_\ell\,;a_k$
  and $a_i\,;a_\ell\,;a_\ell\,;a_k$ is semi-modular]
$\leq[((x\,;y)\cdot(a_i\,;a_j\,;a_j\,;a_k)\cdot(a_i\,;a_\ell\,;a_\ell\,;a_k))\,;z]\cdot(a_i\,;a_\ell)\stackrel{(F1)}{\leq}$
$\leq[((x\,;y)\cdot(a_i\,;a_k))\,;z]\cdot(a_i\,;a_\ell)=(x\otimes y)\otimes z.$

The "$(x\,;y\,;z)\cdot(a_i\,;a_\ell)\leq x\otimes(y\otimes z)=$" part is similar.

Case (ii) is proved in Andréka et al. to appear. Here we quote the proof from there:

$(x\,;y\,;z)\cdot(a_i\,;a_\ell)=$

$\qquad\qquad\qquad\qquad\qquad\qquad\qquad\qquad\text{(13) of Def.3.5}$
$=[((x\,;y)\cdot(a_i\,;a_j\,;a_j\,;a_k))\,;z]\cdot(a_i\,;a_\ell)\quad\leq$

$$\leq [((x\,;y)\cdot(a_i\,;a_j\,;a_j\,;a_k))\cdot(a_i\,;a_\ell\,;z^\vee))\,;z]\cdot(a_i\,;a_\ell) \leq$$
$$\leq [((x\,;y)\cdot(a_i\,;a_j\,;a_j\,;a_k))\cdot(a_i\,;a_\ell\,;(a_k\,;a_\ell)^\vee))\,;z]\cdot(a_i\,;a_\ell) \overset{(\text{F1}^\vee)}{\leq}$$
$$\leq [((x\,;y)\cdot(a_i\,;a_k))\,;z]\cdot(a_i\,;a_\ell) = (x\otimes y)\otimes z.$$

The "$(x\,;y\,;z)\cdot(a_i\,;a_\ell) \leq x\otimes(y\otimes z) =$" part is similar, using (F2$^\vee$).

In case (iii):
$$(x\,;y\,;z)\cdot(a_i\,;a_\ell) =$$
$$= [((x\,;y)\cdot(a_i\,;a_j\,;a_j\,;a_k))\,;z]\cdot(a_i\,;a_\ell) \overset{(16)\text{ of Def.3.5}}{\leq}$$
$$\leq [((x\,;y)\cdot(a_i\,;a_j\,;a_j\,;a_k)\cdot((a_i\,;a_\ell)\triangleleft z))\,;z]\cdot(a_i\,;a_\ell) \leq$$
$$\leq [((x\,;y)\cdot(a_i\,;a_j\,;a_j\,;a_k)\cdot((a_i\,;a_\ell)\triangleleft(a_k\,;a_\ell)))\,;z]\cdot(a_i\,;a_\ell) \overset{(\text{F1}^{\bowtie})}{\leq}$$
$$\leq [((x\,;y)\cdot(a_i\,;a_k))\,;z]\cdot(a_i\,;a_\ell) = (x\otimes y)\otimes z.$$

The "$(x\,;y\,;z)\cdot(a_i\,;a_\ell) \leq x\otimes(y\otimes z) =$" part is similar, using (F2$^{\bowtie}$).

The "$(x\otimes y)\otimes z) \leq (x\,;y\,;z)\cdot(a_i\,;a_\ell)$" and "$x\otimes(y\otimes z) \leq (x\,;y\,;z)\cdot(a_i\,;a_\ell)$" directions are easy by lattice argument and by the monotonicity of $;$.

Now we can repeat the proofs of Lemmas 5.3 and 5.4 in Urquhart 1995. For $x,y \in L_{01}$,

$$(x\otimes c_{12})\otimes(c_{20}\otimes y) \overset{(\text{F3})}{=} (x\otimes(c_{13}\otimes c_{32}))\otimes(c_{20}\otimes y)$$
$$\overset{(5)}{=} ((x\otimes c_{13})\otimes c_{32})\otimes(c_{20}\otimes y)$$
$$\overset{(5)}{=} (x\otimes c_{13})\otimes(c_{32}\otimes(c_{20}\otimes y))$$
$$\overset{(5)}{=} (x\otimes c_{13})\otimes((c_{32}\otimes c_{20})\otimes y)$$
(6) $\overset{(\text{F3})}{=} (x\otimes c_{13})\otimes(c_{30}\otimes y).$

For $x,y,z \in L_{01}$,

$$(x\odot y)\odot z = [[(x\otimes c_{12})\otimes(c_{20}\otimes y)]\otimes c_{12}]\otimes(c_{20}\otimes z)$$
$$\overset{(6)}{=} [[(x\otimes c_{13})\otimes(c_{30}\otimes y)]\otimes c_{12}]\otimes(c_{20}\otimes z)$$
$$\overset{(5)}{=} [(x\otimes c_{13})\otimes[(c_{30}\otimes y)\otimes c_{12}]]\otimes(c_{20}\otimes z)$$
$$\overset{(5)}{=} (x\otimes c_{13})\otimes[[(c_{30}\otimes y)\otimes c_{12}]\otimes(c_{20}\otimes z)]$$
$$\overset{(5)}{=} (x\otimes c_{13})\otimes[[c_{30}\otimes(y\otimes c_{12})]\otimes(c_{20}\otimes z)]$$
$$\overset{(5)}{=} (x\otimes c_{13})\otimes[c_{30}\otimes[(y\otimes c_{12})\otimes(c_{20}\otimes z)]]$$
$$\overset{(6)}{=} (x\otimes c_{12})\otimes[c_{20}\otimes[(y\otimes c_{12})\otimes(c_{20}\otimes z)]]$$
$$= x\odot(y\odot z).$$

□

Lemma 3.9 Let $\mathfrak{A} = \langle +, \cdot, ; \rangle$ be a DL-semigroup.[13] If $a, k \in A$ are such that for every $x \in A$,

$$a + k \geq x, \quad a \cdot k \leq x,$$
$$a\,;k \leq k, \quad k\,;a \leq k$$

then a is semi-modular in $\mathfrak{A}$.

Proof. (Cf. the first part of the proof of Lemma 5.5 in Urquhart 1995.) Assume that $a \geq c$. Then

$$\begin{aligned}
a \cdot (b\,;c) &= a \cdot [((b \cdot a) + (b \cdot k))\,;c] \\
&= a \cdot [((b \cdot a)\,;c) + ((b \cdot k)\,;c)] \\
&\leq a \cdot [((b \cdot a)\,;c) + (k\,;a)] \\
&= [a \cdot ((b \cdot a)\,;c)] + [a \cdot (k\,;a)] \\
&\leq [a \cdot ((b \cdot a)\,;c)] + (a \cdot k) \\
&\leq (a \cdot b)\,;c.
\end{aligned}$$

The case of $a \cdot (c\,;b) \leq c\,;(a \cdot b)$ is similar. □

Now recall that $Queq$ denotes the set of quasiequations of the language of semigroups (using ; for the semigroup operation). For any $q \in Queq$, an other quasiequation $q^{\odot}$ is defined as follows.

In case (I) of Theorem 3.6, let

$$q^{\odot} \stackrel{\text{def}}{=} q.$$

In cases (II)–(IV), let the variables of $q^{\odot}$ are those of q together with new variables a_0, a_1, a_2, a_3, c_{ij} ($i \neq j < 4$). In case (II), add also further new variables t, f and b_{ijk}, for each $\langle i, j, k \rangle$ ($i, j, k < 4$, $|\{i, j, k\}| = 3$). The left-hand side of $q^{\odot}$ is the conjunction of the following equations:

- the equations stating that the variables of q are below $a_0\,;a_1$ (i.e., they belong to L_{01});
- the equations of the left-hand side of q, with the operation symbol ; replaced by the term $\odot$;
-
  - in case (II), frame axioms (F1), (F2), (F3), (F4) for a_0, a_1, a_2, a_3, c_{ij} ($i \neq j < 4$), and the equations

  $$(a_i\,;a_j\,;a_j\,;a_k) + b_{ijk} \geq t, \quad (a_i\,;a_j\,;a_j\,;a_k) \cdot b_{ijk} \leq f,$$
  $$a_i\,;a_j\,;a_j\,;a_k\,;b_{ijk} \leq b_{ijk}, \quad b_{ijk}\,;a_i\,;a_j\,;a_j\,;a_k \leq b_{ijk},$$

  for each $\langle i, j, k \rangle$ ($i, j, k < 4$, $|\{i, j, k\}| = 3$);
  - in case (III), frame axioms (F1$^\smile$), (F2$^\smile$), (F3), for a_0, a_1, a_2, a_3, c_{ij} ($i \neq j < 4$);
  - in case (IV), frame axioms (F1$^{\bowtie}$), (F2$^{\bowtie}$), (F3), for a_0, a_1, a_2, a_3, c_{ij} ($i \neq j < 4$).

---

[13] Associativity of ; is not needed here

The right-hand side of $q^\odot$ is the same as that of q, with ; replaced again by $\odot$.

Lemma 3.10 *If*
  (i) *either $\mathfrak{A}$ is a bounded DL-semigroup and $\mathcal{F}$ is a semi-modular 4-frame in $\mathfrak{A}$*
  (ii) *or $\mathfrak{A}$ is a DL-semigroup with inverse (or conjugates) and $\mathcal{F}$ is a 4-frame in $\mathfrak{A}$*

then for any $q \in Queq$,
$$\langle L_{01}^{\mathcal{F}}, \odot^{\mathcal{F}}\rangle \models q \quad \Longleftrightarrow \quad \mathfrak{A} \models q^\odot.$$

Proof. It is straightforward by the definition of $q^\odot$. □

Notation 3.11 For any set G, let $\mathcal{P}(G)$ denote the powerset of G and let $\mathcal{P}_\omega(G) \stackrel{\text{def}}{=} \{X \subseteq G : X \text{ is finite}\}$. Let $\mathfrak{G} = \langle G, ;\rangle$ be a semigroup and let $X, Y \subseteq G$. Then
$$X; Y \stackrel{\text{def}}{=} \{x; y : x \in X, y \in Y\}$$
$$Cm_\omega(\mathfrak{G}) \stackrel{\text{def}}{=} \langle \mathcal{P}_\omega(G), \cup, \cap, ;\rangle.$$

Now let $\mathfrak{G} = \langle G, ;, ^{-1}, e\rangle$ be a group and let $X, Y \subseteq G$. Then
$$X^\smile \stackrel{\text{def}}{=} \{x^{-1} : x \in X\}$$
$$X \triangleleft Y \stackrel{\text{def}}{=} \{x; y^{-1} : x \in X, y \in Y\}$$
$$X \triangleright Y \stackrel{\text{def}}{=} \{x^{-1}; y : x \in X, y \in Y\}$$
$$Cm_\omega^\smile(\mathfrak{G}) \stackrel{\text{def}}{=} \langle \mathcal{P}_\omega(G), \cup, \cap, ;, ^\smile\rangle \text{ and}$$
$$Cm_\omega^{\bowtie}(\mathfrak{G}) \stackrel{\text{def}}{=} \langle \mathcal{P}_\omega(G), \cup, \cap, ;, \triangleleft, \triangleright\rangle.$$

Then it is easy to check that $Cm_\omega(\mathfrak{G})$ is a DL-semigroup, $Cm_\omega^\smile(\mathfrak{G})$ is a DL-semigroup with inverse and $Cm_\omega^{\bowtie}(\mathfrak{G})$ is a DL-semigroup with conjugates. Moreover, if $\mathfrak{G}$ is finite then $Cm_\omega(\mathfrak{G})$ is bounded. □

Lemma 3.12 *Let $n \in \omega$ and for each $s < 4n$ let $\mathfrak{G}_s$ be a nontrivial finite group. Let $\mathfrak{G} \stackrel{\text{def}}{=} P_{s<4n}\mathfrak{G}_s$. Then there is a semi-modular 4-frame $\mathcal{F}$ in $Cm_\omega(\mathfrak{G})$ such that every finite semigroup with cardinality less than n is embeddable into the semigroup $\langle L_{01}^{\mathcal{F}}, \odot^{\mathcal{F}}\rangle$.*

Proof. For each $s < 4n$ let e_s denote the unit of $\mathfrak{G}_s$, and let $e \stackrel{\text{def}}{=} \langle e_s : s < 4n\rangle$ i.e., e is the unit of $\mathfrak{G}$. For each $s < 4n$ we choose an arbitrary non-unit element b_s from G_s. For each $i < 4$, $j < n$ we define an element a_{ij} of G as follows: for every $s < 4n$ let
$$(a_{ij})_s \stackrel{\text{def}}{=} \begin{cases} b_s, & \text{if } s = i \cdot n + j \\ e_s, & \text{otherwise.} \end{cases}$$

For each $i < 4$, $s < 4n$ let
$$H^i_s \stackrel{\text{def}}{=} \begin{cases} G_s, & \text{if } i \cdot n \le s < i \cdot n + n \\ \{e_s\}, & \text{otherwise,} \end{cases}$$
and let $A_i \stackrel{\text{def}}{=} P_{s<4n} H^i_s$ for each $i < 4$. Then obviously $a_{ij} \in A_i$ ($i < 4$, $j < n$).

For $i \ne j < 4$, $x \subseteq n \times n$ define
$$R_{ij}(x) \stackrel{\text{def}}{=} \{a_{ip}\,; a_{jq}{}^{-1} : \langle p, q \rangle \in x\} \subseteq A_i\,; A_j$$
(where ; denotes the binary group-operation and $^{-1}$ denotes the group-inverse operation of $\mathfrak{G}$). Thus $R_{ij}(x) \in L_{ij}$ holds. Let
$$C_{ij} \stackrel{\text{def}}{=} R_{ij}(Id_n),$$
where $Id_n \stackrel{\text{def}}{=} \{\langle p, p \rangle : p < n\}$.

Claim 3.12.1 *For any $x, y \subseteq n \times n$ and $i, j, k < 4$, $|\{i, j, k\}| = 3$,*
$$R_{ij}(x) \otimes R_{jk}(y) = R_{ik}(x|y).$$
(Here $x|y \stackrel{\text{def}}{=} \{\langle p, q \rangle : \exists r \, \langle p, r \rangle \in x \text{ and } \langle r, q \rangle \in y\}$.)

Proof. Let $x \in (R_{ij}(x)\,; R_{jk}(y)) \cap (A_i\,; A_k)$. Then
$$x = a_{ip}\,; a_{jq}{}^{-1}\,; a_{jr}\,; a_{ks}{}^{-1}$$
for some $\langle p, q \rangle \in x$, $\langle r, s \rangle \in y$, and also $x \in A_i\,; A_k$. Thus $a_{jq}{}^{-1}\,; a_{jr} =$ "unit of A_j" must hold i.e., $q = r$. But then $x = a_{ip}\,; a_{ks}{}^{-1}$ and $\langle p, s \rangle \in x|y$ i.e., $x \in R_{ik}(x|y)$. Conversely, assume that $x \in R_{ik}(x|y)$. This means that $x = a_{ip}\,; a_{ks}{}^{-1}$ for some $\langle p, s \rangle \in x|y$. Thus $x \in A_i\,; A_k$. Let q be such that $\langle p, q \rangle \in x$, $\langle q, s \rangle \in y$. Then $x = a_{ip}\,; a_{ks}{}^{-1} = a_{ip}\,; a_{jq}{}^{-1}\,; a_{jq}\,; a_{ks}{}^{-1} \in R_{ij}(x)\,; R_{jk}(y)$. Hence $x \in R_{ij}(x) \otimes R_{jk}(y)$, and Claim 3.12.1 has been proved. $\square$

Claim 3.12.2 *Let $\mathcal{F} \stackrel{\text{def}}{=} \{A_0, A_1, A_2, A_3\} \cup \{C_{ij} : i \ne j < 4\}$. Then $\mathcal{F}$ is a semi-modular 4-frame in $Cm_\omega(\mathfrak{G})$.*

Proof. Since $\mathfrak{G}$ is finite, $\mathcal{F} \subseteq Cm_\omega(\mathfrak{G})$ obviously holds. We show that frame axiom (F1) is satisfied i.e.,
$$(A_i\,; A_j\,; A_j\,; A_k) \cap (A_i\,; A_l\,; A_l\,; A_k) \subseteq A_i\,; A_k \quad \text{for } \{i, j, k, \ell\} = 4.$$
Let x be in the left-hand side. Then $x \in A_i\,; A_j\,; A_j\,; A_k$. Therefore $x_s = e_s$ for every s with $\ell \cdot n \le s < \ell \cdot n + n$. But $x \in A_i\,; A_\ell\,; A_\ell\,; A_k$ also holds, thus $x_s = e_s$ for every s with $j \cdot n \le s < j \cdot n + n$. Therefore $x \in A_i\,; A_k$.

Frame axioms (F2) and (F4) obviously hold by the definitions.

Frame axiom (F3) says that $C_{ij} \otimes C_{jk} = C_{ik}$ ($i, j, k < 4$, $|\{i, j, k\}| = 3$). Using $C_{pq} = R_{pq}(Id_n)$, by Claim 3.12.1 we get $C_{ij} \otimes C_{jk} = R_{ij}(Id_n) \otimes R_{jk}(Id_n) = R_{ik}(Id_n|Id_n) = R_{ik}(Id_n) = C_{ik}$.

To show that $\mathcal{F}$ is semi-modular in $Cm_\omega(\mathfrak{G})$, let

$$D_{i\ell k} \stackrel{\text{def}}{=} A_i\,;A_\ell\,;A_\ell\,;A_k,$$

and let $\tilde{D}_{i\ell k} \stackrel{\text{def}}{=} G - D_{i\ell k}$. We prove that the conditions of Lemma 3.9 above hold for $D_{i\ell k}$ and $\tilde{D}_{i\ell k}$ in $Cm_\omega(\mathfrak{G})$. Indeed, $D_{i\ell k}, \tilde{D}_{i\ell k} \in Cm_\omega(\mathfrak{G})$ and $D_{i\ell k} \cap \tilde{D}_{i\ell k} = \emptyset$, $D_{i\ell k} \cup \tilde{D}_{i\ell k} = G$ hold. Further, $D_{i\ell k}\,;\tilde{D}_{i\ell k} \leq \tilde{D}_{i\ell k}$ and $\tilde{D}_{i\ell k}\,;D_{i\ell k} \leq \tilde{D}_{i\ell k}$ hold by the definition of the A_i's. □

Now we show that the semigroup $\langle L_{01}^\mathcal{F}, \odot^\mathcal{F}\rangle$ embeds any semigroup of cardinality less than n. Let $\mathfrak{S}$ be a semigroup with $|S| < n$. It is enough to show that $\langle \mathcal{P}(n \times n), |\rangle$ is embedded, since (by taking the Cayley-representation of $\mathfrak{S}$) we have $\mathfrak{S} \hookrightarrow \langle \mathcal{P}(n \times n), |\rangle$.

We take the function

$$R_{01} : \mathcal{P}(n \times n) \longrightarrow L_{01}^\mathcal{F}$$

defined above and show that it is an injective semigroup-homomorphism. First, we show that R_{01} is injective. To this end, it is enough to show that

$$a_{0p}\,;a_{1q}^{-1} = a_{0r}\,;a_{1s}^{-1} \quad \text{implies} \quad \langle p,q\rangle = \langle r,s\rangle.$$

Indeed, assume that $a_{0p}\,;a_{1q}^{-1} = a_{0r}\,;a_{1s}^{-1}$. Then

$$a_{0r}^{-1}\,;a_{0p}\,;a_{1q}^{-1}\,;a_{1s} = e,$$

hence $a_{0r}^{-1}\,;a_{0p} =$ "unit of A_0" and $a_{1q}^{-1}\,;a_{1s} =$ "unit of A_1". Thus $q = s$ and also $p = r$ must hold.

Next we show that R_{01} is a homomorphism i.e., that for any $x, y \subseteq n \times n$,

$$R_{01}(x|y) = R_{01}(x) \odot R_{01}(y).$$

Indeed, this follows almost immediately by Claim 3.12.1 and by $C_{ij} = R_{ij}(Id_n)$ as follows: $R_{01}(x) \odot R_{01}(y) = (R_{01}(x) \otimes C_{12}) \otimes (C_{20} \otimes R_{01}(y)) = (R_{01}(x) \otimes R_{12}(Id_n)) \otimes (R_{20}(Id_n) \otimes R_{01}(y)) = R_{02}(x|Id_n) \otimes R_{21}(Id_n|y) = R_{01}(x|Id_n|Id_n|y) = R_{01}(x|y)$, completing the proof of Lemma 3.12. □

Lemma 3.13 (Lemma 2.7 in Andréka et al. to appear) *Let $\mathfrak{G}$ be an infinite group. Then for every $n \in \omega$ there is a 4-frame $\mathcal{F}$ in $Cm_\omega^\vee(\mathfrak{G})$ $(Cm_\omega^\bowtie(\mathfrak{G}))$ such that every finite semigroup with cardinality less than n is embeddable into the semigroup $\langle L_{01}^\mathcal{F}, \odot^\mathcal{F}\rangle$.*

Proof. The proof of the $Cm_\omega^\vee(\mathfrak{G})$ case can be found in Andréka et al. to appear. The $Cm_\omega^\bowtie(\mathfrak{G})$ case can be proved in a similar way. □

Lemma 3.14 (i) *Let $\mathfrak{A} = \langle A, +, \cdot, ;\rangle$ be a DL-semigroup with a smallest element $0^\mathfrak{A}$. Let $G \subseteq A$ be an antichain, closed under $;^\mathfrak{A}$ and let $\mathfrak{G} \stackrel{\text{def}}{=} \langle G, ;\rangle$. For any finite $X \subseteq G$ define*

$$h(X) \stackrel{\text{def}}{=} \sum_{x \in X} x.$$

Then h embeds $Cm_\omega(\mathfrak{G})$ into $\mathfrak{A}$.

(ii) Let $\mathfrak{A}$ be a DL-semigroup with inverse (or with conjugates) and with a smallest element $0^{\mathfrak{A}}$. Let $G \subseteq A$ be an antichain, closed under $;^{\mathfrak{A}}$ and $\smile^{\mathfrak{A}}$ (or $\triangleright^{\mathfrak{A}}$ and $\triangleleft^{\mathfrak{A}}$), let $e \in G$ be $;^{\mathfrak{A}}$-neutral in G, and let $\mathfrak{G} \stackrel{\text{def}}{=} \langle G, ;, \smile, e \rangle$. Then h above embeds $Cm_\omega^\smile(\mathfrak{G})$ (or $Cm_\omega^{\bowtie}(\mathfrak{G})$) into $\mathfrak{A}$.

Proof. It is easy to check that h is a homomorphism in each case. We prove that h is injective: Assume $X \neq Y \subseteq G$ and $h(X) = h(Y)$. Say, there is some $y \in Y$, $y \notin X$. Then

$$y = y \cdot h(Y) = y \cdot h(X) = \sum_{x \in X}(y \cdot x) = 0^{\mathfrak{A}}.$$

But this contradicts the fact that G is an antichain. □

Now let K any class occurring in Theorem 3.6, and assume that a unary term $c(x)$ of the language of K is given. For any quasiequation p of the language of K of form

$$[(\tau_1 = \sigma_1) \& \ldots \& (\tau_m = \sigma_m)] \Rightarrow (\tau_0 = \sigma_0)$$

an equation $e_c(p)$ of the language of K is defined as follows.

$$e_c(p) \stackrel{\text{def}}{=} [(\tau_0 + c(\tau)) * (\sigma_0 + c(\tau))] + [(\sigma_0 + c(\tau)) * (\tau_0 + c(\tau))] \leq \tau,$$

where $\tau \stackrel{\text{def}}{=} (\tau_1 * \sigma_1) + (\sigma_1 * \tau_1) + \cdots + (\tau_m * \sigma_m) + (\sigma_m * \tau_m)$.

To any $\mathfrak{A} \in \mathsf{K}$, $\bar{a} \in {}^\omega A$, we define

$$B(\mathfrak{A}) \stackrel{\text{def}}{=} \{x + c(\tau)^{\mathfrak{A}}[\bar{a}] : x \in A\}$$

$$x \,;^{\mathfrak{B}} y \stackrel{\text{def}}{=} (x \,;^{\mathfrak{A}} y) + c(\tau)^{\mathfrak{A}}[\bar{a}], \quad \text{for any } x, y \in B(\mathfrak{A}),$$

$$\mathfrak{S}_{\bar{a}}(\mathfrak{A}) \stackrel{\text{def}}{=} \langle B(\mathfrak{A}), ;^{\mathfrak{B}} \rangle.$$

In case (II) of Theorem 3.6, let

$$\mathfrak{B}_{\bar{a}}(\mathfrak{A}) \stackrel{\text{def}}{=} \langle B(\mathfrak{A}), +^{\mathfrak{A}}, \cdot^{\mathfrak{A}}, ;^{\mathfrak{B}} \rangle.$$

In case (III), let

$$x^{\smile \mathfrak{B}} \stackrel{\text{def}}{=} x^{\smile \mathfrak{A}} + c(\tau)^{\mathfrak{A}}[\bar{a}], \quad \text{for any } x \in B(\mathfrak{A}),$$

$$\mathfrak{B}_{\bar{a}}^\smile(\mathfrak{A}) \stackrel{\text{def}}{=} \langle B(\mathfrak{A}), +^{\mathfrak{A}}, \cdot^{\mathfrak{A}}, ;^{\mathfrak{B}}, \smile^{\mathfrak{B}} \rangle.$$

In case (IV), let

$$x \triangleright^{\mathfrak{B}} y \stackrel{\text{def}}{=} (x \triangleright^{\mathfrak{A}} y) + c(\tau)^{\mathfrak{A}}[\bar{a}], \quad \text{and}$$

$$x \triangleleft^{\mathfrak{B}} y \stackrel{\text{def}}{=} (x \triangleleft^{\mathfrak{A}} y) + c(\tau)^{\mathfrak{A}}[\bar{a}], \quad \text{for any } x, y \in B(\mathfrak{A}),$$

$$\mathfrak{B}_{\bar{a}}^{\bowtie}(\mathfrak{A}) \stackrel{\text{def}}{=} \langle B(\mathfrak{A}), +^{\mathfrak{A}}, \cdot^{\mathfrak{A}}, ;^{\mathfrak{B}}, \triangleright^{\mathfrak{B}}, \triangleleft^{\mathfrak{B}} \rangle.$$

Now consider the following properties, concerning the behaviour of the unary term c in class K:

(a) Term c is increasing i.e., $\mathsf{K} \models x \leq c(x)$.
(b$_1$) "Dual"-relativizing with $c(x)$ is a ;-homomorphism for every x i.e.,
$$\mathsf{K} \models (y\,;z) + c(x) = [(y + c(x))\,;(z + c(x))] + c(x).$$
(b$_2$) "Dual"-relativizing with $c(x)$ is a $\smile$-homomorphism for every x i.e.,
$$\mathsf{K} \models y^\smile + c(x) = (y + c(x))^\smile + c(x).$$
(b$_3$) "Dual"-relativizing with $c(x)$ is both a $\triangleright$- and a $\triangleleft$-homomorphism for every x i.e.,
$$\mathsf{K} \models (y \triangleright z) + c(x) = [(y + c(x)) \triangleright (z + c(x))] + c(x) \text{ and}$$
$$\mathsf{K} \models (y \triangleleft z) + c(x) = [(y + c(x)) \triangleleft (z + c(x))] + c(x).$$
(c) For any normal $\mathfrak{A} \in \mathsf{K}$, $c(0^\mathfrak{A}) = 0^\mathfrak{A}$.

Lemma 3.15 *Let K be one of the classes occurring in Theorem 3.6. Assume term c has properties (a) and (b$_1$) above in K, and let $\mathfrak{A} \in \mathsf{K}$, $\bar{a} \in {}^\omega A$. Then (i)–(iii) below hold.*

(i) *In case of (I) of Theorem 3.6,*
  ○ $\mathfrak{S}_{\bar{a}}(\mathfrak{A})$ *is a semigroup;*
  ○ *for any quasiequation $p \in \mathit{Queq}$,*
$$\mathfrak{S}_{\bar{a}}(\mathfrak{A}) \models p \quad \Longrightarrow \quad \mathfrak{A} \models e_c(p)[\bar{a}].$$

(ii) *In case of (II),*
  ○ $\mathfrak{B}_{\bar{a}}(\mathfrak{A})$ *is a bounded DL-semigroup;*
  ○ *for any quasiequation p of the type of $\mathfrak{B}_{\bar{a}}(\mathfrak{A})$,*
$$\mathfrak{B}_{\bar{a}}(\mathfrak{A}) \models p \quad \Longrightarrow \quad \mathfrak{A} \models e_c(p)[\bar{a}].$$

(iii) *In case of (III) (or (IV)), if term c also satisfies (b$_2$) (or (b$_3$)) in K then*
  ○ $\mathfrak{B}^\smile_{\bar{a}}(\mathfrak{A})$ $(\mathfrak{B}^{\bowtie}_{\bar{a}}(\mathfrak{A}))$ *is a DL-semigroup with inverse (with conjugates);*
  ○ *for any quasiequation p of the type of $\mathfrak{B}^\smile_{\bar{a}}(\mathfrak{A})$ $(\mathfrak{B}^{\bowtie}_{\bar{a}}(\mathfrak{A}))$,*
$$\mathfrak{B}^\smile_{\bar{a}}(\mathfrak{A}) \models p \quad \Longrightarrow \quad \mathfrak{A} \models e_c(p)[\bar{a}]$$
$$(\mathfrak{B}^{\bowtie}_{\bar{a}}(\mathfrak{A}) \models p \quad \Longrightarrow \quad \mathfrak{A} \models e_c(p)[\bar{a}]).$$

Proof. The first statements of each item obviously hold by the assumptions on c and on K.

To prove the second group of statements, assume that p is of form $[(\tau_1 = \sigma_1) \& \ldots \& (\tau_m = \sigma_m)] \Rightarrow (\tau_0 = \sigma_0)$. We will use Claims 3.15.1–3.15.2 below.

Claim 3.15.1 $\mathsf{K} \models (\tau_0 + c(\tau) = \sigma_0 + c(\tau)) \Rightarrow e_c(p)$.

Proof of Claim 3.15.1. We let $\mathfrak{A} \in \mathsf{K}$, $\bar{a} \in {}^\omega A$. For every k, if $1 \leq k \leq m$ then
$$((\tau_k * \sigma_k) * (\tau_k * \sigma_k))^{\mathfrak{A}}[\bar{a}] \stackrel{(2)(i) \text{ of Def.3.5}}{\leq} (\tau_k * \sigma_k)^{\mathfrak{A}}[\bar{a}] \stackrel{(1) \text{ of Def.3.5}}{\leq}$$
$$\leq \tau^{\mathfrak{A}}[\bar{a}] \stackrel{(a)}{\leq} c(\tau)^{\mathfrak{A}}[\bar{a}] \stackrel{(1) \text{ of Def.3.5}}{\leq} (\tau_0 + c(\tau))^{\mathfrak{A}}[\bar{a}].$$
Assume that $(\tau_0 + c(\tau))^{\mathfrak{A}}[\bar{a}] = (\sigma_0 + c(\tau))^{\mathfrak{A}}[\bar{a}]$ holds. Then, by (2)(ii) of Definition 3.5,
$$[(\tau_0 + c(\tau)) * (\sigma_0 + c(\tau))]^{\mathfrak{A}}[\bar{a}] \leq ((\tau_k * \sigma_k) * (\tau_k * \sigma_k))^{\mathfrak{A}}[\bar{a}] \stackrel{(2)(i) \text{ of Def.3.5}}{\leq}$$
$$\leq (\tau_k * \sigma_k)^{\mathfrak{A}}[\bar{a}] \stackrel{(1) \text{ of Def.3.5}}{\leq} \tau^{\mathfrak{A}}[\bar{a}].$$
Similarly, $[(\sigma_0 + c(\tau)) * (\tau_0 + c(\tau))]^{\mathfrak{A}}[\bar{a}] \leq \tau^{\mathfrak{A}}[\bar{a}]$ also holds, proving $\mathfrak{A} \models e_c(p)[\bar{a}]$. □

Claim 3.15.2 *For every k, if $1 \leq k \leq m$ then $\mathsf{K} \models \tau_k + c(\tau) = \sigma_k + c(\tau)$.*
Proof of Claim 3.15.2.
$$\mathsf{K} \models \tau_k + c(\tau) \stackrel{(a)}{=} \tau_k + \tau + c(\tau) \stackrel{(1) \text{ of Def.3.5}}{\geq}$$
$$\geq \tau_k + (\tau_k * \sigma_k) + c(\tau) \stackrel{(2)(iii) \text{ of Def.3.5}}{\geq} \sigma_k + c(\tau).$$
The other direction can be proved similarly. □

To prove Lemma 3.15(i), assume that K is as in Theorem 3.6(I), and assume $\mathfrak{A} \not\models e_c(p)[\bar{a}]$ for some $\mathfrak{A} \in \mathsf{K}$, $\bar{a} \in {}^\omega A$. Then, by Claim 3.15.1, $\mathfrak{A} \models (\tau_0 + c(\tau) \neq \sigma_0 + c(\tau))[\bar{a}]$. Let
$$\bar{b} \stackrel{\text{def}}{=} \langle \ldots, a_j + c(\tau)^{\mathfrak{A}}[\bar{a}], \ldots \rangle_{j \in \omega}.$$
Then $\mathfrak{S}_{\bar{a}}(\mathfrak{A}) \models (\tau_0 \neq \sigma_0)[\bar{b}]$ follows. Similarly, by Claim 3.15.2, for any $1 \leq k \leq m$, $\mathfrak{A} \models (\tau_k + c(\tau) = \sigma_k + c(\tau))[\bar{a}]$, which implies $\mathfrak{S}_{\bar{a}}(\mathfrak{A}) \models (\tau_k = \sigma_k)[\bar{b}]$. Therefore $\mathfrak{S}_{\bar{a}}(\mathfrak{A}) \not\models p[\bar{b}]$ holds. Statements (ii) and (iii) can be proved similarly. □

Let FSG denote the class of all finite semigroups.

Proposition 3.16 *(i) Let K as in (I) or (II) of Theorem 3.6. If term c has properties (a) and (b_1) in K then for any $q \in Queq$,*
$$(7) \qquad \mathsf{SG} \models q \implies \mathsf{K} \models e_c(q^\odot).$$
If K is as in (III) or in (IV) then (7) holds only if c also satisfies (b_2) or (b_3) in K, respectively.

(ii) Let K as in (I)(i), (II), (III) or (IV) of Theorem 3.6. If term c has property (c) in K then for any $q \in Queq$,
$$(8) \qquad \mathsf{K} \models e_c(q^\odot) \implies \mathsf{FSG} \models q.$$
If K is as in (I)(ii) then (8) holds with S in place of FSG.

Proof of Proposition 3.16. We prove the cases when class K is as in (I)(i) and in (II) of Theorem 3.6. The other cases are of similar pattern, using the appropriate items of the lemmas above.
Proof of (i): Assume first that K is as in (I)(i). Let $q \in Queq$ and assume $\mathfrak{A} \not\models e_c(q^\odot)[\bar{a}]$ for some $\mathfrak{A} \in \mathsf{K}$, $\bar{a} \in {}^\omega A$. Then, by Lemma 3.15, $\mathfrak{S}_{\bar{a}}(\mathfrak{A})$ is a semigroup and $\mathfrak{S}_{\bar{a}}(\mathfrak{A}) \not\models q^\odot$. But in this case $q^\odot = q$, proving $\mathsf{SG} \not\models q$.

Second, let K be as in (II), and assume $\mathfrak{A} \not\models e_c(q^\odot)[\bar{a}]$ for some $\mathfrak{A} \in \mathsf{K}$, $\bar{a} \in {}^\omega A$, $q \in Queq$. Then, by Lemma 3.15, $\mathfrak{B}_{\bar{a}}(\mathfrak{A})$ is a bounded DL-semigroup and $\mathfrak{B}_{\bar{a}}(\mathfrak{A}) \not\models q^\odot$. Let $\mathcal{F}$ be a 4-frame in $\mathfrak{B}_{\bar{a}}(\mathfrak{A})$. Then, by Lemma 3.10, $\langle L_{01}^{\mathcal{F}}, \odot^{\mathcal{F}} \rangle \not\models q$. Thus, by Lemma 3.8, $\mathsf{SG} \not\models q$.

Proof of (ii): First, let K be as in (I)(i) and assume that there is some finite semigroup $\mathfrak{S}$ with $\mathfrak{S} \not\models q$. Then, by the assumption on K, $\mathfrak{S}$ can be embedded into the ;-reduct of some normal $\mathfrak{A} \in \mathsf{K}$, thus $\mathfrak{A} \not\models q$ also holds. Then, by $q^\odot = q$, $\mathfrak{A} \not\models q^\odot$.

Assume now that $q^\odot$ is of form $[(\tau_1 = \sigma_1) \& \ldots \& (\tau_m = \sigma_m)] \Rightarrow (\tau_0 = \sigma_0)$. Then $\mathfrak{A} \not\models q^\odot$ means that $\tau_k^{\mathfrak{A}}[\bar{a}] = \sigma_k^{\mathfrak{A}}[\bar{a}]$ ($k = 1, \ldots, m$) and $\tau_0^{\mathfrak{A}}[\bar{a}] \neq \sigma_0^{\mathfrak{A}}[\bar{a}]$ hold for some $\bar{a} \in {}^\omega A$. But $\mathfrak{A}$ is normal, thus $\mathfrak{A} \models x * x = 0^{\mathfrak{A}}$. Therefore $\tau^{\mathfrak{A}}[\bar{a}] = 0^{\mathfrak{A}}$ i.e.,

(9) $\qquad$ (right side of $e_c(q^\odot))^{\mathfrak{A}}[\bar{a}] = 0^{\mathfrak{A}}$.

But, since $\mathfrak{A}$ is normal and term c has property (c), $c(\tau)^{\mathfrak{A}}[\bar{a}] = 0^{\mathfrak{A}}$ also holds, thus, by $0^{\mathfrak{A}}$ being the smallest element of $\mathfrak{A}$,

(10) $\qquad$ (left side of $e_c(q^\odot))^{\mathfrak{A}}[\bar{a}] = ((\tau_0 * \sigma_0) + (\sigma_0 * \tau_0))^{\mathfrak{A}}[\bar{a}] \not\leq 0^{\mathfrak{A}}$.

Otherwise,

$$\tau_0^{\mathfrak{A}}[\bar{a}] \stackrel{\text{(2)(iii) of Def.3.5}}{\leq} ((\sigma_0 * \tau_0) + \sigma_0)^{\mathfrak{A}}[\bar{a}] \stackrel{\text{(1) of Def.3.5}}{\leq}$$
$$\leq ((\tau_0 * \sigma_0) + (\sigma_0 * \tau_0) + \sigma_0)^{\mathfrak{A}}[\bar{a}] \stackrel{\text{(1) of Def.3.5}}{\leq} 0^{\mathfrak{A}} + \sigma_0^{\mathfrak{A}}[\bar{a}] = \sigma_0^{\mathfrak{A}}[\bar{a}],$$

and similarly, $\sigma_0^{\mathfrak{A}}[\bar{a}] \leq \tau_0^{\mathfrak{A}}[\bar{a}]$ would hold, contradicting $\sigma_0^{\mathfrak{A}}[\bar{a}] \neq \tau_0^{\mathfrak{A}}[\bar{a}]$. Thus, by (9) and (10), $\mathfrak{A} \not\models e_c(q^\odot)[\bar{a}]$ that is, $\mathsf{K} \not\models e_c(q^\odot)$.

Now let K be as in (II), and assume $\mathsf{FSG} \not\models q$ that is, there is some finite semigroup $\mathfrak{S}$ with $\mathfrak{S} \not\models q$. Assume that $|S| < n$. Then, by the assumption on K, there are nontrivial finite groups $\mathfrak{G}_i$ ($i < 4n$) such that

(11) $\qquad \mathfrak{G} \stackrel{\text{def}}{=} P_{i<4n}\mathfrak{G}_i$ is embeddable into the ;-reduct of some normal $\mathfrak{A} \in \mathsf{K}$ as an antichain.

By Lemma 3.12, there is a semi-modular 4-frame $\mathcal{F}$ in $Cm_\omega(\mathfrak{G})$ such that

(12) $\qquad \mathfrak{S}$ is embeddable into $\langle L_{01}^{\mathcal{F}}, \odot^{\mathcal{F}} \rangle$.

Thus, by (12), $\mathfrak{S} \not\models q$ implies that $\langle L_{01}^{\mathcal{F}}, \odot^{\mathcal{F}} \rangle \not\models q$. Therefore, by Lemma 3.10,

(13) $\qquad Cm_\omega(\mathfrak{G}) \not\models q^\odot$.

But (11), by Lemma 3.14, implies that

(14) $Cm_\omega(\mathfrak{G})$ is embeddable into the $\langle +, \cdot, ; \rangle$-reduct of $\mathfrak{A}$.

Now (13) and (14) together imply $\mathfrak{A} \not\models q^\odot$. From here the proof can be completed as in the (I)(i) case. □

We will use the following generalization of Post's theorem.

Theorem 3.17 *[Cf. Gurevich and Lewis 1984] The quasiequational theory of finite semigroups is recursively inseparable from the set of quasiequations falsifiable in some semigroup that is, any set Q such that*

$$\{q \in Queq : \mathsf{SG} \models q\} \subseteq Q \subseteq \{q \in Queq : \mathsf{FSG} \models q\}$$

is undecidable.[14]

Now we can complete the proof of Theorem 3.6. Let K be as in (I) or as in (II). Let $c(x)$ be the following unary term of the language of K:

$$c(x) \stackrel{\text{def}}{=} x + (1\,;x) + (x\,;1) + (1\,;x\,;1).$$

Then property (a) holds for c in K, because the $+$-reduct of any algebra in K is a $+$-semilattice.

For property (b_1): the distributivity of $;$ over $+$ implies its monotonicity, thus

$$(y\,;z) + c(x) \leq [(y + c(x))\,;(z + c(x))] + c(x)$$

holds. For the other direction, by distributivity and associativity of $;$,

$$[(y + c(x))\,;(z + c(x))] \leq (y\,;z) + (1\,;c(x)) + (c(x)\,;1) \leq (y\,;z) + c(x).$$

For (c): In any normal $\mathfrak{A} \in \mathsf{K}$, $c(0^\mathfrak{A}) = 0^\mathfrak{A}$ obviously holds.

If K is as in (III) then let c be the following.

$$c(x) \stackrel{\text{def}}{=} x + (1\,;x) + (x\,;1) + (1\,;x\,;1) + \\ + x^\smile + (1\,;x)^\smile + (x\,;1)^\smile + (1\,;x\,;1)^\smile.$$

It is obvious that this c also has properties (a) and (c). It is easy to check that (13) of Definition 3.5 and $x^{\smile\smile} \leq x$ together imply that $1\,;x^\smile \leq 1\,;x\,;1$ and $x^\smile\,;1 \leq 1\,;x\,;1$ hold. Therefore c has property (b_1). By distributivity of $\smile$ over $+$ and by $x^{\smile\smile} \leq x$ again, we also have $c(x)^\smile \leq c(x)$, thus, by distributivity again, c has (b_2) as well.

Similarly, it can be proved that if K is as in (IV) and c is

$$c(x) \stackrel{\text{def}}{=} x + (1\,;x) + (x\,;1) + (1\,;x\,;1) + \\ + (1 \triangleleft x) + (x \triangleleft 1) + (1 \triangleright x) + (x \triangleright 1) + \\ + ((1\,;x) \triangleleft 1) + (1 \triangleleft (1\,;x)) + ((x\,;1) \triangleright 1) + (1 \triangleright (x\,;1)) + \\ + ((1 \triangleleft x)\,;1) + ((x \triangleleft 1) \triangleright 1) + (1 \triangleright (x \triangleleft 1)).$$

---

[14] See Theorem 4.3 for stronger results from Gurevich and Lewis 1984

then c has properties (a), (b_1), (b_3), (c).

Now in each cases let $Q \stackrel{\text{def}}{=} \{q \in Queq : \mathsf{K} \models e_c(q^\odot)\}$. Since, by the above, c satisfies the corresponding assumptions, Proposition 3.16 implies that
$$\{q \in Queq : \mathsf{SG} \models q\} \subseteq Q \subseteq \{q \in Queq : \mathsf{FSG} \models q\}^{15}.$$
Thus, by Theorem 3.17,[16] Q is undecidable. Since the translation functions $\odot$ and e_c are recursive, this proves the undecidability of the equational theory $Eq(\mathsf{K})$ of K. □

Now we show how the 'logical' statements of Theorems 3.1 and 3.4 follow from the 'algebraic' results of Theorem 3.6.

Proof of Theorem 3.1. Observe that Boolean symmetric difference $\oplus$ satisfies (2)(i)–(iii) of Definition 3.5. Then, by Fact 1.4, Theorem 3.1(i) and (iv) are straightforward consequences of Theorem 3.6(I)(i) and (II), respectively. We show that Theorem 3.1(ii) also follows from Theorem 3.6(I)(i). Let $\mathfrak{S}$ be a finite semigroup. Then there is a homomorphism h from the free semigroup with two generators onto $\mathfrak{S}$. Then the function
$$f(y) \stackrel{\text{def}}{=} \{x : h(x) = y\} \quad (\text{for all } y \in S)$$
embeds $\mathfrak{S}$ into the complex algebra $\mathfrak{A}$ of the free semigroup with two generators, which is the algebraic counterpart of language models over a two element alphabet. By the assumption on logic $\mathcal{L}$, $\mathfrak{A}$ belongs to $\mathsf{Alg}(\mathcal{L})$. Since $\emptyset \in A$ and $x \oplus^{\mathfrak{A}} x = x$; $^{\mathfrak{A}}\emptyset = \emptyset$; $^{\mathfrak{A}}x = \emptyset$, the assumption of (I)(i) of Theorem 3.6 is satisfied. Theorem 3.1(iii) can be proved similarly, since the range of function f above is finite. □

Proof of Theorem 3.4. It follows from Theorem 3.6(III). □

4 Applications and Connections to Other Approaches

4.1 Non-Classical Logics

As we already mentioned in the proof of Theorem 3.1, Boolean symmetric difference $\oplus$ is a special case of a binary operation $*$ of Definition 3.5 (2)(i)–(iii). However, such an operation is only a "touch" of symmetric difference. E.g. $*$ is not necessarily commutative. Further, $x * x$ is far from being 0-like in general. Indeed, let $\mathfrak{A} = \langle +, * \rangle$ be an arbitrary $+$-semilattice with $*$ satisfying (2)(i)–(iii) of Definition 3.5, and let

$$H \stackrel{\text{def}}{=} \{x \in A : x = y * z \text{ for some } y, z \in A\}$$
$$At(H) \stackrel{\text{def}}{=} \{(x * x) * (x * x) : x \in A\} = \{(x * x) * (x * x) : x \in H\}.$$

Then H is "atomic" i.e., below each element of H there is an element of $At(H)$. This set $At(H)$ of "atoms" can be almost anything, it can form an

---
[15] In case of (I)(ii), with S in place of FSG
[16] In case of (I)(ii), by the assumption on S

arbitrarily large antichain, or there can be a lot of elements below $At(H)$ (but not members of H).

Now let us dualize our assumptions on $*$. Namely, assume that

(2) (i)^d $x * y \leq (x * y) * (x * y)$
(ii)^d $x \leq y * y \implies y * y \leq x * x$
(iii)^d $(x * y) \cdot x \leq y$.

Then what is called *(weak) implication* in the literature (cf. e.g. Nemitz 1965) is a special case of a $*$ satisfying (2)(i)^d–(iii)^d. $\cdot\cdot$-semilattices $\langle A, \cdot, 1, * \rangle$ with greatest element 1 and with $*$ being a weak implication are also called *Brouwerian semilattices* (cf. Köhler 1981).

Brouwerian semilattices are reducts of Heyting algebras (where the usual notation for $*$ is $\to$). Since the class of Heyting algebras are the algebraic counterpart of intuitionistic propositional logic, our results have e.g. the following corollary.

Theorem 4.1 *Let $\mathcal{L}$ be a propositional intuitionistic logic with a binary associative $\square$-type modality that is, assume that (i)–(iii) below hold for $\mathcal{L}$.*[17]

(i) The language of $\mathcal{L}$ consists of connectives $\wedge, \vee, \to, \neg, \bullet$.

(ii) The axioms of $\mathcal{L}$ are

(IPT) all tautologies of intuitionistic propositional logic;
($K^\bullet$) $((p_0 \wedge p_1) \bullet p_2) \leftrightarrow ((p_0 \bullet p_2) \wedge (p_1 \bullet p_2))$
  $(p_0 \bullet (p_1 \wedge p_2)) \leftrightarrow ((p_0 \bullet p_1) \wedge (p_0 \bullet p_2))$.
($A^\bullet$) $((p_0 \bullet p_1) \bullet p_2) \leftrightarrow (p_0 \bullet (p_1 \bullet p_2))$.

(iii) The set of theorems of $\mathcal{L}$ is closed under substitution, modus ponens and the following rule:

$$\frac{\varphi \leftrightarrow \psi}{(\varphi \bullet \chi) \leftrightarrow (\psi \bullet \chi)} \qquad \frac{\varphi \leftrightarrow \psi}{(\chi \bullet \varphi) \leftrightarrow (\chi \bullet \psi)}.$$

Then $\mathcal{L}$ is undecidable.

4.2 "Almost Trivial" Undecidable Logics

So far we proved that a large part of the lattice of the logics considered in this paper consists of undecidable logics. For example, such is any part "between" the logic having all possible Kripke frames with an associative $\circ$ and any logic having only one group frame over some special infinite group. Let us step back and ask ourselves about how well our landscape or lattice of logics is mapped for undecidability. We can conclude that most of our "signs" or "landmarks" seem to converge around those areas where the logics are "compatible with" at least one group frame. Therefore, in order to complete the picture, let us strike out in a completely different

---

[17]For more about intuitionistic modal logics see e.g. Fischer Servi 1977.

direction. Let us choose the "direction" in which logics not compatible with any group frame lie. Theorem 4.2 below shows the undecidability of logics which lie in this new direction namely, their theorems are not valid in any group frame at all. On the other hand, these logics are "almost trivial" in the sense that strong postulates hold for them. This is the sense we used the word "trivial" in formulating Theorem 3.1.

Theorem 4.2 *(i) Let logic $\mathcal{L}$ be the minimal logic with an associative binary modality having the following axiom:*

$$\Diamond\varphi \wedge \neg\varphi \to \Diamond((\varphi \circ \varphi) \wedge \neg\varphi),$$

where $\Diamond\varphi \stackrel{\text{def}}{=} \varphi \vee (\varphi \circ \mathsf{T}) \vee (\mathsf{T} \circ \varphi) \vee (\mathsf{T} \circ \varphi \circ \mathsf{T})$. Then $\mathcal{L}$ is undecidable.

(ii) Let logic $\mathcal{L}$ be the minimal logic with an associative binary modality such that rule

$$\frac{\varphi \circ \varphi \to \varphi}{\varphi \leftrightarrow (\mathsf{T} \circ \varphi \circ \mathsf{T})}$$

is admissible[18] to $\mathcal{L}$. Then $\mathcal{L}$ is undecidable.

We will use another result of Gurevich and Lewis 1984. Let $\mathfrak{S} = \langle S, ; \rangle$ be a semigroup. By a *zero of* $\mathfrak{S}$ we understand some $\mathbf{z} \in S$ such that $\mathfrak{S} \models \forall x \ (x; \mathbf{z} = \mathbf{z}; x = \mathbf{z})$. We say that $\mathfrak{S}$ is a *semigroup with zero* iff $(\exists \mathbf{z} \in S)(\mathbf{z}$ is a zero of $\mathfrak{S})$.

Theorem 4.3 (Gurevich and Lewis 1984)) *The quasiequational theory of finite cancellation[19] semigroups with zero is recursively inseparable from the set of quasiequations falsifiable in some semigroup.*

Corollary 4.4 *The quasiequational theory of all semigroups with zero is recursively inseparable from the set of quasiequations falsifiable in some semigroup.*

Proof of Theorem 4.2. Let $\mathcal{L}$ be any of the logics occurring in the theorem. We show that any semigroup $\mathfrak{S} = \langle S, ; \rangle$ with zero can be embedded into the $\circ$-reduct of some algebra $\mathfrak{A} \in \mathsf{Alg}(\mathcal{L})$ in such a way that zero $\mathbf{z}$ goes to $1^{\mathfrak{A}}$, where for any $X, Y \subseteq S$, $X \circ Y$ is defined as follows.

$$X \circ Y \stackrel{\text{def}}{=} \begin{cases} \{x; y : x \in X, y \in Y\} & \text{if } \mathbf{z} \neq x; y \text{ for some } x \in X, y \in Y \\ S & \text{else} \end{cases}$$

Now let $\mathfrak{A} \stackrel{\text{def}}{=} \langle \mathcal{P}(S), \cup, \cap, -, S, \emptyset, \circ \rangle$. Then it is not hard to check that $\mathfrak{A} \in \mathsf{Alg}(\mathcal{L})$ (in both cases), and

$$f(x) \stackrel{\text{def}}{=} \begin{cases} S & \text{if } x = \mathbf{z} \\ \{x\} & \text{else} \end{cases}$$

---

[18] A rule $\varphi_1, \ldots, \varphi_n \ / \ \varphi$ is called *admissible* to logic $\mathcal{L}$, iff for every model $\mathfrak{M}$ of $\mathcal{L}$ such that $\mathfrak{M} \models \varphi_1 \wedge \cdots \wedge \varphi_n$, we also have $\mathfrak{M} \models \varphi$.
[19] We omit here the definition of *cancellation semigroups*, since it is not needed. It can be found in Gurevich and Lewis 1984.

embeds $\mathfrak{S}$ into the $\circ$-reduct of $\mathfrak{A}$ such that $f(\mathbf{z}) = S = 1^{\mathfrak{A}}$.

Now, by Corollary 4.4, $\mathsf{Alg}(\mathcal{L})$ satisfies the assumption of Theorem 3.6 (I) (ii). Therefore, by Fact 1.4, $\mathcal{L}$ is undecidable. □

4.3 Modal Logics Embedding Lambek Calculus

The *weakest modal logic $\mathcal{L}_{LC}$ embedding Lambek Calculus* has the following connectives: the usual Booleans and binary connectives $\circ$, $\backslash$ and $/$. A Kripke model for $\mathcal{L}_{LC}$ is a triple $\mathfrak{M} = \langle W, C, V \rangle$, where W is a set, $C \subseteq W \times W \times W$ and V is a function from the set of propositional variables into the subsets of W. The truth definitions are as follows (see e.g. Došen 1992, Roorda 1991):

$\mathfrak{M}, w \Vdash \varphi \circ \psi$ iff $(\exists u, v)[C(w, u, v)$ and $\mathfrak{M}, u \Vdash \varphi$ and $\mathfrak{M}, v \Vdash \psi]$,

$\mathfrak{M}, w \Vdash \varphi \backslash \psi$ iff $(\forall u, v)[\text{if } C(u, u, w) \text{ then } (\mathfrak{M}, u \Vdash \neg\varphi \text{ or } \mathfrak{M}, v \Vdash \psi)]$,

$\mathfrak{M}, w \Vdash \varphi / \psi$ iff $(\forall u, v)[\text{if } C(v, w, u) \text{ then } (\mathfrak{M}, u \Vdash \neg\varphi \text{ or } \mathfrak{M}, v \Vdash \psi)]$.

Logic $\mathcal{L}_{LC}$ has all possible Kripke models without any restriction on W and C but satisfying "$(p_0 \circ p_1) \circ p_2 \leftrightarrow p_0 \circ (p_1 \circ p_2)$".

$\mathcal{L}_{LC}$ faithfully embeds Lambek Calculus, see e.g. Buszkowski 1986, Roorda 1991. For more on modal logics having connectives $\circ$, $\backslash$ and $/$, see section 7 of chapter 1, section 5 of chapter 2, and chapter 6.

Theorem 4.5 *(i) $\mathcal{L}_{LC}$ is undecidable.*

(ii) Any extension of $\mathcal{L}_{LC}$ satisfying any one of conditions (i)–(iv) of Theorem 3.1 is undecidable. For example:

- *The extension of $\mathcal{L}_{LC}$ with axiom "$(p_0 \circ p_1) \leftrightarrow (p_1 \circ p_0)$" is undecidable.*
- *Let $\mathcal{L}$ be any extension of $\mathcal{L}_{LC}$ with further axioms (which can involve other arbitrary new logical connectives). If the $\circ$-fragment of $\mathcal{L}$ is valid in all finite semigroup frames then $\mathcal{L}$ is undecidable.*
- *The so called resource-sensitive logics DL, DX, DD, DE, DU, DF in Roorda 1991, 1992 are all undecidable. Moreover, they remain undecidable if we add axioms to them which are true in all language models.*

Proof. All the statements follow from Theorem 3.1. □

4.4 Two Dimensional Logics of Pratt

Theorem 4.6 *The following two dimensional logics (using the terminology of Pratt 1994) are undecidable (i.e., they are varieties having undecidable equational theories): Boolean Monoids, Boolean Semirings (**BSR**), Residuated Boolean Monoids (**RBM**), Relation Algebras (**RA**), Boolean Semirings with transitive closure (**BSRT**), Relation Algebras with transi-*

tive closure (**RAT**), Residuated Boolean Monoids with transitive closure (**BACT**).

Consider the decidable[20] *variety of Idempotent Semirings* **ISR**. *Add a new binary operation* $*$, *and add axioms (2)(i)–(iii) of Definition 3.5 above. Denote the so obtained variety by* **ISR***. *Then Eq(***ISR****) is undecidable.*

Proof. All the statements follow from Theorem 3.1. □

5 Undecidable Logics with a Binary Conjugated Modality

So far we have seen that associativity causes undecidability of many logics. Now we discuss an other axiom having the same effect. This axiom, the *Euclidean* law, can be formulated with the help of new connectives.

Definition 5.1 $\mathcal{L}$ is a *logic with a binary conjugated modality* iff (i–iii) below hold.

(i) $\mathcal{L}$ is a logic with a binary modality $\circ$.

(ii) The language of $\mathcal{L}$ also includes the binary connectives $\triangleright$ and $\triangleleft$ (the *conjugates* of $\circ$).

(iii) $\mathcal{L}$ also has the following axioms.

$$(p_0 \wedge (p_1 \triangleleft p_2)) \to (p_1 \wedge (p_0 \circ p_2)) \triangleleft p_2$$
$$(p_0 \wedge (p_1 \circ p_2)) \to (p_1 \wedge (p_0 \triangleleft p_2)) \circ p_2$$
$$(p_0 \wedge (p_1 \triangleright p_2)) \to p_1 \triangleright (p_2 \wedge (p_1 \circ p_0))$$
$$(p_0 \wedge (p_1 \circ p_2)) \to p_1 \circ (p_2 \wedge (p_1 \triangleright p_0)). \quad \square$$

As a consequence, conjugates $\triangleright$ and $\triangleleft$ also distribute over $\vee$ (thus they are monotonous as well). It is easier to 'see' conjugates from the semantical point of view. If $\mathfrak{M} = \langle W, C, V \rangle$ is a model then the new clauses in the truth definition are

$\mathfrak{M}, w \Vdash (\varphi \triangleright \psi)$ iff $(\exists u, v) [C(v, u, w), \mathfrak{M}, u \Vdash \varphi \text{ and } \mathfrak{M}, v \Vdash \psi]$
$\mathfrak{M}, w \Vdash (\varphi \triangleleft \psi)$ iff $(\exists u, v) [C(u, w, v), \mathfrak{M}, u \Vdash \varphi \text{ and } \mathfrak{M}, v \Vdash \psi]$.

We note that conjugates are duals of connectives \ and / (introduced in subsection 4.3 above) that is, $x \triangleright y = -(x \setminus -y)$ and $x \triangleleft y = -(x / -y)$.

Of course, the former theorems extend for logics with a binary conjugated modality, since already the $\circ$-fragments of the logics in question are undecidable.

Corollary 5.2 *(i) The minimal logic with an associative binary conjugated modality is undecidable.*

(ii) Any extension of the minimal logic with an associative binary conjugated modality, satisfying any of the conditions of Theorem 3.1, is undecidable.

---

[20] Cf. Theorem 6.3 below.

A logic $\mathcal{L}$ with a binary conjugated modality is called *Euclidean* iff $\mathcal{L}$ also has a constant connective $\iota\delta$, and $\mathcal{L}$ contains the following axioms:

(E) $((p_0 \triangleright p_1) \circ p_2) \to (p_0 \triangleright (p_1 \circ p_2))$
(U) $(p_0 \circ \iota\delta) \leftrightarrow p_0$
    $(\iota\delta \circ p_0) \leftrightarrow p_0$.

Axiom (E) is 'of similar character' as one half of associativity, in the sense that the corresponding frame conditions look similar, as the picture below shows.

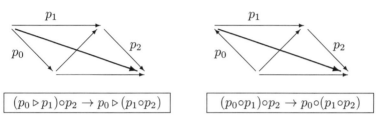

A frame for a logic with a binary conjugated modality is called *symmetric* iff formula "$(p_0 \triangleright \iota\delta) \leftrightarrow p_0$" is valid in it. If all the frames of a logic $\mathcal{L}$ are symmetric then $\mathcal{L}$ is called symmetric as well. A symmetric frame is called *totally symmetric* iff it also validates formula "$p_0 \to (p_0 \circ p_0)$".

Let $\mathcal{L}_{EU}$ denote the minimal normal Euclidean logic with a binary conjugated modality.

Theorem 5.3 *(i) $\mathcal{L}_{EU}$ is undecidable.*
(ii) Any extension of $\mathcal{L}_{EU}$ + "$\circ$ is commutative" whose $\langle \circ, \triangleright, \triangleleft, \iota\delta \rangle$-fragment[21] is valid in all totally symmetric frames is undecidable.
(iii) Any extension of $\mathcal{L}_{EU}$ whose $\langle \circ, \triangleright, \triangleleft, \iota\delta \rangle$-fragment is valid in the group frames[22] over arbitrarily large symmetric (permutation) groups is undecidable.
(iv) Any extension of symmetric $\mathcal{L}_{EU}$ whose $\langle \circ, \triangleright, \triangleleft, \iota\delta \rangle$-fragment is valid in the group frames over arbitrarily large Boolean groups[23] is undecidable.
(v) Any extension of $\mathcal{L}_{EU}$ + "$\circ$ is commutative" whose $\langle \circ, \triangleright, \triangleleft, \iota\delta \rangle$-fragment is valid in the group frames over arbitrarily large Boolean groups is undecidable.

Proof. The statements are proved in Németi et al. 1995. □

---
[21] I.e., those theorems which include only $\circ$, $\triangleright$, $\triangleleft$, $\iota\delta$ and the Booleans
[22] The notion of a group frame can be extended in a natural way to evaluate $\triangleright$ and $\triangleleft$.
[23] A group $\mathfrak{G} = \langle G, ; , ^{-1}, e \rangle$ is called *Boolean* iff $\mathfrak{G} \models x;x = e$ (i.e., $\mathfrak{G}$ is of order 2). Every Boolean group is commutative.

For further results concerning undecidable Euclidean logics see Németi et al. 1995.

Next we state the solutions of some open problems from Jipsen 1992, concerning varieties which are algebraic counterparts of Euclidean logics. For this, we use the notation of Jipsen 1992 without recalling it.

Corollary 5.4 *All the classes listed in Problem 3.61 of Jipsen 1992 have undecidable equational theories. Namely,*

(i) ERM, IERM, CERM, RM, IRM, CRM, ICRM, ARM, SERM, ISERM *and;*

(ii) EUR, IEUR, SEUR, ISEUR, CEUR, ICEUR *have undecidable equational theories.*

Proof. All the statements of (i) follow from Theorem 3.6. The statements of (ii) follow from Theorem 5.3. □

6 What Can We Save from Associativity? — Decidable Versions

So far we have seen that associativity kills decidability in many logics. Is there a way out of this difficulty? Can we have at least some associativity and decidability together? Below we outline some positive solutions.

Consider the following weaker but quite a powerful and useful version of associativity.

(WA) $((\varphi \wedge \iota\delta) \circ \top) \circ \top \leftrightarrow (\varphi \wedge \iota\delta) \circ (\top \circ \top)$,

where $\top$ abbreviates the formula "$p_0 \to p_0$". (This weakening of associativity was investigated first by Maddux 1978 in connection with Relation Algebras.)

Let AL_{1-5} and AL_{1-8} denote the arrow logics defined by axioms (A1)–(A5) and (A1)–(A8) of chapter 1, respectively (cf. also van Benthem 1994).

Theorem 6.1 *(i)* AL_{1-5} *is decidable.*

(ii) $AL_{1-5} + (WA)$ *is decidable.*

(iii) $AL_{1-8} + (WA)$ *is decidable.*

Proof. It is proved in Németi 1987 (cf. also Mikulás et al. 1995) that the equational theories of the algebraic counterparts of the logics in Theorem 6.1 are decidable by showing that these classes all have the *finite algebra property*. This property says that if an equation fails in an algebra of the class in question then it must already fail in a finite algebra. The filtration given in chapter 2 (Lemma 3.5) can also be used to prove this theorem. □

Now let us investigate more expressive logics with a binary modality, namely add to arrow logic new logical connectives, like the *difference operator* D, *universal modality* E, the *"graded"* or "stratified" modalities

$\langle n\text{-times}\rangle$ $(n \in \omega)$, [24] and the *Kleene-star* $*$. The truth definitions of these new connectives in a model $\mathfrak{M} = \langle W, C, \ldots, V \rangle$ are as follows.

$\mathfrak{M}, w \Vdash D\varphi$ iff $(\exists v \in W)[w \neq v$ and $\mathfrak{M}, v \Vdash \varphi]$,
$\mathfrak{M}, w \Vdash \langle n\text{-times}\rangle \varphi$ iff $|\{v \in W : \mathfrak{M}, v \Vdash \varphi\}| \geq n$,
$\mathfrak{M}, w \Vdash \varphi^*$ iff w can be "C-decomposed" into some finite sequence of worlds satisfying φ.

The universal modality E is defined by $E\varphi \stackrel{\text{def}}{=} \varphi \vee D\varphi$. In Sain 1983 it was proved that D increases the program verifying powers of dynamic logic and of some temporal logics of computation. $\langle n\text{-times}\rangle$ was introduced by Sain in 1985 and it was applied to reasoning about actions in Pásztor 1992. Sain 1988 pointed out that for $n < 3$, $\langle n\text{-times}\rangle$ is expressible from D (but to express D we need both $\langle 1\text{-times}\rangle$ and $\langle 2\text{-times}\rangle$).

Theorem 6.2 *Any logic in Theorem 6.1 remains decidable if we add D, or E, or $\langle n\text{-times}\rangle$, or $*$ to it.*

Proof. The statements concerning D alone are proved in Andréka et al. 1994b, see also Mikulás 1995, Marx et al. 1995. The $\langle n\text{-times}\rangle$ cases are treated in Mikulás 1995, Mikulás and Németi 1995. For the statements including Kleene-star but without (WA) see van Benthem 1994, for the weakly associative case see Mikulás et al. 1995. The additions of $*$ (and E) are treated in chapter 5; these decidability results can easily be obtained using the "star lemma" from that chapter (cf. also Marx 1995). □

The significance of Theorem 6.2 is that the logic which is obtained from $AL_{1-8} + $ (WA) by adding connectives D and $\langle n\text{-times}\rangle$ $(n \in \omega)$, is very expressive. It is decidable, and in some directions it is more expressive than the undecidable associative logic $AL_{1-8} + $ (A) (the logical equivalent of Relation Algebras), e.g. already $\langle 4\text{-times}\rangle$ is not expressible in the latter.

But what if we want to keep full associativity and still want some decidability? The theorem below shows in an algebraic setting that it is possible by weakening the Boolean part.

Theorem 6.3 *The following varieties all have decidable equational theories:*

*(i) Idempotent semirings (**ISR**) (i.e., algebras $\langle A, +, 0, ; \rangle$ with $+$ semilattice with smallest element 0 and $;$ associative and distributing over $+$),*

(ii) DL-semigroups,

(iii) bounded DL-semigroups,

---

[24] See e.g. Sain 1984, Sain 1988, Gargov et al. 1987, Ohlbach 1993, van der Hoek 1992 and the references therein.

(iv) bounded DL-monoids (i.e., bounded DL-semigroups with ;-neutral element id).

Proof. The proofs can be found in Andréka and Kurucz 1995. □

However, Fact 1.4 cannot be used to formulate the "logical" statements corresponding to Theorem 6.3 above, since in all cases connective $\to$ is missing (and it is not expressible from the others). For algebraization of such logics, the reader is kindly referred to e.g. Blok and Pigozzi 1992.

Finally we discuss another direction, in which one can obtain an associative but decidable logic, namely by omitting axiom (K°). We say that $\mathcal{L}$ is a *classical logic with a binary modality* iff $\mathcal{L}$ contains (PT) as axiom and the set of validities of $\mathcal{L}$ is closed under substitution, modus ponens and replacement of equivalents.[25]

Theorem 6.4 *The minimal associative classical logic with a binary modality is decidable.*

Proof. It is a consequence of a general result of Pigozzi 1974, saying that the "join" of two disjoint decidable equational theories is decidable. □

More about associative decidable logics without the axiom (K°) can be found in chapter 4.

Acknowledgment. We are especially grateful to Steven Givant for extensive help. Thanks are due to Peter Jipsen, Maarten Marx, Szabolcs Mikulás and Vaughan Pratt.

References

Andréka, H., S. Givant, and I. Németi. to appear. Decision Problems for Equational Theories of Relation Algebras. *Memoirs of the American Mathematical Society.*

Andréka, H., and Á Kurucz. 1995. Some Decidable Classes of Algebras and Logics. Preprint. Math. Inst. Hungar. Acad. Sci., Budapest.

Andréka, H., Á. Kurucz, I. Németi, and I. Sain. 1994a. Applying algebraic logic; A general methodology. In *Algebraic Logic and the Methodology of Applying it*, ed. H. Andréka, I. Németi, and I. Sain, 1–72. A shortened version appeared as *Applying Algebraic Logic to Logic* in 'Algebraic Methodology and Software Technology (AMAST'93)', (eds: M.Nivat, C.Rattray, T.Rus and G.Scollo), in series *Workshops in Computing*, Springer-Verlag, 1994, 7–28.

Andréka, H., Sz. Mikulás, and I. Németi. 1994b. You Can Decide Differently: Decidability of Relativized Representable Relation Algebras with the Difference Operator. Preprint. Math. Inst. Hungar. Acad. Sci., Budapest.

van Benthem, J. 1991. *Language in Action (Categories, Lambdas and Dynamic Logic).* Studies in Logic, Vol. 130. North-Holland, Amsterdam.

---

[25] Such logics are called "classical" in the usual mono-modal setting, cf. Segerberg 1971.

van Benthem, J. 1994. A Note on Dynamic Arrow Logics. In *Logic and Information Flow*, ed. J. van Eijck and A. Visser. 15–29. Massachusetts: MIT Press.

Blok, W. J., and D. Pigozzi. 1989. Algebraizable logics. *Memoirs of the American Mathematical Society* 77, 396:vi+78 pp.

Blok, W. J., and D. Pigozzi. 1992. Algebraic Semantics for Universal Horn Logic without Equality. In *Universal Algebra and Quasigroup Theory*, ed. A. Romanowska and J. D. H. Smith. 1–56. Berlin: Heldermann Verlag.

Burris, S., and H. Sankappanavar. 1981. *A Course in Universal Algebra*. Graduate Texts in Mathematics. Springer–Verlag, New York.

Buszkowski, W. 1986. Completeness Results for Lambek Syntactic Calculus. *Z. Math. Logik Grundlag. Math.* 32:13–28.

Davis, M. 1977. Unsolvable Problems. In *Handbook of Mathematical Logic*, ed. J. Barwise. 567–594. North-Holland, Amsterdam.

Došen, K. 1992. A Brief Survey of Frames for the Lambek Calculus. *Z. Math. Logik Grundlag. Math.*

Fischer Servi, G. 1977. On Modal Logics with an Intuitionistic Base. *Studia Logica* 36:141–149.

Gargov, G., S. Passy, and T. Tinchev. 1987. Modal environment for Boolean speculations. In *Mathematical Logic and its Applications*, ed. D. Skorde, 253–263. Plenum Press, New York.

Gurevich, Y., and H. R. Lewis. 1984. The Word Problem for Cancellation Semigroups with Zero. *Journal of Symbolic Logic* 49(1):184–191.

Henkin, L., J. D. Monk, and A. Tarski. 1985. *Cylindric Algebras, Part II*. North-Holland, Amsterdam.

van der Hoek, W. 1992. *Modalities for Reasoning about Knowledge and Quantities*. Doctoral dissertation, Department of Mathematics and Computer Science, Free University, Amsterdam.

Jipsen, P. 1992. *Computer Aided Investigations of Relation Algebras*. Doctoral dissertation, Vanderbilt University, Nashville, Tennessee.

Jónsson, B., and A. Tarski. 1951. Boolean algebras with operators. Parts I and II. *American Journal of Mathematics* 73:891–939. and 74:127–162, 1952.

Köhler, P. 1981. Brouwerian Semilattices. *Transactions of the American Mathematical Society* 158:103–126.

Kurucz, Á., I. Németi, I. Sain, and A. Simon. 1995. Decidable and undecidable modal logics with a binary modality. *Journal of Logic, Language and Information* 4:191–206.

Lipshitz, L. 1974. The Undecidability of the Word Problems for Projective Geometries and Modular Lattices. *Transactions of the American Mathematical Society* 193:171–180.

Maddux, R. 1978. Topics in Relation Algebras. Doctoral Dissertation, Berkeley.

Marx, M. 1995. *Algebraic Relativization and Arrow Logic*. Doctoral dissertation, Institute for Logic, Language and Computation, University of Amsterdam. ILLC Dissertation Series 1995-3.

Marx, M., Sz. Mikulás and I. Németi. 1995. Taming Logic. *Journal of Logic, Language and Information* 4:207–226.

McCarthy, J., and P. J. Hayes. 1969. Some Philosophical Problems from the Standpoint of Artificial Intelligence. In *Machine Intelligence 4*, ed. B. Meltzer and D. Michie. 463–502. Edinburgh University Press.

Mikulás, Sz. 1995. *Taming Logics*. Doctoral dissertation, Institute for Logic, Language and Computation, University of Amsterdam. ILLC Dissertation Series 95-12.

Mikulás, Sz., and I. Németi. 1995. Relativized Algebras of Relations with Generalized Quantifiers (Positive Results). Manuscript. Math. Inst. Hungar. Acad. Sci., Budapest.

Mikulás, Sz., I. Németi, and I. Sain. 1995. Decidable Logics of the Dynamic Trend, and Relativized Relation Algebras. In *Logic Colloquium '92*, ed. L. Csirmaz, D. M. Gabbay, and M. de Rijke, 165–176. Studies in Logic, Language and Computation. Stanford, California. CSLI Publications.

Németi, I. 1980. Some Constructions of Cylindric Algebra Theory Applied to Dynamic Algebras of Programs. *Comp. Linguist. Comp. Lang.* 14:43–65.

Németi, I. 1987. Decidability of relation algebras with weakened associativity. *Proc. Amer. Math. Soc.* 100, 2:340–344.

Németi, I. 1991. Algebraizations of quantifier logics: an overview (version 11.2). Preprint. Math. Inst. Hungar. Acad. Sci., Budapest. A short version without proofs appeared in *Studia Logica*, 50(3/4):485–569, 1991.

Németi, I., I. Sain, and A. Simon. 1995. Undecidability of the equational theories of some classes of residuated Boolean algebras with operators. *Bulletin of the IGPL* 3(1):93–105.

Nemitz, W. C. 1965. Implicative Semi-lattices. *Transactions of the American Mathematical Society* 117:128–142.

von Neumann, J. 1960. *Continuous Geometry (Ed.: J. Halpern)*. Princeton: Princeton University Press.

Ohlbach, H. J. 1993. Translation Methods for Non-classical Logics: An Overview. *Bulletin of the IGPL* 1(1):69–90.

Pásztor, A. 1992. An Infinite Hierarchy of Program Verification Methods. In *Many Sorted Logic and its Applications*, ed. J. V. Tucker and K. Meinke. John Wiley and Sons Ltd. Also available as Florida International University, School of Computer Science, Miami, Technical Report, 1987.

Pigozzi, D. 1974. The Join of Equational Theories. *Colloquium Mathematicum* 30(1):15–25.

Pixley, A. F. 1971. The Ternary Discriminator Function in Universal Algebra. *Math. Ann.* 191:167–180.

Pratt, V. 1994. A Roadmap of Some Two Dimensional Logics. In *Logic and Information Flow*, ed. J. van Eijck and A. Visser. 149–162. Massachusetts: MIT Press.

Roorda, D. 1991. *Resource Logics. Proof-Theoretical Investigations*. Doctoral dissertation, Institute for Logic, Language and Computation, University of Amsterdam.

Roorda, D. 1992. Lambek Calculus and Boolean Connectives: on the Road. OTS Working Paper OTS-WP-CL-92-004. Research Inst. Language and Speech, Utrecht University.

Sain, I. 1978. On Dynamic Algebras and on Boolean Algebras with Operators; Their Generalizations. Math. Inst. Budapest, manuscript.

Sain, I. 1982. Strong Amalgamation and Epimorphisms of Cylindric Algebras and Boolean Algebras with Operators. Preprint. Math. Inst. Hungar. Acad. Sci., Budapest.

Sain, I. 1983. Successor Axioms for Time Increase the Program Verifying Power of Full Computational Induction. Preprint 23/1983. Math. Inst. Hungar. Acad. Sci., Budapest.

Sain, I. 1984. How Can the Difference Operator Improve Modal Logic of Programs? Manuscript of a lecture in D. Scott's seminar, Pittsburgh.

Sain, I. 1988. Is Sometimes 'Some Other Time' Better than 'Sometimes'? *Studia Logica* 47(3):279–301. Longer version appeared as Preprint No.50/1985, Math. Inst. Budapest.

Segerberg, K. 1971. *An Essay in Classical Modal Logic*. Filosofiska Studier 13. University of Uppsala.

Urquhart, A. 1995. Decision Problems for Distributive Lattice-ordered Semigroups. *Algebra Universalis* 33(4):399–418.

Venema, Y. 1991. *Many–Dimensional Modal Logic*. Doctoral dissertation, Institute for Logic, Language and Computation, Universiteit van Amsterdam.

4

Associativity Does Not Imply Undecidability without the Axiom of Modal Distribution

VIKTOR GYURIS

ABSTRACT. We show that non-distributive arrow logic (i.e., in which the modal axiom K fails) is decidable even in the presence of the associativity axiom of ○. Decidability will still hold if we allow addition of several further axioms which involve no Boolean connectives. In the proof we will use the methodology described in Andréka et al. 1994 of applying algebraic tools in solving logical problems. As a consequence of a general theorem of Don Pigozzi 1974, the class of algebras corresponding to non-distributive arrow logic is decidable.

By *Arrow Logic* we understand a propositional multi-modal logic with a binary modality ○, a unary one[1] ˇ and a constant one $\iota\delta$.

From chapter 2 we know that if we put several natural restrictions on the frames of arrow logic the result is a decidable logic provided that the associativity axiom of ○ is not among the restrictions. Furthermore, the authors suggest two additional axioms that can be interpreted as 'cautious' associativity and they still do not imply undecidability.

On the other hand, in chapter 3, it is shown that associativity of ○ forces the logic to be undecidable. Moreover, it is proved that adding almost any further axioms and rules leaves the logic undecidable. For that reason they call that logic essentially undecidable.

The above results might suggest that the associativity axiom alone is to blame for the undecidability. Here we will show that the picture is a little bit more detailed.

---

[1]Editors note: in this chapter the converse operation is denoted by ˇ, instead of ⊗.

Arrow Logic and Multi-Modal Logic
M. Marx, L. Pólos and M. Masuch, eds.
Copyright © 1996, CSLI Publications.

In all versions of arrow logic which appear in this book, the axiom of Modal Distribution (briefly MD) is automatically assumed. Weakening minimal arrow logic, we get what we will call non-distributive arrow logic (see Definition 1 below) in which MD is not automatically satisfied. Theorem 2 shows that if we restrict non-distributive arrow logic by the associativity axiom of $\circ$, then the result is still decidable. Furthermore, some additional restrictions can be added, and the logic remains decidable.

The price we pay for decidability is that our non-distributive arrow logic is not a modal logic since MD is not valid.

Definition 1 Let Ax_0 be the following finite set of formula schemes:

(1) $\qquad (\varphi \circ \psi) \circ \chi \;\leftrightarrow\; \varphi \circ (\psi \circ \chi),$

(2) $\qquad \varphi^{\smile\smile} \;\leftrightarrow\; \varphi,$

(3) $\qquad (\varphi \circ \psi)^{\smile} \;\leftrightarrow\; \psi^{\smile} \circ \varphi^{\smile},$

(4) $\qquad \iota\delta \;\leftrightarrow\; \iota\delta^{\smile},$

(5) $\qquad \iota\delta \circ \varphi \;\leftrightarrow\; \varphi.$

For any $S \subseteq Ax_0$, we define the non-distributive arrow logic $\mathcal{L}_{\vdash_S}$ to be the pair $\langle F, \vdash_S \rangle$ where

- F is the language of arrow logic, i.e., F is the smallest set H satisfying:
  - all the propositional variables are in H,
  - $\iota\delta \in H$,
  - if $\varphi, \psi \in H$ then $(\varphi \vee \psi), (\neg\varphi), (\varphi \circ \psi), (\varphi^{\smile}) \in H$,
- $\vdash_S$ is an inference system whose axioms are S together with the axioms of propositional logic, and whose inference rules are Modus Ponens and the substitution rules

$$\frac{\varphi \leftrightarrow \varphi', \psi \leftrightarrow \psi'}{\varphi \circ \psi \leftrightarrow \varphi' \circ \psi'}, \qquad \frac{\varphi \leftrightarrow \varphi'}{\varphi^{\smile} \leftrightarrow \varphi'^{\smile}},$$

for $\circ$ and $^{\smile}$, respectively.

Theorem 2 *Suppose that $\mathcal{L}_{\vdash_S} = \langle F, \vdash_S \rangle$ is a non-distributive arrow logic for some set $S \subseteq Ax_0$. Then $\mathcal{L}_{\vdash_S}$ is decidable.*

In the proof of the theorem above, we will use the methodology described in Andréka et al. 1994. Namely, we will use a kind of duality theorem (a variant of Theorem B1 in Andréka et al. 1994) according to which, for proving that $\mathcal{L}_{\vdash_S}$ is decidable, it is enough to prove that the equational theory of the class $\mathrm{Alg}_m(\mathcal{L}_{\vdash_S})$ of algebras associated to $\mathcal{L}_{\vdash_S}$ according to Definition 3.1.6 of Andréka et al. 1994 is decidable.[2]

Then we will show that the equational theory of $\mathrm{Alg}_m(\mathcal{L}_{\vdash_S})$ is a 'disjoint union' of two decidable theories and so, according to a result of Don Pigozzi 1974, it is decidable.

---

[2] Editors note: see also the discussion just before Fact 1.4 in chapter 3.

For the proof of Theorem 2, we recall some definitions from Andréka et al. 1994. By a logic (or a logical system) we mean a 4-tuple $\langle F, M, mng, \models \rangle$ where F is a set (of *formulas*), M is a class (of *models*), mng, called the *meaning function*, is a function with domain $F \times M$, and $\models \,\subseteq (F \times M)$ (*validity relation*), such that the condition

$$(mng(\varphi, \mathcal{M}) = mng(\psi, \mathcal{M}) \text{ and } \mathcal{M} \models \varphi) \text{ implies } \mathcal{M} \models \psi$$

is satisfied. For more detail, see Andréka et al. 1994: Definition 2.1.3. Sometimes we concentrate on deductive systems $\mathcal{L}_\vdash = \langle F, \vdash \rangle$ associated to a logic $\mathcal{L}$, and call it a logic (this happened in our Definition 1). Andréka et al. 1994 calls $\mathcal{L}_\vdash$ a logic in the syntactical sense or syntactic logic for short (see pp.8–9 there). It is shown there that, given a syntactic logic $\mathcal{L}_\vdash = \langle F, \vdash \rangle$, one can associate a(n artificial) logic $\mathcal{L} = \langle F, M, mng, \models \rangle$ in the "4-tuple sense" above, as follows. We put

- $M = \{T \subseteq F \mid T \text{ is closed under } \vdash\}$,
- $mng(\varphi, T) = mng_T(\varphi) = \{\psi \in F \mid T \vdash (\varphi \leftrightarrow \psi)\}$, for every $\varphi \in F$ and $T \in M$,
- $T \models \varphi$ if $\varphi \in T$ for every $\varphi \in F$ and $T \in M$.

The semantical consequence relation induced by $\models$ coincides with the original $\vdash$, i.e., for any $\Gamma \subseteq F$,

$$\Gamma \models \varphi \quad \text{iff} \quad \Gamma \vdash \varphi.$$

Therefore, the constructed logic has a strongly complete, strongly sound inference system, namely $\vdash$.

Proof. (Theorem 2) From the non-distributive arrow logic $\mathcal{L}_{\vdash_S}$, we construct the logic $\mathcal{L}_S$ as described above. $\mathcal{L}_S$ is a (strongly) nice logic in the sense of Andréka et al. 1994 Definition 3.1.1. In Andréka et al. 1994, the decidability problem of nice logics is reduced to the problem of deciding the validity of formulas of the form $\tau_1 \leftrightarrow \tau_2$ where $\tau_1, \tau_2 \in F$.

The algebraic counterpart of $\mathcal{L}_S$ is

$$\mathrm{Alg}_m(\mathcal{L}_S) = \mathbf{I}\{mng_T(F) \mid T \in M\},$$

where $mng_T(F)$ is the image of the formula algebra F under the homomorphism mng_T. (See Definition 3.1.7 of Andréka et al. 1994.)

As an immediate consequence of Corollary 3.2.2 of Andréka et al. 1994, we have that for any $\tau_1, \tau_2 \in F$

$$\models \tau_1 \leftrightarrow \tau_2 \quad \text{iff} \quad \mathrm{Alg}_m(\mathcal{L}_S) \models \tau_1 = \tau_2.$$

By Theorem 3.2.3 of Andréka et al. 1994, the quasi-variety generated by $\mathrm{Alg}_m(\mathcal{L}_S)$ is axiomatizable by the following (finite) set Ax of quasi-equations:

- $\tau(x_1, \ldots, x_n) = 1$ if $\tau(\varphi_1, \ldots, \varphi_n)$ is an axiom of $\vdash$ for some natural number n,

- $(x_1 = 1, (x_1 \to x_2) = 1) \Rightarrow x_2 = 1$,
  $((x_1 \leftrightarrow x_2) = 1, (x_3 \leftrightarrow x_4) = 1) \Rightarrow (x_1 \circ x_3 \leftrightarrow x_2 \circ x_4) = 1$,
  $(x_1 \leftrightarrow x_2) = 1 \Rightarrow (x_1^{\smile} \leftrightarrow x_2^{\smile}) = 1$,
- $(x_1 \leftrightarrow x_2) = 1 \Rightarrow x_1 = x_2$,
  $x_1 = x_2 \Rightarrow (x_1 \leftrightarrow x_2) = 1$.

For every axiom $\tau_1 \leftrightarrow \tau_2$ of $\vdash_S$, its equational translation $\tau_1 = \tau_2$ is derivable from Ax. Denote the set of all Boolean axioms together with the equational translations of axioms of S by Ax'. From Ax, the set Ax' can be derived, but this holds in the other direction as well. (To verify the latter statement, we need to use only that in Boolean Algebras it holds that $x_1 \leftrightarrow x_2 = 1$ if and only if $x_1 = x_2$.)

That shows that the quasi-variety V generated by Ax is actually a variety (generated by Ax').

It remains to check that the equational theory of V is decidable. Ax' is the union of two disjoint sets: the set of Boolean axioms and the set S' of axioms coming from S. The equational theory of the variety generated by the Boolean axioms (the class of Boolean Algebras) is decidable. In Proposition 3 below we show that the same holds for the equational theory of the variety generated by S'. The set of Boolean connectives and the set of those connectives that appear in the axioms in S' are disjoint. Using Theorem 1 of Pigozzi 1974 we conclude that V is decidable. This completes the proof. □

Proposition 3 *Suppose S' is a subset of the following set of equations*
(1) $x \circ (y \circ z) = (x \circ y) \circ z$,
(2) $x^{\smile\smile} = x$,
(3) $(x \circ y)^{\smile} = y^{\smile} \circ x^{\smile}$,
(4) $\iota\delta^{\smile} = \iota\delta$,
(5a, b) $\iota\delta \circ x = x, x \circ \iota\delta = x$.
Then the equational theory of the variety generated by S' is decidable.

Proof. We give a decision algorithm, and we show that the decision given by the algorithm answers the problem of equality of two terms.

A complexity measure m is defined in the set of terms by induction: $m(x) = m(\iota\delta) = 1$ for all variables x, $m(\tau_1 \circ \tau_2) = m(\tau_1) + 2m(\tau_2)$, $m(\tau^{\smile}) = 2^{m(\tau)}$. Any axiom of S' can be considered as a term manipulation rule. If a term matches the pattern that appears on the left side of the equation then it can be rewritten using the pattern on the right side of the axiom. E.g. $(\tau_1 \circ \tau_2)^{\smile} \Rightarrow \tau_2^{\smile} \circ \tau_1^{\smile}$. When a rewrite rule of S' is applied to a term, then the complexity of the term decreases. Hence, applying the rewrite rules to a term τ, in less than $m(\tau)$ steps we arrive at a point where no rewrite rule can be applied.

We say that the *normal form of a term* τ_1 is τ_2 if τ_2 has minimal complexity among all the terms that can be reached from τ_1 by multiple use of the rewrite rules of S'. We say that a *term is of normal form* if it coincides with its normal form. Clearly, the transformation to normal form is a computable function.

The decision algorithm is the following. Transform both input terms to normal form and check if they coincide. We need to show only that for any pair τ_1, τ_2 of distinct terms of normal form there is an algebra $\mathcal{A}$ with $\tau_1^{\mathcal{A}} \neq \tau_2^{\mathcal{A}}$ and $\mathcal{A} \models S'$.

First we show it when S' consists of all five equations. If a term of normal form contains $\iota\delta$ as its subterm, then it cannot be in the scope of $\circ$ or $\check{}$ (see (5a,b) and (4)). So the only term of normal form with $\iota\delta$ as a subterm is $\iota\delta$ itself. If $\tau_1^{\vee}$ is a subterm of τ then the main operand of τ_1 cannot be $\check{}$ or $\circ$ (axioms (2) and (3)). So τ_1 is a variable. This shows that τ is of normal form if and only if it is $\iota\delta$ or of the form $(\ldots((\tau_0 \circ \tau_1) \circ \tau_2) \circ \ldots \circ \tau_{n-1})$ where τ_i ($i \in n$) is either x or $x^{\vee}$ for some variable x. Let Σ be a set, A be the set of all finite words over $\Sigma \times \Sigma$. If the universe of the algebra $\mathcal{A}$ is A, its operations are the concatenation and the total conversion (e.g. $([a,b][c,d])^{\vee} = [d,c][b,a]$), and $\iota\delta$ is evaluated to the empty word, then $\tau_1^{\mathcal{A}} \neq \tau_2^{\mathcal{A}}$ whenever τ_1 and τ_2 are distinct terms of normal form.

If S' is all the equations but (5a,b) then a term of normal form is $(\ldots((\tau_0 \circ \tau_1) \circ \tau_2) \circ \ldots \circ \tau_{n-1})$ where τ_i ($i \in n$) is $\iota\delta$, x or $x^{\vee}$ for some variable x. $\mathcal{A}'$ can be chosen to be the same algebra as above with the modification that $\iota\delta$ is evaluated to $[\sigma, \sigma]$ for some $\sigma \in \Sigma$.

If S' is all the equations but (4) then the first case holds since (4) can be derived from S':

$$\iota\delta^{\vee} \stackrel{5}{=} \iota\delta \circ \iota\delta^{\vee} \stackrel{2}{=} \iota\delta^{\vee\vee} \circ \iota\delta^{\vee} \stackrel{3}{=} (\iota\delta \circ \iota\delta^{\vee})^{\vee} \stackrel{5}{=} \iota\delta^{\vee\vee} \stackrel{2}{=} \iota\delta.$$

Suppose that S' is all the equations but (3). A term of normal form is either $\iota\delta$, or does not contain $\iota\delta$ as its subterm. By Theorem 1 of Pigozzi 1974 the equational theory of (1) and (2) is decidable. (We used that the equational theories of the varieties generated by (1) and (2) independently are decidable.) Since any model of (1) and (2) can be expanded by a constant $\iota\delta$ that satisfies (4) and (5), we have proved this case.

Suppose that S' is all the equations but (2). A term of normal form is either $\iota\delta$ or $(\ldots((\tau_0 \circ \tau_1) \circ \tau_2) \circ \ldots \circ \tau_{n-1})$ where τ_i ($i \in n$) is $x^{\vee\cdots\vee}$ for some variable x. Let $f: \Sigma \to \Sigma$ be an arbitrary function. We put $\mathcal{A}''$ to be the algebra described in the first case ($\mathcal{A}$) with the modification that $\check{}$ is the conversion of a word and f-shifting of its letters (e.g. $([a,b][c,d])^{\vee} = [f(c), f(d)][f(a), f(b)])$.

Suppose that S' is all the equations but (1). A term of normal form is either $\iota\delta$ or a member of the free term algebra of type $\{\circ\}$ with generators $\{x, x^{\vee} : x \text{ is a variable}\}$. Under a *binary directed tree* we understand a finite binary tree with a distinguished vertex (called root). For each vertex, the two outgoing edges are labeled by 'left' and 'right'. The *mirroring* of a binary directed tree is changing every label from 'left' to 'right' and from 'right' to 'left'. The *sum* of the trees T_1 and T_2 is T if the subtree on the 'left' edge of the root is T_1 while that on the 'right' is T_2. The set of finite trees on an infinite set with the two operations introduced above forms an algebra that is a model of S' and any two distinct normal formed terms are evaluated into different trees.

We leave the verification of the claim for the other choices of S' to the reader. □

With the proof of this proposition we have completed the proof of Theorem 2. To obtain the arrow logics to which our theorem applies, we stripped off all modal clothes, i.e., the distribution axioms and the necessitation rules:

$$\frac{\varphi}{\neg((\neg\varphi)^{\smile})}, \quad \frac{\varphi}{\neg(\neg\varphi \circ \neg\psi)}, \quad \frac{\varphi}{\neg(\neg\psi \circ \neg\varphi)}.$$

There are several interesting intermediate systems to which the "disjoint union" proof-technique used here does not (immediately) apply, and for which the decidability question is open. We just mention some possibilities.

- add necessitation and distribution for $\smile$;
- add necessitation for $\circ$;
- add the monotonicity rule (i.e., one important half of distribution) for $\circ$:

$$\frac{\varphi \to \varphi'}{\varphi \circ \psi \to \varphi' \circ \psi}, \quad \frac{\varphi \to \varphi'}{\psi \circ \varphi \to \psi \circ \varphi'};$$

- any combination of these.

We end this chapter with establishing the connection between two versions of the modal distribution axiom. Though, Proposition 4 is folklore, at least for the unary case, we state and prove it because of its usefulness.

Proposition 4 *In every non-distributive arrow logic the Axiom K of minimal modal logic implies the Axiom of MD. These axioms can be stated in the following form (see chapter 1).*

Axiom K:
$$\varphi \bullet (\psi_1 \to \psi_2) \leftrightarrow (\varphi \bullet \psi_1) \to (\varphi \bullet \psi_2),$$
$$(\varphi_1 \to \varphi_2) \bullet \psi \leftrightarrow (\varphi_1 \bullet \psi) \to (\varphi_2 \bullet \psi),$$
$$(\varphi \to \psi)^* \leftrightarrow \varphi^* \to \psi^*.$$

Axiom of MD:
$$\varphi \circ (\psi_1 \vee \psi_2) \leftrightarrow (\varphi \circ \psi_1) \vee (\varphi \circ \psi_2),$$
$$(\varphi_1 \vee \varphi_2) \circ \psi \leftrightarrow (\varphi_1 \circ \psi) \vee (\varphi_2 \circ \psi),$$
$$(\varphi \vee \psi)^{\smile} \leftrightarrow \varphi^{\smile} \vee \psi^{\smile}.$$

Here $\varphi \bullet \psi$ and φ^* abbreviates the dual of $\circ$ and $\check{}$ respectively, that is, $\varphi \bullet \psi$ is $\neg(\neg\varphi \circ \neg\psi)$ and φ^* is $\neg(\neg\varphi)^{\vee}$.

Proof. Suppose $\Diamond$ is a unary connective (might be derived from the basic connectives) and $\Box$ is its derived dual connective, i.e., $\Box\varphi = \neg\Diamond\neg\varphi$. Let $\tau(\varphi,\psi)$ be the term $\Box(\varphi \to \psi) \leftrightarrow (\Box\varphi \to \Box\psi)$ and $\sigma(\varphi,\psi)$ be the term $\Diamond(\varphi \vee \psi) \leftrightarrow (\Diamond\varphi \vee \Diamond\psi)$.

We show that τ implies σ. Suppose that τ holds in a non-distributive arrow logic. Then we derive

$\Diamond(\neg\varphi \wedge \psi) \leftrightarrow (\neg\Diamond\varphi \wedge \Diamond\psi)$	since it is $\tau(\neg\varphi, \neg\psi)$,	(1)
$\Diamond(\neg\varphi \wedge \psi) \leftrightarrow (\neg\Diamond\varphi \wedge \Diamond(\varphi \vee \psi))$	since it is $\tau(\neg\varphi, \neg(\varphi \vee \psi))$,	(2)
$(\neg\Diamond\varphi \wedge \Diamond\psi) \leftrightarrow (\neg\Diamond\varphi \wedge \Diamond(\varphi \vee \psi))$	from (1) and (2),	(3)
$(\neg\Diamond(\varphi \vee \psi) \wedge \Diamond\varphi) \leftrightarrow (\neg\Diamond(\varphi \vee \psi) \wedge \Diamond(\varphi \vee \psi))$		(4)
	by substituting $\varphi \vee \psi$ into φ and φ into ψ in (3),	
$\Diamond\varphi \to \Diamond(\varphi \vee \psi)$	by (4),	(5)
$\sigma(\varphi, \psi)$	by (3) and (5).	(6)

If we apply the result proved above to the cases

- $\Diamond\varphi = (\neg\psi \circ \varphi)$ where $\psi \in F$,
- $\Diamond\varphi = (\varphi \circ \neg\psi)$ where $\psi \in F$,
- $\Diamond\varphi = \varphi^{\vee}$,

we get the proof of Proposition 4. $\square$

References

Andréka, H., Á. Kurucz, I. Németi, and I. Sain. 1994. Applying algebraic logic; A general methodology. In *Algebraic Logic and the Methodology of Applying it*, ed. H. Andréka, I. Németi, and I. Sain, 1–72. A shortened version appeared as *Applying Algebraic Logic to Logic* in 'Algebraic Methodology and Software Technology (AMAST'93)', (eds: M.Nivat, C.Rattray, T.Rus and G.Scollo), in series *Workshops in Computing*, Springer-Verlag, 1994, 7–28.

Pigozzi, D. 1974. The Join of Equational Theories. *Colloquium Mathematicum* 30:15–25.

5
Dynamic Arrow Logic
MAARTEN MARX

Arrow logic is similar to Propositional Dynamic Logic (PDL) (cf., Harel 1984) in the sense that both logics are designed to reason about transitions or programs. The difference is that in PDL we can only reason about the *effect* of programs because formulas can only be interpreted at *states*. In arrow logic we can only interpret formulas at *transitions*. We will change arrow logic such that a comparison of the two systems becomes possible. Two changes are needed. First, the key operator of PDL, the Kleene * (which we use here for transitive closure), is missing from the arrow logics discussed so far, and second, we need to be able to reason about states as well in arrow logic.

Van Benthem 1994 expands the language of arrow logic with the Kleene star (*) and indicates how (weak) completeness and decidability can be obtained for various systems, where the formulas are interpreted on Kripke style arrow-frames. In the first part of this paper we give a general lemma, arising from a filtration argument, from which these and more results follow almost immediately. We also show that completeness and decidability can in several cases be obtained with respect to pair-frames as well.

In the second part of this paper we will look at a two-sorted arrow logic, especially designed for reasoning both about states and transitions. This logic is also introduced in van Benthem 1994, but only in the context of arrow-frames. Here we will give a complete axiomatization with respect to locally square pair-frames and show that this logic is decidable.

We end with a comparison between this logic and PDL on the one hand and a similar system called *Peirce Algebras* (Brink et al. 1994) on the other. The two sorted logic of local squares is term definable equivalent with that of Peirce Algebras. The system discussed here can be viewed as a very nice behaving superclass of Peirce Algebras. In sharp contrast with

---

The author is supported by UK EPSRC grant No. GR/K54946.

Arrow Logic and Multi-Modal Logic
M. Marx, L. Pólos and M. Masuch, eds.
Copyright © 1996, CSLI Publications.

(Representable) Peirce Algebras, this larger class is finitely axiomatizable, and its set of validities is decidable.

Connection with Other Contributions to this Volume. We use the terminology and notation developed in chapters 1 and 2. This chapter can be see as an extension of the results on pair-frames in chapter 2 in the research direction outlined in the last section therein.

1 Arrow Logic with the Kleene star

We start by defining the meaning of Kleene * on arrow-frames, using a special accessibility relation. After that, we prove our main result, roughly saying that, if a finitely axiomatizable canonical arrow logic admits filtrations, we have a weakly complete finite derivation system for its expansion with *. Using the filtration and axiomatization results from chapter 2, this gives us several complete and decidable *-expansions of pair arrow logics.

Some of our results could be derived using the ideas and results in van Benthem 1994. The disadvantage of the type of proof presented there is that for every different arrow logic, one has to adjust the argument considerably. Moreover, most of its work is in proving the finite frame property for the logic without *. Since we know this already for several arrow logics, we tried to find a proof which separates the latter issue from the difficulties arising from adding *. The result (lemma 1.1) is widely applicable, and provides a means of reducing the problem of axiomatizing an arrow logic with * to the problem of finding an appropriate filtration for that logic without *.

Meaning of the Kleene star. The intuitive meaning of the Kleene star on arrow-frames is given as follows:

$M, x \Vdash \phi^*$ *iff* $M, x \Vdash \phi$ or x can be finitely C-decomposed into ϕ-arrows.

Note that the star is not a modality, because it does not distribute over disjunction. It is *normal* and *monotonic*. We make its meaning precise, using an accessibility relation between points and sets of points. For that, we use the concept of a *mountain*. Let $\mathcal{F} = \langle W, C, R, I \rangle$ be an arrow-frame. $\mathcal{F}$-mountains are defined inductively:

- for all $x \in W$, the tuple $\langle x, \emptyset, \{x\} \rangle$ is an $\mathcal{F}$-mountain;
- if $Cxyz$, and $\langle y, Y_1, B_1 \rangle$ and $\langle z, Y_2, B_2 \rangle$ are $\mathcal{F}$-mountains, then $\langle x, (\{\langle xyz \rangle\} \cup Y_1 \cup Y_2), (B_1 \cup B_2) \rangle$ is an $\mathcal{F}$-mountain;
- these are all the $\mathcal{F}$-mountains.

Every mountain $\langle x, Y, B \rangle$ represents a finite decomposition of x into arrows from B. We define the accessibility relation for * on a frame $\mathcal{F} = \langle W, C, R, I \rangle$ as a relation $S^{\mathcal{F}} \subseteq W \times \mathcal{P}(W)$:

$$S^{\mathcal{F}}(x, B) \textit{ iff } (\exists Y \subseteq C) : \langle x, Y, B \rangle \text{ is an } \mathcal{F}\text{-mountain.}$$

This relation is determined by the domain of $\mathcal{F}$ and the relation C. Now the meaning of ϕ^* can be defined as follows:

$$M = \langle \mathcal{F}, v \rangle, x \Vdash \phi^* \textit{ iff } (\exists B \subseteq W) : S^{\mathcal{F}}(x, B) \text{ \& } B \subseteq [\![\phi]\!]_M.$$

Let K be a class of frames $\mathcal{F} = \langle W, C, \ldots \rangle$ where $C \subseteq W \times W \times W$. So on these frames we can interpret composition and maybe more modalities. We do not care what the other modalities are. As in chapter 2 we denote the arrow logic of this class by AL(K). Note that the type of K uniquely determines the language of this arrow logic (i.e. for every $n+1$ ary relation R there is an n place modality which gets its meaning in the standard way from R).

If $AL(\mathsf{K})$ is an arrow logic of any type containing at least the composition operator, $AL(\mathsf{K})+^*$ denotes its expansion with *. Note that we do not change the class of models, only the set of formulas and the truth definition.

Compactness. Let ϕ^n stand for the disjunction of all formulas containing n copies of ϕ separated by ∘'s and brackets. (E.g., $\phi^3 = \phi \circ (\phi \circ \phi) \vee (\phi \circ \phi) \circ \phi$.) Intuitively, ϕ^* is equivalent to the infinite disjunction of all ϕ^n. Clearly, $\{\neg \phi^n : n < \omega\} \models \neg \phi^*$, but ϕ^* is always satisfiable together with any finite set of $\neg \phi^n$'s. This shows that the logic is *not compact*, and that means that we can only hope for *weak* completeness in the presence of star.

Obtaining Completeness. Van Benthem 1994 proposed the following axioms and rule for *, which are easily seen to be valid.[1]

(1*) $p \to p^*$
(2*) $p^* \circ p^* \to p^*$

*-rule $\quad \dfrac{\vdash \phi \to \psi, \ \vdash \psi \circ \psi \to \psi}{\vdash \phi^* \to \psi}$

The next lemma shows that these often suffice for weakly axiomatizing *-expansions of arrow logics.

We call an arrow logic AL(K) *canonical* if the canonical frame of the logic is a member of K. For the definition of *filtration* and related notions we refer to chapter 2.

Lemma 1.1 (Star Lemma) *Let $AL(\mathsf{K})$ be an arrow logic of any type containing at least the composition operator. Suppose $AL(\mathsf{K})$ is canonical and that there exists a derivation system $\Omega(\Sigma)$ which is sound and complete with respect to K. If $AL(\mathsf{K})$ admits filtrations, with the restriction that the filtration for the accessibility relation C for the composition operator is minimal, then the set of validities of $AL(\mathsf{K})+^*$ is axiomatizable by the derivation system obtained from $\Omega(\Sigma)$ by adding the axioms $(1^*), (2^*)$ and*

---

[1] At the frame level, (1*) and (2*) correspond to the conditions $\forall x S^{\mathcal{F}} x \{x\}$ and $\forall x y z \forall B_1 B_2 ((Cxyz \wedge S^{\mathcal{F}} y B_1 \wedge S^{\mathcal{F}} z B_2) \to S^{\mathcal{F}} x (B_1 \cup B_2))$, respectively.

the *-rule. Moreover, if $AL(\mathsf{K})$ is finitely axiomatizable, then $AL(\mathsf{K})+^*$ is decidable.

Remark 1.2 The restriction on the relation C is not crucial. If we do not know that the filtration for C is minimal, it is sufficient that the closure set $CL(X)$ (through which we filtrate) satisfies ($\phi^* \in CL(X) \Rightarrow \phi^* \circ \phi^* \in CL(X)$). Since the filtrations used in chapter 2 for arrow-frames are minimal for C, this weaker version of the lemma is sufficient here.

Proof. (Lemma 1.1) Let $AL(\mathsf{K})$ be a canonical arrow logic which admits filtrations with the mentioned restriction, and which is axiomatizable by $\Omega(\Sigma)$. Let $\Omega(\Sigma^*)$ be the derivation system obtained from $\Omega(\Sigma)$ by adding the *-axioms and rule. We have to show that $\Omega(\Sigma^*)$ forms a weakly complete axiomatization of $AL(\mathsf{K})+^*$. So assume $\Omega(\Sigma^*) \not\vdash \phi$. Let $\mathsf{M} = \langle \mathcal{F}, \mathsf{v} \rangle$ be the canonical model of the derivation system $\Omega(\Sigma^*)$. Because $AL(\mathsf{K})$ is canonical, and we assumed that the derivation system is complete for $AL(\mathsf{K})$, the frame $\mathcal{F}$ is a member of K. The problem with this model (due to the failure of compactness) is that it includes ultrafilters containing a formula ψ^* without any "witness" ψ^n. Hence, such an ultrafilter cannot be "finitely C-decomposed into ψ arrows". We will solve this problem by creating a finite model.

Let X^ϕ be the set of subformulas of ϕ. This set will contain formulas of the form ψ^*. From the point of view of the old logic, these mean nothing, so we can just regard them as propositional variables[2]. Then X^ϕ can be viewed as a (finite) set of formulas in the old (*-free) language. By assumption, we can filtrate the canonical model through a set $CL(\phi) \supseteq X^\phi$, obtaining a finite model $\mathsf{M}^* = \langle \mathcal{F}^*, \mathsf{v}^* \rangle$, in which the relation C is defined *minimally*. Define v^* only on the "real" propositional variables. Recall that we used F to denote the domain of the frame $\mathcal{F}$, and that a "minimal filtration" meant that

$$C^*\overline{x},\overline{y},\overline{z} \text{ iff } (\exists uvw \in F) : Cuvw \ \& \ \overline{x} = \overline{u} \ \& \ \overline{y} = \overline{v} \ \& \ \overline{z} = \overline{w}.$$

Note that we also use a superscript * to denote the filtration. (In this proof, a superscript * *only* denotes the Kleene star if it is attached to a formula.) By assumption, $\mathcal{F}^* \in \mathsf{K}$ and $\mathcal{F}^*$ is finite. We are almost ready if we can prove the truth-lemma:

Claim 1 (Truth Lemma)

(**T**) $\qquad (\forall \psi \in CL(\phi))(\forall w \in F) : \psi \in \overline{w} \iff \mathsf{M}^*, \overline{w} \Vdash \psi.$

Proof of Claim. By induction on the complexity of ψ. By assumption, we can perform the inductive proof for all connectives except * (because **min** and **max** are satisfied for all the "old" modalities). For *, we need

---

[2] Note that the only important thing is the outermost * subformula (i.e., $(p \circ q^*)^*$ is regarded as a propositional variable).

additional work. In the sequel, we use δ^w to denote the $CL(\phi)$-formula which uniquely describes the point $\overline{w}$ in the filtration[3]. For every formula ψ^* in the closure set, we define a formula which describes the set of points in the filtration where ψ^* holds by the truth definition. Define this formula as:
$$\psi^\# \stackrel{\text{def}}{=} \bigvee \{\delta^w : M^*, \overline{w} \Vdash \psi^*\}.$$
Because F^* is finite, $\psi^\#$ is a formula.

Claim 2 $(\forall \psi^* \in CL(\phi)) : \Omega(\Sigma^*) \vdash \psi^* \leftrightarrow \psi^\#$.

Proof of Claim. Suppose that $\psi^* \in CL(\phi)$. We start with the easy side. $\Omega(\Sigma^*) \vdash \psi^\# \to \psi^*$. Suppose that $\psi^\# \in w$, for an arbitrary $w \in F$. We have to show that $\psi^* \in w$. By the definition of $\psi^\#$ and the truth definition, we should show that (1) below holds (this is sufficient, because $\psi^* \in \overline{w} \iff \psi^* \in w$).

(1) $[(\exists Y \subseteq C^*) : \langle \overline{w}, Y, B \rangle$ is an $\mathcal{F}^*$-mountain $]$ & $B \subseteq \llbracket \psi \rrbracket_{M^*} \Rightarrow \psi^* \in \overline{w}$.

We do an induction on the "height of the mountains", measured by the cardinality of the set Y. So we do a double induction. Call the induction hypothesis for the truth-lemma IH1, and the one for this claim IH2.

$|Y| = 0$. Suppose $\langle \overline{w}, \emptyset, B \rangle$ is an $\mathcal{F}^*$-mountain and $B \subseteq \llbracket \psi \rrbracket_{M^*}$. Then, by definition, $B = \{\overline{w}\}$, whence $M^*, \overline{w} \Vdash \psi$. But then, by IH1, $\psi \in \overline{w}$, and by axiom (1^*), $\psi^* \in \overline{w}$.

$|Y| = n+1$. We may assume that (1) holds for all smaller Y. Suppose $\langle \overline{w}, Y, B \rangle$ is an $\mathcal{F}^*$-mountain and $B \subseteq \llbracket \psi \rrbracket_{M^*}$. By definition of mountains, we can find $\overline{y}, \overline{z}$ such that $C^*\overline{w}, \overline{y}, \overline{z}$, and $\langle \overline{y}, Y_1, B_1 \rangle$ and $\langle \overline{z}, Y_2, B_2 \rangle$ are $\mathcal{F}^*$-mountains with $Y = (Y_1 \cup Y_2 \cup \langle \overline{w}, \overline{y}, \overline{z} \rangle)$, $B = (B_1 \cup B_2)$, and Y_1 and Y_2 are strictly smaller than Y. So IH2 implies that $\psi^* \in \overline{y}$ and $\psi^* \in \overline{z}$. Now we use our assumption of a minimal definition for C and compute

$C^*\overline{w}, \overline{y}, \overline{z}$ & $\psi^* \in \overline{y}$ & $\psi^* \in \overline{z} \iff$ (assumption)
$(\exists u, v, x) : Cuvx$ & $\overline{u} = \overline{w}, \overline{v} = \overline{y}$ & $\overline{x} = \overline{z}$ &
   & $\psi^* \in \overline{y}$ & $\psi^* \in \overline{z} \Rightarrow$ (because $\psi^* \in CL(\phi)$)
$\psi^* \in v$ & $\psi^* \in x \Rightarrow$ (by definition of C)
$\psi^* \circ \psi^* \in u \Rightarrow$ (axiom (2^*))
$\psi^* \in u \Rightarrow$ (because $\psi^* \in CL(\phi)$ and $\overline{u} = \overline{w}$)
$\psi^* \in \overline{w}$.

This finishes the first half of the claim. The crucial step for the other half is the next claim.

Claim 3 For all δ^w, δ^v belonging to $\psi^\#$, $\Omega(\Sigma^*) \vdash \delta^w \circ \delta^v \to \psi^\#$.

Proof of Claim. Suppose otherwise. Then $\delta^w \circ \delta^v \wedge \neg \psi^\#$ is consistent. Let u be a maximally consistent extension of $\delta^w \circ \delta^v \wedge \neg \psi^\#$ (that is, an

---
[3]See Hughes and Creswell 1984 p.137 how to define such a formula. Note that, $u \equiv_{CL(\phi)} v \iff \vdash \delta^u \leftrightarrow \delta^v$ and $\delta^w \in v \iff \overline{w} = \overline{v}$.

ultrafilter containing $\delta^w \circ \delta^v \wedge \neg \psi^\#$). Then the canonical frame $\mathcal{F}$ has w', v' with $Cuw'v'$ & $\delta^w \in w'$ & $\delta^v \in v'$. Hence in the filtration, $C^*\overline{u}, \overline{w}, \overline{v}$. By definition of $\psi^\#$, $\overline{w} \Vdash \psi^*$ & $\overline{v} \Vdash \psi^*$, hence $\overline{u} \Vdash \psi^* \circ \psi^*$. Thus, $\overline{u} \models \psi^*$ and δ^u belongs to $\psi^\#$, whence $\vdash \delta^u \to \psi^\#$ and $\psi^\# \in u$, a contradiction. ◀

To prove $\Omega(\Sigma^*) \vdash \psi^* \to \psi^\#$, we now proceed as follows. We have $\vdash \psi \to \psi^\#$, since $\vdash \psi \leftrightarrow \bigvee\{\delta^w : \psi \in w\}$ by propositional logic and the fact that $\psi \in CL(\phi)$. By claim 2, using distribution of $\circ$ over $\vee$: $\vdash \psi^\# \circ \psi^\# \to \psi^\#$. So, by the $*$-rule, $\vdash \psi^* \to \psi^\#$. ◀

Now we are ready to finish the proof of the truth-lemma with the inductive step for $*$. Suppose that $\psi^* \in CL(\phi)$. Then

$$\begin{aligned} \psi^* \in \overline{w} &\overset{\text{def}}{\Longleftrightarrow} \psi^* \in w \\ &\Longleftrightarrow (\text{claim 2}) \; \psi^\# \in w \\ &\Longleftrightarrow \delta^{\overline{w}} \in w \; \& \; \mathrm{M}^*, \overline{w} \Vdash \psi^* \\ &\overset{\text{def}}{\Longleftrightarrow} \mathrm{M}^*, \overline{w} \Vdash \psi^*. \end{aligned}$$

◀

We now conclude our main proof. Henceforth, we may assume that (**T**) holds for all formulas in $CL(\phi)$. Now, since $\Omega(\Sigma^*) \not\vdash \phi$, there exists a $w \in F$ (the universe of the canonical frame) such that $\phi \notin w$: whence, by (**T**), $\mathrm{M}^*, \overline{w} \not\Vdash \phi$. Since M^* is an $AL(\mathsf{K})+^*$ model, we have $AL(\mathsf{K})+^* \not\models \phi$. This proves weak completeness. Because M^* is finite, we have also shown that $AL(\mathsf{K})+^*$ has the finite frame property. If the old logic is finitely axiomatizable, the expansion is finitely axiomatizable, so it is also decidable. □

Applications of the Star Lemma. We can apply this lemma for instance to the arrow logics of pair-frames $AL(\mathsf{KP}_H)$ for $H \subseteq \{R, S\}$ (see chapter 2).

Theorem 1.3 *Let $H \subseteq \{R, S\}$. The logic $AL(\mathsf{KP}_H)+^*$ admits a weakly complete axiomatization by the $AL(\mathsf{KP}_H)$ derivation system extended with axioms (1^*) and (2^*) and the $*$-rule.*

Theorem 1.4 *Let $H \subseteq \{R, S\}$. The logic $AL(\mathsf{KP}_H)+^*$ is decidable.*

We prove the two theorems together.

Proof. Let $H \subseteq \{R, S\}$. Lemma 2.5 in chapter 2 states that the class of arrow-frames KA_H equals the class of zigzagmorphic images of the class of pair-frames KP_H. It follows that $AL(\mathsf{KP}_H)+^*$ is equivalent to $AL(\mathsf{KA}_H)+^*$, because the meaning of $*$ is determined by the accessibility relation C of the composition operator. But then, it suffices to prove the theorem for the arrow logics $AL(\mathsf{KA}_H)+^*$. These logics are canonical, finitely axiomatizable, and — by the proof of lemma 3.5 in chapter 2 — they allow filtrations in the restricted sense of lemma 1.1. Now apply that lemma. □

Once $T \in H$, the old arrow logic becomes undecidable, hence its corresponding class of frames cannot allow filtrations, and the above proof does not go through.

Expansion with the Universal Modality. A debatable aspect of the above theorem is that its derivation system is not orthodox in the sense of Definition 18 in chapter 1, because we added an *inference rule*. With the universal modality $\lozenge$ in the language, we can replace the ∗-rule with axiom (3^*):

(3^*) $\qquad \Box((p \to q) \land (q \circ q \to q)) \to (p^* \to q).$

If $\Box$ is the dual of the universal modality, this axiom is valid on pair-frames, and we can derive the ∗-rule from it. The next theorem states that we can also axiomatize this expansion of $AL(\mathsf{KP}_H)+^*$.

Theorem 1.5 *Let $H \subseteq \{R, S\}$. $AL(\mathsf{KP}_H)+\{^*, \lozenge\}$ admits a weakly complete axiomatization by the following derivation system:*

1. *The $AL(\mathsf{KP}_H)$ derivation system;*
2. *an S5 axiomatization for $\lozenge$ (cf. e.g., Hughes and Creswell 1984);*
3. *axioms $\otimes p \to \lozenge p$ and $p \circ q \to \lozenge p \land \lozenge q$ for the interaction of $\lozenge$ with the old connectives[4];*
4. *axioms $(1^*), (2^*)$ and (3^*) for the Kleene star.*

Theorem 1.6 *Let $H \subseteq \{R, S\}$. $AL(\mathsf{KP}_H)+\{^*, \lozenge\}$ is decidable.*

We prove the two theorems together.
Proof. We give a proof-sketch. Take the canonical model $M = \langle \mathcal{F}, v \rangle$ of the derivation system given in the theorem. Filtrate it as described in the proof of lemma 3.5 in chapter 2. Define the relation U^* for the $\lozenge$ in the filtration by $U^*\overline{x}, \overline{y} \stackrel{\text{def}}{\iff} (\lozenge\psi \in \overline{x} \iff \lozenge\psi \in \overline{y})$. Then (by Theorem 8.7 in Hughes and Creswell 1984), this is a filtration and U^* is an equivalence relation. Since both the relations for composition and converse are defined minimally, the axioms in (iii) hold. Define the star in the filtration, and use the Star Lemma to prove the truth-lemma. Suppose the point which falsified the non-derivable formula was $\overline{w}$. Generate the subframe from $\overline{w}$. Here the formula is still falsified, and U^* will be the universal relation. Clearly this frame is finite. Because its frame validates all the arrow-logical axioms, it is a zigzagmorphic image of a pair-frame in which $\lozenge$ is the universal modality. Since the meaning of star is determined by the relation C for the composition operator, the formula will still fail in this pair-frame. □

---

[4]Note that $p^* \to \lozenge p$ is derivable.

2 Two-Sorted Arrow Logic

One can also develop a two sorted arrow logic of pair-frames, as proposed by van Benthem (1994) in the context of abstract arrow-frames. This logic can reason about two domains: both states and transitions. The new language has appropriate modalities to reason within the two domains, and to reason about connections between them. In de Rijke 1993, the importance of many-sorted modal logics is stressed. All applications mentioned there can also be performed in our framework, without occurring undecidability.

The Connectives. The proposed logic has a two-sorted language of state assertions (with meta-variables $\phi, \psi, \ldots$) and transition formulas or *programs* (with meta-variables $\pi_1, \pi_2, \ldots$). Two new connectives, taken from propositional dynamic logic, provide the *connection* between the two sorts:

$$M, x \Vdash \langle \pi \rangle \phi \quad \textit{iff} \quad (\exists y) : M, \langle x, y \rangle \Vdash \pi \ \& \ M, y \Vdash \phi,$$
$$M, \langle x, y \rangle \Vdash \phi? \quad \textit{iff} \quad x = y \ \& \ M, x \Vdash \phi.$$

Van Benthem proposed three simpler connectives[5] from which these two can be defined (see 2.3 below).

$$M, \langle x, y \rangle \Vdash \mathbf{L}\phi \quad \textit{iff} \quad M, x \Vdash \phi,$$
$$M, \langle x, y \rangle \Vdash \mathbf{R}\phi \quad \textit{iff} \quad M, y \Vdash \phi,$$
$$M, x \Vdash \mathbf{D}\pi \quad \textit{iff} \quad M, \langle x, x \rangle \Vdash \pi.$$

Schematically, we can represent the language as follows (see van Benthem 1991, for an explanation of the concepts *mode* and *projection*):

$$\begin{array}{ccc} & \rightarrow \text{ modes } \rightarrow & \\ & \mathbf{L}, \mathbf{R}, ? & \\ \text{propositional logic} & & \text{arrow logic} \\ \text{interpreted on states} & & \text{interpreted on transitions} \\ & \leftarrow \text{ projections } \leftarrow & \\ & \mathbf{D}, \langle . \rangle \phi & \end{array}$$

The **D** and ? connectives make most sense in reflexive pair-frames. If we have a symmetric domain as well, $\mathbf{L}\phi$ and $\mathbf{R}\phi$ become interdefinable. Hence, from now on we will work in this class of frames.

Definition 2.1 State assertions ST and programs PR are the smallest sets satisfying:

- $\{q_i : i < \omega\} \subseteq \mathsf{ST}$, and
- if $\phi, \psi \in \mathsf{ST}$ and $\pi \in \mathsf{PR}$, then $\neg \phi, (\phi \vee \psi), \mathbf{D}\pi \in \mathsf{ST}$.
- $\{p_i : i < \omega\} \cup \{\iota \delta\} \subseteq \mathsf{PR}$, and

---

[5] In the terminology of van Benthem 1991, **D** is the only permutation-invariant projection which is a Boolean homomorphism, and **L** and **R** are the only such modes which are Boolean homomorphisms.

- if $\pi_1, \pi_2 \in \mathsf{PR}$ and $\phi \in \mathsf{ST}$, then $-\pi_1$, $(\pi_1 \cup \pi_2)$, $\otimes \pi_1$, $(\pi_1 \circ \pi_2)$, $\mathbf{L}\phi \in \mathsf{PR}$.

Here, "$-$" denotes the negation of a program, "$\cup$" the disjunction of two programs (for conjunction, we use "$\cap$"). For the Boolean top, we use "$\top$", for the arrow logical one, we use "1". We will use "$\rightarrow$" for material implication in both sorts. Now we can define *propositional dynamic arrow logic* of locally square pair-frames.

Definition 2.2 DAL_{pair} is a triple $\langle \mathsf{Fml}, \mathsf{Mod}, \Vdash \rangle$ in which:

- $\mathsf{Fml} = \mathsf{ST} \cup \mathsf{PR}$.
- $M = \langle Ar, Po, \mathsf{v}^{\mathsf{PR}}, \mathsf{v}^{\mathsf{ST}} \rangle$ is a DAL_{pair} model if Ar is a reflexive and symmetric binary relation with base Po, and

$$\mathsf{v}^{\mathsf{PR}} : \{p_i : i < \omega\} \longrightarrow \mathcal{P}(Ar)$$

and

$$\mathsf{v}^{\mathsf{ST}} : \{q_i : i < \omega\} \longrightarrow \mathcal{P}(Po)$$

are valuation functions for the propositional variables in PR and ST, respectively. Mod is the class of all such models.

- $\Vdash$ gives meaning to the formulas in every model. ST-formulas are interpreted on the set Po (of states) and PR-formulas on the set Ar (of arrows) as one would expect for the given connectives. For the new connectives, $\Vdash$ was defined above.

The next proposition shows that DAL_{pair} is strong enough to capture the mode and projection from PDL.

Proposition 2.3 *On reflexive and symmetric pair-frames, the languages $\{\circ, \otimes, \iota\delta, \mathbf{L}, \mathbf{D}\}$ and $\{\circ, \otimes, \iota\delta, ?, \langle . \rangle.\}$ are equally expressive.*

Proof. First, we express $\mathbf{R}, ?$ and $\langle . \rangle.$ in DAL_{pair}, just as in van Benthem 1994. We need symmetry of the universe for the first, and reflexivity and symmetry for the last clause.

$$\begin{aligned} \mathbf{R}\phi &\leftrightarrow \otimes(\mathbf{L}\phi), \\ \phi? &\leftrightarrow \iota\delta \cap \mathbf{L}\phi, \\ \langle \pi \rangle \phi &\leftrightarrow \mathbf{D}(\pi \circ \mathbf{L}\phi). \end{aligned}$$

On the other hand, with ? and $\langle . \rangle.$ as primitives, one can express $\mathbf{L}$ and $\mathbf{D}$ as follows:

$$\begin{aligned} \mathbf{L}\phi &\leftrightarrow \phi? \circ 1, \\ \mathbf{D}\pi &\leftrightarrow \langle \iota\delta \cap \pi \rangle \top. \end{aligned}$$

For these definitions, only reflexivity is needed. $\square$

Decidability

Theorem 2.4 DAL_{pair} *is decidable.*

Proof. We can give a direct proof, adjusting our previous filtration. An

easier way is provided by the fact that in $AL(\mathsf{KP}_{RS})$, one can encode all
the *state assertions* of DAL_{pair}, viewed as programs which hold only at
identity arrows. Define the following inductive translation function ($°$)
from DAL_{pair} formulas to arrow logical formulas:

$$\begin{array}{rclrcl}
p_i^\circ & = & p_i & q_i^\circ & = & q_i \cap \iota\delta \\
(\neg \pi)^\circ & = & \neg(\pi^\circ) & (\neg \phi)^\circ & = & \neg(\phi^\circ) \cap \iota\delta \\
(\pi_1 \cup \pi_2)^\circ & = & \pi_1^\circ \cup \pi_2^\circ & (\phi \vee \psi)^\circ & = & (\phi^\circ \cup \psi^\circ) \cap \iota\delta \\
(\mathbf{L}\phi)^\circ & = & (\iota\delta \cap \phi^\circ) \circ 1 & (\mathbf{D}\pi)^\circ & = & \pi^\circ \cap \iota\delta \\
\iota\delta^\circ & = & \iota\delta \\
(\otimes \pi)^\circ & = & \otimes(\pi^\circ) \\
(\pi_1 \circ \pi_2)^\circ & = & (\pi_1^\circ \circ \pi_2^\circ)
\end{array}$$

An easy induction shows that a formula is DAL_{pair} valid if and only if
its translation is $AL(\mathsf{KP}_{RS})$ valid. (Cf. Brink et al. 1994 for a similar
translation). But then, we can decide DAL_{pair} formulas in the decidable
logic $AL(\mathsf{KP}_{RS})$. □

Completeness

Next, we provide a complete axiomatization for DAL_{pair}. We use a similar
strategy as in the completeness proof in chapter 2: we first define abstract
DAL frames, then we restrict that class to a suitable class K_{dal}, and show
that every frame from K_{dal} is a zigzagmorphic image of a DAL_{pair} frame.

Definition 2.5 (i) $\mathcal{F} = \langle Ar, Po, C, R, I, l, d\rangle$ is a *DAL frame* if (1) Ar is
a set (of arrows), (2) Po is a set (of begin and end points of arrows), (3)
$\langle Ar, C, R, I\rangle$ is an arrow-frame, (4) $l : Ar \longrightarrow Po$ is a function (providing
each arrow with its starting point), and (5) $d : Po \longrightarrow Ar$ is a function
(providing each point with the identity arrow on that point). The meaning
of the two new connectives on these frames is:

$$\begin{array}{rl}
(\forall x \in Ar): & \mathrm{M}, x \Vdash \mathbf{L}\phi \quad \textit{iff} \quad \mathrm{M}, l(x) \Vdash \phi, \\
(\forall w \in Po): & \mathrm{M}, w \Vdash \mathbf{D}\pi \quad \textit{iff} \quad \mathrm{M}, d(w) \Vdash \pi.
\end{array}$$

(ii) K_{dal} is the class of all DAL frames which satisfy:

(D_0) conditions (C_1) – (C_{15}) from Proposition 2.1 in chapter 2[6],
(D_1) l and d are total functions,
(D_2) $(\forall w \in Po) : I(d(w))$,
(D_3) $(\forall w \in Po) : w = l(d(w))$,
(D_4) $(\forall x \in Ar) : Ix \Rightarrow x = d(l(x))$,
(D_5) $(\forall xy \in Ar) : Cxyx \ \& \ Iy \Rightarrow l(x) = l(y)$.

(iii) Let $V \subseteq U \times U$ be a symmetric and reflexive relation. Define
$\mathcal{FR}(V) = \langle V, V_0, C_V, R_V, I_V, l_V, d_V\rangle$, in which $V_0 = \mathsf{Base}(V)$, the relations C_V, R_V, I_V are defined as in Definition 1.1 in chapter 2, and

---

[6]These are the requirements on $\langle Ar, C, R, I\rangle$ which suffices for a representation as a
reflexive and symmetric pair-frame.

$(\forall \langle u, v \rangle \in V) : l_V(\langle u, v \rangle) = u$ and $(\forall w \in V_0) : d_V(w) = \langle w, w \rangle$. These frames are called DAL_{pair} frames.

Theorem 2.6 DAL_{pair} *is strongly completely axiomatizable by adding the following axioms to the basic derivation system:*[7]

(DA_0) all $AL(\mathsf{KP}_{RS})$ axioms, (cf. Proposition 2.1 in chapter 2)
(DA_1) **L** and **D** distribute over negation,
(DA_2) **D**$\iota\delta$,
(DA_3) **DL**$\phi \leftrightarrow \phi$,
(DA_4) $\iota\delta \to (\mathbf{LD}\pi \leftrightarrow \pi)$,
(DA_5) $(\iota\delta \cap \mathbf{L}\phi) \circ 1 \leftrightarrow \mathbf{L}\phi$.

Proof. Soundness is easy to check. As for completeness, it is easy to see that any DAL frame satisfies the axioms iff it satisfies conditions $(D_0) - (D_5)$ (because all axioms are Sahlqvist formulas). So, by the same argument as in chapter 2, it suffices to show that any frame $\mathcal{F} \in \mathsf{K}_{dal}$ is a zigzagmorphic image of some frame $\mathcal{FR}(V)$, for V a reflexive and symmetric relation. The relevant "two-sorted zigzagmorphism" works as follows: Let

$$\mathcal{F} = \langle Ar^F, Po^F, C^F, R^F, I^F, l^F, d^F \rangle$$

and

$$\mathcal{G} = \langle Ar^G, Po^G, C^G, R^G, I^G, l^G, d^G \rangle$$

be in K_{dal}. The functions $p : Ar^F \longrightarrow Ar^G$ and $p^* : Po^F \longrightarrow Po^G$ constitute a *zigzagmorphism* if (1) p is a zigzagmorphism for the C, R, I part, (2) p^* is surjective, and (3) $(\forall w \in Po^F) : p(d^F(w)) = d^G(p^*(w))$ and $(\forall x \in Ar^F) : p^*(l^F(x)) = l^G(p(x))$.

Let $\mathcal{F} = \langle Ar, Po, C, R, I, l, d \rangle \in \mathsf{K}_{dal}$. By lemma 2.5 in chapter 2, the reduct $\langle Ar, C, R, I \rangle$ is a zigzagmorphic image of some locally square pair-frame $\langle V, C_V, R_V, I_V \rangle$, say by the function $p : V \longrightarrow Ar$. Take the frame $\mathcal{FR}(V)$ and define $p^* : V_0 \longrightarrow Po$ as $p^*(w) = l(p(d_V(w)))$.

Claim. The functions p and p^* form a zigzagmorphism from $\mathcal{FR}(V)$ onto $\mathcal{F}$.

Proof of Claim. We have to show that (1) p^* is surjective, (2) $p(d_V(w)) = d(p^*(w))$ and (3) $p^*(l_V(x)) = l(p(x))$. Let us compute.

(1) Suppose $w \in Po$, then $d(w) \in Ar$. Because p is surjective and (by condition (D_2)) $I(d(w))$, there is some $x \in V$ such that $p(x) = d(w)$ & $I_V(x)$. Since $I_V(x)$, also $x = d_V(x_0)$, whence

$$p^*(x_0) \stackrel{\text{def}}{=} l(p(d_V(x_0))) = l(p(x)) = l(d(w)) \stackrel{(D_3)}{=} w.$$

(2) Let $w \in V_0$. Then, by definition, $I_V(d_V(w))$, whence by assumption,

---

[7]I.e., we add (UG) rules for **L** and **D** and distribution axioms $\mathbf{L}(q \vee q') \leftrightarrow (\mathbf{L}q \cup \mathbf{L}q')$ and $\mathbf{D}(p \cup p') \leftrightarrow (\mathbf{D}p \vee \mathbf{D}p')$ to the basic derivation system for arrow logic defined in chapter 1.

$I(p(d_V(w)))$, so by (D_4):
$$p(d_V(w)) = d(l(p(d_V(w)))) \stackrel{\text{def}}{=} d(p^*(w)).$$

(3) Let $x \in V$. Then, by definition, $C_V x, d_V(l_V(x)), x$ and $I_V d_V(l_V(x))$. Since p is a zigzagmorphism: $Cp(x), p(d_V(l_V(x))), p(x)$ and $Ip(d_V(l_V(x)))$. Then, (D_5) implies that
$$l(p(x)) = l(p(d_V(l_V(x)))) \stackrel{\text{def}}{=} p^*(l_V(x)).$$
□

Definability and Interpolation

Theorem 2.7 DAL_{pair} *enjoys Craig interpolation and Beth definability.*

Proof. Since DAL_{pair} is equivalent to the logic of the Sahlqvist definable class of two-sorted arrow-frames K_{dal} it is — by the zigzagproduct lemma in chapter 2 — sufficient to show that this class is closed under taking binary zigzag products. That this class is closed under that construction is easy to see and left to the reader. □

3 Conclusion

We have seen that making arrow logic two-sorted can be done without losing any of the positive properties of the one-sorted system. This is also the conclusion of van Benthem 1994 in the context of Kripke frames. What is new here is that the logic can be given a natural pair-frame semantics, which is finitely axiomatizable. This logic also behaves well from the computational point of view: it is decidable and it has the interpolation property.

Connections with Other Systems

Propositional dynamic logic. We briefly compare DAL_{pair} with propositional dynamic logic (PDL). As we have seen, the Kleene star can be added to the arrow logical part of DAL_{pair}, which yields a PDL-like system over locally square pair-frames. For this comparison, define that subclass of K_{dal} in which *composition is associative*, namely: $\mathsf{K}_{dal}^{ass} \stackrel{\text{def}}{=} \{\mathcal{F} \in \mathsf{K}_{dal} : \mathcal{F} \models \pi_1 \circ (\pi_2 \circ \pi_3) \leftrightarrow (\pi_1 \circ \pi_2) \circ \pi_3\}$. Note that K_{dal}^{ass} validates all RA-axioms, so it inherits all negative properties of RA.

Proposition 3.1 (i) *Every $*$-free PDL formula which is valid in PDL, is also valid in* K_{dal}^{ass}.
(ii) *All $*$-free PDL axioms, except $\langle \pi_1 \rangle \langle \pi_2 \rangle \phi \to \langle \pi_1 \circ \pi_2 \rangle \phi$, are valid in* K_{dal}.

Proof. The validities follow from a straightforward computation. The following DAL_{pair} model is a counterexample for $\langle \pi_1 \rangle \langle \pi_2 \rangle \phi \to \langle \pi_1 \circ \pi_2 \rangle \phi$. Its domain consists of the set $(^2\{u,v\} \cup {}^2\{v,w\})$. Let $v^{\mathsf{ST}}(q) = \{w\}$,

$v^{\mathsf{PR}}(p_1) = \{\langle u,v\rangle\}$, and $v^{\mathsf{PR}}(p_2) = \{\langle v,w\rangle\}$ Then $u \Vdash \langle p_1\rangle\langle p_2\rangle q$, but $u \not\Vdash \langle p_1 \circ p_2\rangle q$.

(2)
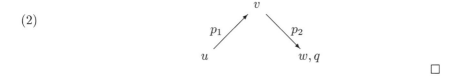

□

Weak Peirce Algebras
Peirce Algebras are discussed in Brink et al. 1994 and in de Rijke 1993. They have several applications in computer science and knowledge engineering. De Rijke (1993 p.104) asks for an "arrow version of Peirce algebras which is sufficiently expressive for applications but is still decidable". The logic DAL_{pair} might be an answer to this question[8]. To conclude, we present a weakened version of Peirce Algebras with some nice properties that Peirce Algebras lack.

Peirce Algebras. We copy the definition from de Rijke 1993. A *Peirce Algebra* is a two sorted algebra $\langle \mathfrak{R}, \mathfrak{B}, :, ^c\rangle$ in which $\mathfrak{R} \in$ RA and $\mathfrak{B} \in$ BA. The binary operator ":" is a function from $R \times B$ to B, called *Peirce product*, and the unary operator "^{c}" is a function from B to R. The operators which form the connections between the two sorts have to satisfy $(P_1) - (P_8)$ below.

(P_1) $\pi : (\phi \vee \psi) = (\pi : \phi) \vee (\pi : \psi)$ (P_5) $0 : \phi = 0$
(P_2) $(\pi_1 \vee \pi_2) : \phi = (\pi_1 : \phi) \vee (\pi_2 : \phi)$ (P_6) $\pi \otimes : \neg(\pi : \phi) \leq \neg\phi$
(P_3) $\pi_1 : (\pi_2 : \phi) = (\pi_1 \circ \pi_2) : \phi$ (P_7) $\phi^c : 1 = \phi$
(P_4) $\iota\delta : \phi = \phi$ (P_8) $(\pi : 1)^c = \pi \circ 1$.

The intended models are subalgebras of direct products of two-sorted algebras $\langle (\mathcal{F}_{pair}(U \times U))^+, \mathfrak{P}(U), \langle . \rangle., \mathbf{L}\rangle$ for some set U.

Peirce Algebras and DAL_{pair}. As we have seen, $\langle . \rangle.$ and $\mathbf{D}$ are interdefinable in DAL_{pair}. So the intended models of Peirce Algebras are that subclass of DAL_{pair} models in which the set of pairs is a full Cartesian square. Axiomatically, the only difference between *the relational part* of Peirce Algebras and that of DAL_{pair} is that in Peirce Algebras composition is associative, while in DAL_{pair} it is only *weakly associative*. The next proposition tells us that this is the only important difference between DAL_{pair} and Peirce Algebras. Here are the trivial translations from the

---
[8]If DAL_{pair} is still not expressive enough, we can add for instance the difference operator or the Kleene star. It follows from the previous results that, except if we add both, decidability remains.

above axioms into the DAL_{pair} language.

(P_1') $\langle \pi \rangle(\phi \vee \psi) \leftrightarrow \langle \pi \rangle \phi \vee \langle \pi \rangle \psi$ (P_5') $\langle 0 \rangle \phi \leftrightarrow \bot$
(P_2') $\langle \pi_1 \cup \pi_2 \rangle \phi \leftrightarrow \langle \pi_1 \rangle \phi \vee \langle \pi_2 \rangle \phi$ (P_6') $\langle \otimes \pi \rangle \neg \langle \pi \rangle \phi \to \neg \phi$
(P_3') $\langle \pi_1 \rangle \langle \pi_2 \rangle \phi \leftrightarrow \langle \pi_1 \circ \pi_2 \rangle \phi$ (P_7') $\langle \mathbf{L} \phi \rangle \top \leftrightarrow \phi$
(P_4') $\langle \iota \delta \rangle \phi \leftrightarrow \phi$ (P_8') $\mathbf{L} \langle \pi \rangle \top \leftrightarrow \pi \circ 1$.

Proposition 3.2
(i) $DAL_{pair} \models (P_1'), (P_2'), (P_{3\leftarrow}'), (P_4'), (P_5'), (P_6'), (P_7'), (P_{8\leftarrow}')$.
(ii) $DAL_{pair} \not\models (P_{3\to}'), (P_{8\to}')$.
(iii) $\mathsf{K}_{dal}^{ass} \models (P_1') - (P_8')$.

Proof. (i). By direct calculation. (ii). For $(P_{3\to}')$, this was proved in 3.1. We give a counterexample to $\mathbf{L} \langle \pi \rangle \top \to \pi \circ 1$. Let the domain of model M be the reflexive and symmetric closure of $\{\langle u, v \rangle, \langle u, w \rangle\}$ and set $v^{\mathsf{PR}}(p) = \{\langle u, v \rangle\}$. Then $\langle u, w \rangle \Vdash \mathbf{L} \langle p \rangle \top$, but $\langle u, w \rangle \not\Vdash p \circ 1$.

(3)

(iii). By direct calculation. □

Thus, K_{dal}^{ass} is at least as strong as the logic of Peirce Algebras. Conversely, consider the class of *Representable Weak Peirce Algebras* (RWPA), whose relational component consists only of subalgebras of complex algebras of locally square pair-frames. (I.e., an RWPA is an algebra of the form $\langle \mathfrak{R}, \mathfrak{B}, \langle . \rangle ., \mathbf{L} \rangle$ with $\mathfrak{R} \leq (\mathcal{F}_{pair}(V))^+$ and $\mathfrak{B} \leq \mathfrak{P}(V_0)$, for a reflexive and symmetric relation V with base V_0.) By proposition 2.3, the algebraic version of DAL_{pair} is (term-definably equivalent to) the class RWPA. All earlier positive results for DAL_{pair}, which do not hold in Peirce Algebras, carry over to RWPA. So the strategy of obtaining positive results by widening the class of models to the "logical core", also works for Peirce Algebras.

References

van Benthem, J. 1991. *Language in Action (Categories, Lambdas and Dynamic Logic)*. Studies in Logic, Vol. 130. North-Holland, Amsterdam.

van Benthem, J. 1994. A Note on Dynamic Arrow Logics. In *Logic and Information Flow*, ed. J. van Eijck and A. Visser. 15–29. MIT Press, Cambridge (Mass.).

Brink, Ch., K. Britz, and R. Schmidt. 1994. Peirce Algebras. *Formal Aspects of Computing* 6(3):339–358.

Harel, D. 1984. Dynamic Logic. In *Handbook of Philosophical Logic Volume II*, ed. D.M. Gabbay and F. Guenther. 497–604. Dordrecht: Reidel.

Hughes, G., and M. Creswell. 1984. *A Companion to Modal Logic*. Methuen.

de Rijke, M. 1993. *Extending Modal Logic*. Doctoral dissertation, Institute for Logic, Language and Computation, Universiteit van Amsterdam. ILLC Dissertation Series 1993–4.

6
Complete Calculus for Conjugated Arrow Logic

SZABOLCS MIKULÁS

ABSTRACT. We will give a strongly complete Hilbert-style inference system for a variant of arrow logic. Namely, we will consider a logic whose connectives, beside the Booleans, are identity, the binary modality •, and its two conjugate modalities ▶ and ◀. The models for this logic are Kripke models with one ternary accessibility relation corresponding to the modality •. A Hilbert-style inference system will be defined; and we will prove the completeness of this calculus with respect to the above semantics. We will also prove that this logic is decidable for it has the finite model property. The completeness proof uses an algebraic representation theorem, which will be proved in the paper as well.

1 Introduction

Arrow logic is treated for instance in van Benthem 1991 and chapter 1. That version of arrow logic contains all the Boolean connectives, and three modal operators: identity, converse and composition. The proposed models for that logic are Kripke models with three accessibility relations corresponding to the modalities above.

We will investigate a variant of arrow logic, called *conjugated arrow logic*, CARL for short, which is given by forgetting the modality converse and adding two new binary (or dyadic) modalities: ▶ and ◀. The models for CARL are Kripke models with one ternary accessibility relation corresponding to composition •. The other two modalities are conjugate modalities of • as, in temporal logic, *sometimes-in-the-past* is a conjugate of *sometimes-in-the-future*. Since • is a binary modality, the accessibility relation inter-

---

Research is supported by the Hungarian National Foundation for Scientific Research grant Nos. F17452 and T16448.

Arrow Logic and Multi-Modal Logic
M. Marx, L. Pólos and M. Masuch, eds.
Copyright © 1996, CSLI Publications.

preting it is a ternary relation. Thus we can permute the arguments of the relation in many ways; the interpretations of ▶ and ◀ are given by two such permutations. The idea that the binary connectives •, (and the duals of) ▶ and ◀ can be considered as (dyadic) modalities is, e.g., in Dirk Roorda's dissertation on resource logics Roorda 1991. However, his semantics uses three accessibility relations, one for each modality. See also Kosta Došen's survey paper Došen 1992. The connectives ▶ and ◀ are interesting for several reasons. Their presence is equivalent with that of the *residuals* \ and / of • as we will see later. (The syntactic nature of ▶ and ◀ and their relationship to • are investigated, e.g., in Jónsson and Tsinakis 1992.) The residuals \ and / are the well-known connectives of *categorial grammars*, e.g., of the *Lambek calculus*. These are also connectives of many *substructural logics* and Girard's *linear logic*. Further motivation is that \ and / can be considered as some kinds of implication as well: $\varphi \backslash \psi$ is *preimplication* meaning *had φ then ψ* and φ/ψ is *postimplication φ if-ever ψ*, see Vaughan Pratt's paper 1990. Further, identity is interpreted using the above accessibility relation too: identity holds at a world iff it is the composition of itself.

We will define a Hilbert-style inference system and show its strong completeness for CARL. We will also prove by filtration that CARL has the finite model property, hence it is decidable.

One direction of further investigations would be adding more and more connectives to our logic; another interesting direction is to consider more "concrete" semantics for our logics, e.g., where the set of possible worlds is a binary relation and the accessibility relation is relational composition. Actually, Maarten Marx proved in Marx 1995 completeness for the pair version of CARL, i.e., where the models are arbitrary binary relations and the accessibility relation is relational composition restricted to the universe of the model. Similar investigations are, e.g., in chapter 2 and Andréka and Mikulás 1994. We could also restrict the class of models by requiring that some further axioms should hold, as suggested, e.g., in van Benthem 1994.

2 Who is CARL?

In this section we give the precise definition of CARL.

Definition 2.1 (Conjugated Arrow Logic) Let P be a set, called the set of *parameters*, and $Cn = \{\wedge, \neg, \bullet, \blacktriangleright, \blacktriangleleft, \iota\delta\}$. The set of *formulas* of CARL is built up from P using the elements of Cn as connectives, where $\iota\delta$ is a 0-ary and $\neg$ is a unary connective and the others are binary connectives, in the usual way. The notion of *subformula* (of a formula) is assumed to be known. We will also use the well-known derived connectives: $\rightarrow$, $\leftrightarrow$, $\vee$, and the formula schemes: *False*, $\bot$, and *True*, $\top$.

A *Kripke frame* for CARL is an ordered pair $\langle W, C \rangle$, where W, called the set of *possible worlds*, is a non-empty set and C, called *accessibility relation*, is a ternary relation on W, i.e., $C \subseteq W \times W \times W$.

A *Kripke model* for CARL is a frame enriched with a valuation v. More precisely, it is an ordered triple $\langle W, C, v \rangle$, where $v : P \longrightarrow \mathcal{P}(W)$, i.e., v associates to every parameter a subset of W.

(Local) Truth of a formula φ at a world $w \in W$ in a model $\langle W, C, v \rangle$, denoted by $w \Vdash_v \varphi$, is defined by recursion as follows.

- If $p \in P$, then $w \Vdash_v p \stackrel{\text{def}}{\iff} w \in v(p)$.
- $w \Vdash_v \varphi \wedge \psi \stackrel{\text{def}}{\iff} w \Vdash_v \varphi \ \& \ w \Vdash_v \psi$.
- $w \Vdash_v \neg \varphi \stackrel{\text{def}}{\iff}$ not $w \Vdash_v \varphi$ (also denoted as $w \not\Vdash_v \varphi$).
- $w \Vdash_v \varphi \bullet \psi \stackrel{\text{def}}{\iff} (\exists w_1, w_2 \in W) C w_1 w_2 w \ \& \ w_1 \Vdash_v \varphi \ \& \ w_2 \Vdash_v \psi$.
- $w \Vdash_v \varphi \blacktriangleright \psi \stackrel{\text{def}}{\iff} (\exists w_1, w_2 \in W) C w_1 w w_2 \ \& \ w_1 \Vdash_v \varphi \ \& \ w_2 \Vdash_v \psi$.
- $w \Vdash_v \varphi \blacktriangleleft \psi \stackrel{\text{def}}{\iff} (\exists w_1, w_2 \in W) C w w_2 w_1 \ \& \ w_1 \Vdash_v \varphi \ \& \ w_2 \Vdash_v \psi$.
- $w \Vdash_v \iota\delta \stackrel{\text{def}}{\iff} Cwww$.

(Global) Truth in a model and *validity* in a frame are defined in the usual way. That is,

- $\langle W, C, v \rangle \models \varphi \stackrel{\text{def}}{\iff}$ for every world $w \in W$, $w \Vdash_v \varphi$
- $\langle W, C \rangle \models \varphi \stackrel{\text{def}}{\iff}$ for every valuation v, $\langle W, C, v \rangle \models \varphi$.

We say that a formula φ is a *semantical consequence* of the set Γ of formulas, in symbols $\Gamma \models \varphi$, iff for every model $\langle W, C, v \rangle$,

$$\langle W, C, v \rangle \models \Gamma \Rightarrow \langle W, C, v \rangle \models \varphi$$

where $\langle W, C, v \rangle \models \Gamma$ abbreviates that, for every $\psi \in \Gamma$, $\langle W, C, v \rangle \models \psi$.

We recall the definitions of the *residuals*, denoted by $\backslash$ and $/$, of $\bullet$:

- $w \Vdash_v \varphi \backslash \psi \stackrel{\text{def}}{\iff} (\forall w_1, w_2 \in W)(C w_1 w w_2 \ \& \ w_1 \Vdash_v \varphi \Rightarrow w_2 \Vdash_v \psi)$
- $w \Vdash_v \varphi / \psi \stackrel{\text{def}}{\iff} (\forall w_1, w_2 \in W)(C w w_2 w_1 \ \& \ w_2 \Vdash_v \psi \Rightarrow w_1 \Vdash_v \varphi)$.

Then the following four formulas are semantically valid:

$$\varphi \backslash \psi \ \leftrightarrow \ \neg(\varphi \blacktriangleright \neg\psi) \qquad \varphi / \psi \ \leftrightarrow \ \neg(\neg\varphi \blacktriangleleft \psi)$$
$$\varphi \blacktriangleright \psi \ \leftrightarrow \ \neg(\varphi \backslash \neg\psi) \qquad \varphi \blacktriangleleft \psi \ \leftrightarrow \ \neg(\neg\varphi / \psi).$$

3 Some Logic

The main result of this paper is the following strong completeness theorem, which we will prove later using an algebraic representation theorem (Theorem 4.3).

Theorem 3.1 (Strong Completeness) *There is a Hilbert-style inference system $\vdash$ which is strongly complete and strongly sound with respect to*

CARL. *That is, for every set Γ of formulas and formula φ,*

$$\Gamma \vDash \varphi \iff \Gamma \vdash \varphi.$$

In fact, the inference system $\vdash$ defined in Definition 3.4 below has the above property.

We will prove the following two theorems as well.

Theorem 3.2 (Decidability) CARL *is decidable, i.e., the set of valid formulas is a decidable set.*

Theorem 3.3 (Finite Model Property) CARL *has the finite model property, i.e., for every formula φ, if φ is not valid, then there is a finite Kripke model which refutes it.*

Definition 3.4 (Inference System $\vdash$) The Hilbert-style inference system $\vdash$ is given by the following axiom schemes and inference rules, where capital Greek letters denote formula schemes (metavariables which can be substituted by formulas) and $\bot$ stands for the formula scheme *False*.

Axiom Schemes.
- (i) axiom schemes for classical propositional logic
- (ii) $\bot \bullet \Phi \leftrightarrow \Phi \bullet \bot \leftrightarrow \bot$
- (iii) $\Phi \bullet (\Psi \vee \Theta) \leftrightarrow (\Phi \bullet \Psi) \vee (\Phi \bullet \Theta)$
- (iv) $(\Phi \vee \Psi) \bullet \Theta \leftrightarrow (\Phi \bullet \Theta) \vee (\Psi \bullet \Theta)$
- (v) $\Phi \wedge \iota \delta \to \Phi \bullet \Phi$.

Inference Rules.

(vi) $\dfrac{(\Phi \bullet \Psi) \wedge \Theta \leftrightarrow \bot}{(\Phi \blacktriangleright \Theta) \wedge \Psi \leftrightarrow \bot}$

(vii) $\dfrac{(\Phi \bullet \Psi) \wedge \Theta \leftrightarrow \bot}{(\Theta \blacktriangleleft \Psi) \wedge \Phi \leftrightarrow \bot}$

(viii) $\dfrac{\Phi \to \Psi \quad \Phi}{\Psi}$

(ix) $\dfrac{\Phi \leftrightarrow \Psi \quad \Theta \leftrightarrow \Lambda}{(\Phi \bullet \Theta) \leftrightarrow (\Psi \bullet \Lambda)}$

where double bar indicates that we have both the downward and the upward rules. The definition of derivability of a formula φ from a set Γ of formulas, $\Gamma \vdash \varphi$, is the usual.

The following formula is an equivalent version of (v) (in the presence of (i), (iii) and (iv)), and we will use it sometimes:

$$\Phi \wedge \Psi \wedge \iota\delta \to \Phi \bullet \Psi.$$

Remark 3.5 Instead of the conjugates $\blacktriangleright$ and $\blacktriangleleft$ we could choose the residuals $\backslash$ and $/$. Then the above three theorems still hold. This is true because the conjugates and the residuals are definable by each other as we mentioned after Definition 2.1. Thus we can replace each occurrence of $\varphi \blacktriangleright \psi$ by its definition $\neg(\varphi \backslash \neg \psi)$ and similarly for $\blacktriangleleft$.

Clearly, the conjugates and the residuals can be present at the same time too. Then we have to add the definitions of the residuals as new axioms to the inference system in Definition 3.4.

Further, we could define a very weak *converse* too:

$$\varphi^{\smile} \stackrel{\text{def}}{\iff} (\varphi \blacktriangleright \iota\delta) \wedge (\iota\delta \blacktriangleleft \varphi).$$

4 Some Algebra

In this section, we define the algebraic counterparts both of syntax (the class **KA** of algebras) and of semantics (the class **RKA** of algebras), and prove that they are identical (Theorem 4.3).

Note that by **RKA** we will denote a class of algebras, and we will use it as the abbreviation of 'representable Kripke algebra' too. Given a class **K** of algebras, we denote by **IK** and by **SK** the class of isomorphic copies and subalgebras of members of **K**, respectively.

Definition 4.1 (Representable Kripke Algebra) The class of *representable Kripke algebras* is defined as

$$\mathsf{RKA} \stackrel{\text{def}}{=} \mathbf{S}\{\langle \mathcal{P}(W), \cap, \sim, \circ^C, \triangleright^C, \triangleleft^C, Id^C \rangle : W \text{ a set } \& \; C \subseteq W \times W \times W\}$$

where $\cap$ is intersection, $\sim$ is set-theoretic complementation with respect to W, i.e., $\sim a = W \smallsetminus a$, and

$$a \circ^C b \stackrel{\text{def}}{=} \{z \in W : (\exists x, y \in W) Cxyz \; \& \; x \in a \; \& \; y \in b\}$$
$$a \triangleright^C b \stackrel{\text{def}}{=} \{z \in W : (\exists x, y \in W) Cxzy \; \& \; x \in a \; \& \; y \in b\}$$
$$a \triangleleft^C b \stackrel{\text{def}}{=} \{z \in W : (\exists x, y \in W) Czyx \; \& \; x \in a \; \& \; y \in b\}$$
$$Id^C \stackrel{\text{def}}{=} \{z \in W : Czzz\}$$

for all elements a, b.

Definition 4.2 (Kripke Algebra) The class of (abstract) *Kripke algebras*, denoted by **KA**, is defined as the class of algebras similar to **RKA**'s satisfying the axioms below. More precisely, every element of **KA** has the form

$$\langle A, \wedge, \neg, \bullet, \blacktriangleright, \blacktriangleleft, \iota\delta \rangle$$

where A is a non-empty set, $\iota\delta$ is a constant, $\neg$ is a unary operation on A, and $\wedge, \bullet, \blacktriangleright, \blacktriangleleft$ are binary operations on A; and the following set Ax of quasi-equations is valid in **KA**:

(1) Boolean axioms
(2) $0 \bullet x = x \bullet 0 = 0$
(3) $x \bullet (y \vee z) = (x \bullet y) \vee (x \bullet z)$
(4) $(x \vee y) \bullet z = (x \bullet z) \vee (y \bullet z)$
(5) $x \wedge \iota\delta \leq x \bullet x$
(6) $(x \bullet y) \wedge z = 0 \iff (x \blacktriangleright z) \wedge y = 0$
(7) $(x \bullet y) \wedge z = 0 \iff (z \blacktriangleleft y) \wedge x = 0$,

where 0 is Boolean zero, $x \vee y$ abbreviates $\neg(\neg x \wedge \neg y)$ and $x \leq y$ abbreviates $x \wedge y = x$.

Again, an axiom equivalent to (5) is

$$x \wedge y \wedge \iota\delta \leq x \bullet y.$$

The main algebraic result of this paper is the following theorem that we will prove in the following section as well.

Theorem 4.3 (Representation Theorem) *For every algebra $\mathcal{A}$,*

$$\mathcal{A} \in \mathsf{KA} \iff \mathcal{A} \in \mathsf{IRKA}.$$

Moreover, if $\mathcal{A}$ is a finite algebra, then there are a finite set W and a relation $C \subseteq W \times W \times W$ such that $\mathcal{A}$ is isomorphic to a subalgebra of the RKA

$$\langle \mathcal{P}(W), \cap, \sim, \circ^C, \triangleright^C, \triangleleft^C, Id^C \rangle.$$

5 Proofs

To prove Theorems 3.1 and 4.3 we need some lemmas (the proof of which we postpone till the end of this section) and some definitions.

Let Γ be an arbitrary set of formulas of CARL; then for any formulas φ and ψ, we set

$$\varphi \equiv_\Gamma \psi \stackrel{\text{def}}{\iff} \Gamma \vdash \varphi \leftrightarrow \psi.$$

The *formula algebra* $\mathcal{F}$ is defined as

$$\mathcal{F} \stackrel{\text{def}}{=} \langle F, \wedge, \neg, \bullet, \blacktriangleright, \blacktriangleleft, \iota\delta \rangle,$$

where F denotes the set of formulas of CARL.

Lemma 5.1 $\equiv_\Gamma$ *is a congruence relation on $\mathcal{F}$.*

Let $\mathcal{F}_\Gamma$ be the factor algebra of $\mathcal{F}$ by $\equiv_\Gamma$, i.e., $\mathcal{F}_\Gamma \stackrel{\text{def}}{=} \mathcal{F}/\equiv_\Gamma$. Further, for any formula φ, we let $\overline{\varphi} \stackrel{\text{def}}{=} \{\psi : \varphi \equiv_\Gamma \psi\}$. Let $\top$ denote the formula *True*.

Lemma 5.2 1. *For every formula φ, $\Gamma \vdash \varphi \iff \overline{\varphi} = \overline{\top}$.*
2. $\mathcal{F}_\Gamma \models Ax$.

Lemma 5.3 *Let W be a non-empty set, $\mathcal{A}$ be a subalgebra of the full RKA $\langle \mathcal{P}(W), \cap, \sim, \circ^C, \triangleright^C, \triangleleft^C, Id^C \rangle$, and v be a valuation. Then*

$$\langle \mathcal{A}, v \rangle \models \varphi = 1 \iff \langle W, C, v \rangle \models \varphi,$$

where 1 denotes Boolean unit.

Proof. (Theorem 3.1) Soundness is easy to check.

For the completeness direction, we assume that $\Gamma \not\vdash \varphi$. Then, by Lemma 5.2, $\overline{\varphi} \neq \overline{\top}$. Now, let n be that valuation which associates its equivalence class to each formula, i.e., $n(\varphi) = \overline{\varphi}$ for every φ. Then $\langle \mathcal{F}_\Gamma, n \rangle \not\models \varphi = \top$.

By Lemma 5.2, for every $\psi \in \Gamma$, $\langle \mathcal{F}_\Gamma, n \rangle \vDash \psi = \top$. Since, by Lemma 5.2, $\mathcal{F}_\Gamma \vDash Ax$, Theorem 4.3 says that $\mathcal{F}_\Gamma \in \mathsf{IRKA}$. Thus we have an $\mathcal{A} \in \mathsf{RKA}$ and a valuation v such that $\langle \mathcal{A}, v \rangle \not\vDash \varphi = 1$, while for every $\psi \in \Gamma$, $\langle \mathcal{A}, v \rangle \vDash \psi = 1$. Whence, by Lemma 5.3, there is a Kripke model $\mathcal{M}$ such that $\mathcal{M} \not\vDash \varphi$, while for every $\psi \in \Gamma$, $\mathcal{M} \vDash \psi$, i.e., $\Gamma \not\vDash \varphi$, which was to be proved. □

Next we prove that CARL is decidable.
Proof. (Theorem 3.2) By the Completeness Theorem 3.1, $\{\varphi : \vDash \varphi\} = \{\varphi : \vdash \varphi\}$; thus the set of valid formulas form a recursively enumerable set.

By Theorem 3.3, for every formula, we have a finite model that refutes it. Since we can enumerate the set of finite models and the semantic value of a formula is computable, the set $\{\varphi : \not\vDash \varphi\}$ is recursively enumerable.

Since both $\{\varphi : \vDash \varphi\}$ and its complement are recursively enumerable, $\{\varphi : \vDash \varphi\}$ is a decidable set. □

Now we prove the Representation Theorem. We will use an idea of Németi 1992.
Proof. (Theorem 4.3) It is easy to verify that $\mathsf{RKA} \vDash Ax$, so we will omit it.

For the other direction, let us assume that $\mathcal{A} \in \mathsf{KA}$. Then we will represent this $\mathcal{A}$ as an algebra $\mathcal{B}$ whose operations are almost good. Indeed, there will be a ternary relation C on the set of ultrafilters $Uf(\mathcal{A})$ such that the unary and binary operations can be defined via this C. Namely, if we let
$$CFGH \stackrel{\text{def}}{\Longleftrightarrow} F \bullet G \subseteq H,$$
for every $F, G, H \in Uf(\mathcal{A})$, and for every $a \in A$,
$$rep(a) \stackrel{\text{def}}{=} \{F \in Uf(\mathcal{A}) : a \in F\},$$
then rep is almost an isomorphism. The only problem is that there may be ultrafilters F such that $F \bullet F \subseteq F$ but $\iota\delta \notin F$, i.e., $rep(\iota\delta) \neq \{F : CFFF\}$. If we split these "bad" ultrafilters into two parts carefully, then we will get a $\mathcal{B}' \in \mathsf{RKA}$ isomorphic to the original algebra $\mathcal{A}$.

First we state and prove a lemma, where $Uf(\mathcal{A})$ denotes the set of Boolean ultrafilters of the algebra $\mathcal{A}$, and for any subsets F and G of A (the universe of $\mathcal{A}$),
$$F \bullet G \stackrel{\text{def}}{=} \{a \in A : (\exists f \in F)(\exists g \in G) a = f \bullet g\}.$$
$F \blacktriangleright G$ and $F \blacktriangleleft G$ are defined similarly.

Lemma 5.4 *Let $\mathcal{A} \in \mathsf{KA}$ and F_0, G_0 be subsets of A with the finite intersection property (i.e., $x, y \in F_0 \Rightarrow x \wedge y \neq 0$). Then*
$$(\exists H \in Uf(\mathcal{A})) F_0 \bullet G_0 \subseteq H \Rightarrow$$
$$(\exists F \supseteq F_0)(\exists G \supseteq G_0) F \in Uf(\mathcal{A}) \,\&\, G \in Uf(\mathcal{A}) \,\&\, F \bullet G \subseteq H.$$

Proof. Let F_0, G_0 and H be as in the formulation of the lemma. Assume that F_0 is not an ultrafilter, i.e., $\exists x(x \notin F_0 \ \& \ \neg x \notin F_0)$. Let F' be the filter generated by $F_0 \cup \{x\}$ and F'' be the filter generated by $F_0 \cup \{\neg x\}$. Assume that $F' \bullet G_0 \not\subseteq H$ and $F'' \bullet G_0 \not\subseteq H$, i.e., $\exists f_1 \exists f_2 (\exists f, f' \in F_0)(\exists g, g' \in G_0)$

$$x \wedge f \leq f_1 \ \& \ f_1 \bullet g \notin H \ \& \ \neg x \wedge f' \leq f_2 \ \& \ f_2 \bullet g' \notin H.$$

Then, since H is upward closed, we have, by (4), $(x \wedge f) \bullet g \notin H$. Similarly, by (3) and (4), we get $(x \wedge f \wedge f') \bullet (g \wedge g') \notin H$. By the same argument, $(\neg x \wedge f \wedge f') \bullet (g \wedge g') \notin H$. Putting together,

$$(f \wedge f') \bullet (g \wedge g') \stackrel{(1)}{=} ((x \wedge f \wedge f') \vee (\neg x \wedge f \wedge f')) \bullet (g \wedge g') \stackrel{(4)}{=}$$
$$= ((x \wedge f \wedge f') \bullet (g \wedge g')) \vee ((\neg x \wedge f \wedge f') \bullet (g \wedge g')) \notin H,$$

a contradiction, since $f \wedge f' \in F_0$ and $g \wedge g' \in G_0$ and $F_0 \bullet G_0 \subseteq H$. So either $F' \bullet G_0 \subseteq H$ or $F'' \bullet G_0 \subseteq H$. Note that if $F' \bullet G_0 \subseteq H$, then $0 \notin F'$, by (2), so F' is a proper filter. Using recursion, one can extend F_0 to an ultrafilter F with the property $F \bullet G_0 \subseteq H$. Then, in the same way, G_0 can be extended to an ultrafilter G such that $F \bullet G \subseteq H$. Thus the lemma has been proved. □

As we mentioned before, let for every $a \in A$,

$$rep(a) \stackrel{\text{def}}{=} \{F \in Uf(\mathcal{A}) : a \in F\},$$

and for every $F, G, H \in Uf(\mathcal{A})$,

$$CFGH \stackrel{\text{def}}{\iff} F \bullet G \subseteq H.$$

Let $rep''A \stackrel{\text{def}}{=} \{rep(a) : a \in A\}$ and

$$\mathcal{B} \stackrel{\text{def}}{=} \langle rep''A, \cap, \sim, \circ^C, \triangleright^C, \triangleleft^C, rep(\iota\delta) \rangle$$

where $\sim$ is set-theoretic complementation with respect to $Uf(\mathcal{A})$, and $\circ^C$, $\triangleright^C$, $\triangleleft^C$ are defined by the frame $\langle Uf(\mathcal{A}), C \rangle$, i.e.,

$$rep(a) \circ^C rep(b) \stackrel{\text{def}}{=} \{F \in Uf(\mathcal{A}) : (\exists G \in rep(a))(\exists H \in rep(b))CGHF\}$$
$$rep(a) \triangleright^C rep(b) \stackrel{\text{def}}{=} \{F \in Uf(\mathcal{A}) : (\exists G \in rep(a))(\exists H \in rep(b))CGFH\}$$
$$rep(a) \triangleleft^C rep(b) \stackrel{\text{def}}{=} \{F \in Uf(\mathcal{A}) : (\exists G \in rep(a))(\exists H \in rep(b))CFHG\}.$$

Now we will show that $rep : \mathcal{A} \longrightarrow \mathcal{B}$ is an isomorphism. rep is a Boolean isomorphism by Stone's representation theorem.

$$rep(a) \circ^C rep(b) =$$
$$= \{F \in Uf(\mathcal{A}) : (\exists G \in rep(a))(\exists H \in rep(b))CGHF\} =$$
$$= \{F \in Uf(\mathcal{A}) : \exists G \exists H (a \in G \ \& \ b \in H \ \& \ G \bullet H \subseteq F)\} \stackrel{(a)}{=}$$
$$= \{F \in Uf(\mathcal{A}) : a \bullet b \in F\}$$
$$= rep(a \bullet b).$$

$(a) : (\subseteq) : a \in G \ \& \ b \in H \Rightarrow a \bullet b \in F.$

($\supseteq$) : If $a \bullet b \in F$, then, by Lemma 5.4, we can construct two ultrafilters G and H such that $a \in G$, $b \in H$ and $G \bullet H \subseteq F$.

$$rep(a) \rhd^C rep(b) =$$
$$= \{F \in Uf(\mathcal{A}) : (\exists G \in rep(a))(\exists H \in rep(b))CGFH\} =$$
$$= \{F \in Uf(\mathcal{A}) : \exists G \exists H(a \in G \,\&\, b \in H \,\&\, G \bullet F \subseteq H)\} \stackrel{(b)}{=}$$
$$= \{F \in Uf(\mathcal{A}) : a \blacktriangleright b \in F\} =$$
$$= rep(a \blacktriangleright b).$$

(b) : ($\subseteq$) : Assume $\neg(a \blacktriangleright b) \in F$. Then $a \bullet (\neg(a \blacktriangleright b)) \in H$, so $(a \bullet (\neg(a \blacktriangleright b))) \wedge b \neq 0$. Thus, by (6), $(a \blacktriangleright b) \wedge \neg(a \blacktriangleright b) \neq 0$, a contradiction. So $a \blacktriangleright b \in F$.

($\supseteq$) : Let $a \blacktriangleright b \in F$. Let $G_0 = \{x \in A : x \geq a\}$ and $H_0 = \{x \in A : x \geq b\}$. First we show that $(G_0 \bullet F) \cup H_0$ can be extended to an ultrafilter H. Let

$$\uparrow (G_0 \bullet F) \stackrel{\text{def}}{=} \{x \in A : (\exists g \in G_0)(\exists f \in F) x \geq g \bullet f\}$$

and $x_1, x_2 \in \,\uparrow (G_0 \bullet F)$. Then, using monotonicity of $\bullet$, $x_1 \wedge x_2 \geq (g_1 \bullet f_1) \wedge (g_2 \bullet f_2) \geq (a \bullet f_1) \wedge (a \bullet f_2) \geq a \bullet (f_1 \wedge f_2) \in G \bullet F$ whence $x_1 \wedge x_2 \in \,\uparrow (G_0 \bullet F)$. So $\uparrow (G_0 \bullet F)$ is $\wedge$-closed, and so is H_0. Now we show that $(\uparrow (G_0 \bullet F)) \cup H_0$ has the finite intersection property, so it can be extended to an ultrafilter H. Assume to the contrary that $x \in \,\uparrow (G_0 \bullet F)$, $h \in H_0$ and $x \wedge h = 0$, then $0 = x \wedge h \geq (g \bullet f) \wedge b \geq (a \bullet f) \wedge b$, i.e., $0 = (a \bullet f) \wedge b$. Then, by (6), $0 = (a \blacktriangleright b) \wedge f \in F$, a contradiction. Clearly, $b \in H$ and $G_0 \bullet F \subseteq H$. Then, by Lemma 5.4, we can extend G_0 to an ultrafilter G such that $G \bullet F \subseteq H$, as desired.

Similar argument shows that $rep(a) \lhd^C rep(b) = rep(a \blacktriangleleft b)$. Since $rep(\iota\delta)$ is the identity constant in $\mathcal{B}$, $\mathcal{A} \cong \mathcal{B}$.

Later we will need the following fact.

(∗) $\qquad (\forall F \in Uf(\mathcal{A}))\iota\delta \in F \Rightarrow CFFF$.

This holds, since $x, y, \iota\delta \in F$ implies, by (5), $x \bullet y \geq x \wedge y \wedge \iota\delta \in F$, i.e., $F \bullet F \subseteq F$.

Now we define a Kripke frame $\langle W', C' \rangle$ and the corresponding RKA $\mathcal{B}'$, which will turn out to be isomorphic to our original algebra $\mathcal{A}$. First we split the "bad" ultrafilters on $\mathcal{A}$ into two parts. Let

$$D \stackrel{\text{def}}{=} \{F \in Uf(\mathcal{A}) : \iota\delta \notin F \,\&\, CFFF\}.$$

Let for every $F \in D$, F_1 and F_2 be two distinct elements (of our set-theoretic universe) not occurring in $Uf(\mathcal{A})$, and

$$s(F) \stackrel{\text{def}}{=} \begin{cases} \{F_1, F_2\} & \text{if } F \in D \\ \{F\} & \text{if } F \notin D. \end{cases}$$

We assume also that for different F's from D the F_i's are completely different. If $F \notin D$, then by both F_1 and F_2 we mean F. Let

$$W' \stackrel{\text{def}}{=} \{F_1, F_2 : F \in Uf(\mathcal{A})\}$$

and

$$C' = (\bigcup \{s(F) \times s(G) \times s(H) : CFGH\}) \smallsetminus \{\langle F_i, F_i, F_i \rangle : F \in D, i \in \{1, 2\}\}.$$

Now we define a representation function Rep as

$$Rep(b) \stackrel{\text{def}}{=} \{F_1, F_2 : F \in b\}$$

for every $b \in B$. Then let B' be the Rep-image of B, i.e., $B' \stackrel{\text{def}}{=} Rep''B$ and

$$\mathcal{B}' \stackrel{\text{def}}{=} \langle B', \cap, \sim, \circ^{C'}, \triangleright^{C'}, \triangleleft^{C'}, Id^{C'} \rangle,$$

where $\sim$ is set-theoretic complementation with respect to W' and the other operations are defined as in Definition 4.1 (using W' and C' instead of W and C, respectively). Clearly, $\mathcal{B}'$ is an RKA; so it remains to show that $Rep : \mathcal{B} \longrightarrow \mathcal{B}'$ is an isomorphism, for then $\mathcal{A} \cong \mathcal{B}'$, by $\mathcal{A} \cong \mathcal{B}$.

Clearly, Rep is a Boolean isomorphism. For the identity we have:

$$\begin{aligned}
Rep(rep(\iota\delta)) &= \bigcup\{s(F) : F \in rep(\iota\delta)\} \stackrel{(c)}{=} \\
&= \bigcup\{s(F) : \iota\delta \in F \,\&\, CFFF\} = \\
&= \{F : F \in Uf(\mathcal{A}) \smallsetminus D \,\&\, \iota\delta \in F \,\&\, CFFF\} \cup \\
&\quad \{F_1, F_2 : F \in D \,\&\, \iota\delta \in F \,\&\, CFFF\} \stackrel{(d)}{=} \\
&= \{F : F \in W' \,\&\, \iota\delta \in F \,\&\, C'FFF\} \stackrel{(e)}{=} \\
&= \{F : F \in W' \,\&\, C'FFF\} = \\
&= Id^{C'}.
\end{aligned}$$

(c): $F \in rep(\iota\delta)$ implies $\iota\delta \in F$, so, by $(*)$, $CFFF$.

(d): By the definition of D, $F \in D$ implies not $CFFF$; then apply the definition of C'.

(e): $C'FFF$ implies $F \in Uf(\mathcal{A}) \smallsetminus D$ and $CFFF$, whence $\iota\delta \in F$, by the definition of D.

In checking that Rep preserves $\circ$, $\triangleright$ and $\triangleleft$, we will use the following lemma.

Lemma 5.5 $(\forall a, b \in B)(\forall z \in W')$

$$(\exists x \in Rep(a))(\exists y \in Rep(b))C'xyz \iff$$
$$(\exists F \in a)(\exists G \in b)(\exists H \in Uf(\mathcal{A}))z \in s(H) \,\&\, CFGH.$$

The same holds for $C'xzy$ and $C'zyx$.

Proof. ($\Rightarrow$): Assume $C'xyz$. Let $F \in a$, $G \in b$ with $x \in s(F)$ and $y \in s(G)$, and let $H \in Uf(\mathcal{A})$ such that $z \in s(H)$. Then $\langle x, y, z \rangle \in s(F) \times s(G) \times s(H)$, thus, by the definition of C', $CFGH$ holds.

($\Leftarrow$): We have two cases.

CASE 1: $F = G = H \in D$. Then $H_1 \neq H_2$, so we can choose $x, y \in s(H)$

with $x = y \neq z$. Now, $\langle x,y,z\rangle \in s(H) \times s(H) \times s(H)$ whence $C'xyz$.
CASE 2: not Case 1. Then let $x \in s(F)$, $y \in s(G)$. Again, $\langle x,y,z\rangle \in s(F) \times s(G) \times s(H)$. Thus $C'xyz$ holds, which finishes the proof of the lemma. □

$Rep(a) \circ^{C'} Rep(b) =$
$= \{z \in W' : (\exists x \in Rep(a))(\exists y \in Rep(b))C'xyz\} \stackrel{\text{L.5.5}}{=}$
$= \{z \in W' : (\exists F \in a)(\exists G \in b)(\exists H \in Uf(\mathcal{A}))z \in s(H) \ \& \ CFGH\} =$
$= \bigcup\{s(H) : H \in Uf(\mathcal{A}) \ \& \ (\exists F \in a)(\exists G \in b)CFGH\} =$
$= \bigcup\{s(H) : H \in a \circ^C b\} =$
$= Rep(a \circ^C b).$

$Rep(a) \triangleright^{C'} Rep(b) =$
$= \{z \in W' : (\exists x \in Rep(a))(\exists y \in Rep(b))C'xzy\} \stackrel{\text{L.5.5}}{=}$
$= \{z \in W' : (\exists F \in a)(\exists G \in b)(\exists H \in Uf(\mathcal{A}))z \in s(H) \ \& \ CFHG\} =$
$= \bigcup\{s(H) : H \in Uf(\mathcal{A}) \ \& \ (\exists F \in a)(\exists G \in b)CFHG\} =$
$= \bigcup\{s(H) : H \in a \triangleright^C b\} =$
$= Rep(a \triangleright^C b).$

The case of $\triangleleft$ is similar.

Thus $Rep : \mathcal{B} \longrightarrow \mathcal{B}'$ is an isomorphism, whence $\mathcal{A} \cong \mathcal{B} \cong \mathcal{B}' \in$ RKA, i.e., $\mathcal{A} \in$ IRKA.

If $|A| < \omega$, then $|Uf(\mathcal{A})| < \omega$, and so $|B'| < \omega$. Thus finite algebras are represented on finite bases. So we have proved Theorem 4.3. □

In the following proof of the finite model property, we will follow the strategy in the proofs of Theorems 3.6.2 and 3.6.3 in Roorda 1991. There, Dirk Roorda proves finite model property for logics with dyadic modalities using filtration.

Proof. (Theorem 3.3) Assume that $\not\vdash \varphi$ and let $\langle W, C, v\rangle$ be any model refuting φ. Let Σ be the set consisting of $\iota\delta$ and the subformulas of φ. We define the equivalence relation $\approx$ on W as

$$x \approx y \stackrel{\text{def}}{\iff} (\forall \psi \in \Sigma)(x \Vdash_v \psi \iff y \Vdash_v \psi).$$

We choose an arbitrary but fixed element w' from every equivalence class $\{x \in W : x \approx w\}$. Let

$$W' \stackrel{\text{def}}{=} \{w' : w \in W\}$$

and

$$v'(p_i) \stackrel{\text{def}}{=} \{w' : w \in v(p_i)\}$$

for every $p_i \in P$. Clearly, W' is finite and v' is a valuation. Let $C' \subseteq$

$W' \times W' \times W'$ be defined as

$$C'x'y'w' \stackrel{\text{def}}{\Longleftrightarrow}$$
$$[((\forall \psi_1 \bullet \psi_2 \in \Sigma)(x' \Vdash_v \psi_1 \ \& \ y' \Vdash_v \psi_2 \Rightarrow w' \Vdash_v \psi_1 \bullet \psi_2) \ \&$$
$$(\forall \psi_1 \blacktriangleright \psi_2 \in \Sigma)(x' \Vdash_v \psi_1 \ \& \ w' \Vdash_v \psi_2 \Rightarrow y' \Vdash_v \psi_1 \blacktriangleright \psi_2) \ \&$$
$$(\forall \psi_1 \blacktriangleleft \psi_2 \in \Sigma)(w' \Vdash_v \psi_1 \ \& \ y' \Vdash_v \psi_2 \Rightarrow x' \Vdash_v \psi_1 \blacktriangleleft \psi_2)) \text{ or}$$
$$(x' = y' = w' \ \& \ Cw'w'w')].$$

Let (min) be the following formula:

$$(w' \in W' \ \& \ x, y \in W \ \& \ Cxyw' \Rightarrow C'x'y'w') \ \&$$
$$(w' \in W' \ \& \ x, y \in W \ \& \ Cxw'y \Rightarrow C'x'w'y') \ \&$$
$$(w' \in W' \ \& \ x, y \in W \ \& \ Cw'yx \Rightarrow C'w'y'x')$$

where x' denotes the distinguished element of the equivalence class of x, and similarly for y.

By (max) we mean the following formula:

$$((C'x'y'w' \ \& \ \psi_1 \bullet \psi_2 \in \Sigma \ \& \ x' \Vdash_v \psi_1 \ \& \ y' \Vdash_v \psi_2) \Rightarrow w' \Vdash_v \psi_1 \bullet \psi_2) \ \&$$
$$((C'x'y'w' \ \& \ \psi_1 \blacktriangleright \psi_2 \in \Sigma \ \& \ x' \Vdash_v \psi_1 \ \& \ w' \Vdash_v \psi_2) \Rightarrow y' \Vdash_v \psi_1 \blacktriangleright \psi_2) \ \&$$
$$((C'x'y'w' \ \& \ \psi_1 \blacktriangleleft \psi_2 \in \Sigma \ \& \ w' \Vdash_v \psi_1 \ \& \ y' \Vdash_v \psi_2) \Rightarrow x' \Vdash_v \psi_1 \blacktriangleleft \psi_2).$$

Lemma 5.6 *Let C' be defined as above. Then* (min) *and* (max) *hold.*

Proof. Let $w' \in W' \ \& \ x, y \in W \ \& \ Cxyw'$. If $\psi_1 \bullet \psi_2 \in \Sigma$ is arbitrary and $x' \Vdash_v \psi_1 \ \& \ y' \Vdash_v \psi_2$, then $x \Vdash_v \psi_1 \ \& \ y \Vdash_v \psi_2$, thus $w' \Vdash_v \psi_1 \bullet \psi_2$, by $Cxyw'$. If $\psi_1 \blacktriangleright \psi_2 \in \Sigma$ is arbitrary and $x' \Vdash_v \psi_1 \ \& \ w' \Vdash_v \psi_2$, then $x \Vdash_v \psi_1$, thus $y \Vdash_v \psi_1 \blacktriangleright \psi_2$, by $Cxyw'$. So $y' \Vdash_v \psi_1 \blacktriangleright \psi_2$. Similarly, if $\psi_1 \blacktriangleleft \psi_2 \in \Sigma$ is arbitrary and $w' \Vdash_v \psi_1 \ \& \ y' \Vdash_v \psi_2$, then $x' \Vdash_v \psi_1 \blacktriangleleft \psi_2$. Then, by the definition of C', we have $C'x'y'w'$.

Similar arguments prove the other two implications of (min). One can easily prove (max) using case distinction according to whether $x' = y' = w'$ holds. $\square$

Now we define the relation $\Vdash'_{v'}$ between the set of possible worlds W' of the model $\langle W', C', v' \rangle$ and the set of formulas. The definition of $\Vdash'_{v'}$ is the usual for parameters, Boolean connectives, and for $\bullet$, $\blacktriangleright$ and $\blacktriangleleft$, cf. Definition 2.1; but for the identity we have:

$$w' \Vdash'_{v'} \iota \delta \stackrel{\text{def}}{\Longleftrightarrow} w' \Vdash_v \iota \delta.$$

Then we can prove the following lemma.

Lemma 5.7 $(\forall \psi \in \Sigma)(\forall w \in W)$

$$w \Vdash_v \psi \iff w' \Vdash'_{v'} \psi.$$

Proof. Proving the lemma we will use without mentioning that, since $w' \approx w$, $w \Vdash_v \psi \iff w' \Vdash_v \psi$, and that $W' \subseteq W$. We will prove by induction and refer to the induction hypothesis by 'i.h.'.

$$(\forall p_i \in P)(w' \in v(p_i) \iff w' \in v'(p_i))$$

COMPLETE CALCULUS FOR CONJUGATED ARROW LOGIC / 137

$$w \Vdash_v \neg\psi_1 \iff w' \not\Vdash_v \psi_1 \overset{\text{i.h.}}{\iff} w' \not\Vdash_{v'}' \psi_1 \iff w' \Vdash_{v'}' \neg\psi_1$$

$$w \Vdash_v \psi_1 \wedge \psi_2 \iff w' \Vdash_v \psi_1 \ \& \ w' \Vdash_v \psi_2 \overset{\text{i.h.}}{\iff}$$
$$\iff w' \Vdash_{v'}' \psi_1 \ \& \ w' \Vdash_{v'}' \psi_2 \iff$$
$$\iff w' \Vdash_{v'}' \psi_1 \wedge \psi_2$$

$$w \Vdash_v \psi_1 \bullet \psi_2 \Rightarrow (\exists x, y \in W) Cxyw' \ \& \ x \Vdash_v \psi_1 \ \& \ y \Vdash_v \psi_2 \overset{(\min)}{\Rightarrow}$$
$$\Rightarrow (\exists x, y \in W) C'x'y'w' \ \& \ x \Vdash_v \psi_1 \ \& \ y \Vdash_v \psi_2 \overset{\text{i.h.}}{\Rightarrow}$$
$$\Rightarrow (\exists x', y' \in W') C'x'y'w' \ \& \ x' \Vdash_{v'}' \psi_1 \ \& \ y' \Vdash_{v'}' \psi_2 \Rightarrow$$
$$\Rightarrow w' \Vdash_{v'}' \psi_1 \bullet \psi_2$$

$$w' \Vdash_{v'}' \psi_1 \bullet \psi_2 \Rightarrow (\exists x', y' \in W') C'x'y'w' \ \& \ x' \Vdash_{v'}' \psi_1 \ \& \ y' \Vdash_{v'}' \psi_2 \overset{\text{i.h.}}{\Rightarrow}$$
$$\Rightarrow (\exists x', y' \in W') C'x'y'w' \ \& \ x' \Vdash_v \psi_1 \ \& \ y' \Vdash_v \psi_2 \overset{(\max)}{\Rightarrow}$$
$$\Rightarrow w \Vdash_v \psi_1 \bullet \psi_2$$

$$w \Vdash_v \psi_1 \blacktriangleright \psi_2 \Rightarrow (\exists x, y \in W) Cxw'y \ \& \ x \Vdash_v \psi_1 \ \& \ y \Vdash_v \psi_2 \overset{(\min)}{\Rightarrow}$$
$$\Rightarrow (\exists x, y \in W) C'x'w'y' \ \& \ x \Vdash_v \psi_1 \ \& \ y \Vdash_v \psi_2 \overset{\text{i.h.}}{\Rightarrow}$$
$$\Rightarrow (\exists x', y' \in W') C'x'w'y' \ \& \ x' \Vdash_{v'}' \psi_1 \ \& \ y' \Vdash_{v'}' \psi_2 \Rightarrow$$
$$\Rightarrow w' \Vdash_{v'}' \psi_1 \blacktriangleright \psi_2$$

$$w' \Vdash_{v'}' \psi_1 \blacktriangleright \psi_2 \Rightarrow (\exists x', y' \in W') C'x'w'y' \ \& \ x' \Vdash_{v'}' \psi_1 \ \& \ y' \Vdash_{v'}' \psi_2 \overset{\text{i.h.}}{\Rightarrow}$$
$$\Rightarrow (\exists x', y' \in W') C'x'w'y' \ \& \ x' \Vdash_v \psi_1 \ \& \ y' \Vdash_v \psi_2 \overset{(\max)}{\Rightarrow}$$
$$\Rightarrow w \Vdash_v \psi_1 \blacktriangleright \psi_2$$

and similar argument proves the case of ◀. Finally,

$$w' \Vdash_{v'}' \iota\delta \iff w' \Vdash_v \iota\delta$$

by definition. □

So far so good, but we would like to have

$$w' \Vdash_{v'}' \iota\delta \iff C'w'w'w'.$$

To achieve it we will apply the same trick as in the proof of the Representation Theorem 4.3.

Let
$$\mathcal{T} \overset{\text{def}}{=} \langle \mathcal{P}(W'), \cap, \sim, \circ^{C'}, \triangleright^{C'}, \triangleleft^{C'}, \{w' : w' \Vdash_{v'}' \iota\delta\} \rangle.$$

Note that we do not know whether $\mathcal{T} \in \mathsf{RKA}$. It is easy to check that for all formula ψ
$$\langle \mathcal{T}, v' \rangle \vDash \psi = 1 \iff \langle W', C', v' \rangle \vDash \psi,$$
and so $\langle \mathcal{T}, v' \rangle \nvDash \varphi = 1$. First we note that

$$w' \Vdash_{v'}' \iota\delta \Rightarrow w' \in Id^{C'}.$$

Indeed,
$$w' \Vdash_{v'}' \iota\delta \iff w' \Vdash_v \iota\delta \iff Cw'w'w' \Rightarrow C'w'w'w' \iff w' \in Id^{C'}.$$

In the proof of the Representation Theorem 4.3 we met exactly the same problem. There we had an algebra $\mathcal{B}$ whose identity was not the "real" one. Then we could split the "bad" elements of the base of $\mathcal{B}$ and then we got a $\mathcal{B}' \in \mathsf{RKA}$ isomorphic to $\mathcal{B}$. Now, if we do the same trick with $\mathcal{T}$ instead of $\mathcal{B}$, then we get a $\mathcal{T}' \in \mathsf{RKA}$ isomorphic to $\mathcal{T}$, where W''' and C''' are defined using W' and C' precisely in the same way as in the representation proof. (There a certain W' and C' were defined using some W and C.) Note that since W' was finite, so is W'''. Then $\mathcal{T}'$ is the subalgebra of the finitely based full RKA

$$\langle \mathcal{P}(W'''), \cap, \sim, \circ^{C'''}, \triangleright^{C'''}, \triangleleft^{C'''}, Id^{C'''} \rangle.$$

By $\mathcal{T} \cong \mathcal{T}'$, we have

$$\langle \mathcal{T}', v'' \rangle \not\models \varphi = 1,$$

where v'' is the valuation determined by v', i.e., $v''(\varphi)$ is the image of $v'(\varphi)$ along the isomorphism. Thus, by Lemma 5.3,

$$\langle W''', C''', v'' \rangle \not\models \varphi.$$

That is we constructed a finite model refuting the non-valid formula φ. □

Now we will prove the lemmas that were used in the proof of the Completeness Theorem 3.1.

Proof. (Lemma 5.1) By propositional axioms and Modus Ponens, $\equiv_\Gamma$ is an equivalence relation.

Now assume that $\Gamma \vdash A \leftrightarrow B$ & $\Gamma \vdash C \leftrightarrow D$. Then, by propositional calculus again, $\Gamma \vdash \neg A \leftrightarrow \neg B$ and $\Gamma \vdash A \wedge C \leftrightarrow B \wedge D$. By the substitution rule (ix) for $\bullet$, $\Gamma \vdash A \bullet C \leftrightarrow B \bullet D$. For $\blacktriangleright$ we have

$\Gamma \vdash \neg((B \blacktriangleright D) \wedge (\neg(B \blacktriangleright D))) \quad \Leftrightarrow \quad$ (propositional calculus)
$\Gamma \vdash (B \blacktriangleright D) \wedge (\neg(B \blacktriangleright D)) \leftrightarrow \bot \quad \Leftrightarrow \quad (vi)$
$\Gamma \vdash (B \bullet (\neg(B \blacktriangleright D))) \wedge D \leftrightarrow \bot \quad \Leftrightarrow \quad$ (propositional calculus and (ix))
$\Gamma \vdash (A \bullet (\neg(B \blacktriangleright D))) \wedge C \leftrightarrow \bot \quad \Leftrightarrow \quad (vi)$
$\Gamma \vdash (A \blacktriangleright C) \wedge (\neg(B \blacktriangleright D)) \leftrightarrow \bot \quad \Leftrightarrow \quad$ (propositional calculus)
$\Gamma \vdash A \blacktriangleright C \rightarrow B \blacktriangleright D.$

By symmetry, $\Gamma \vdash B \blacktriangleright D \rightarrow A \blacktriangleright C$.

The proof for $\blacktriangleleft$ is similar. Thus we have proved the lemma. □

Proof. (Lemma 5.2) 1 is true by propositional calculus.

The proof of 2 is as follows. Let $x_1 = x_2$ be an equation of Ax, v be an arbitrary valuation and $v(x_1) = \overline{p_1}$ and $v(x_2) = \overline{p_2}$. Then, since $p_1 \leftrightarrow p_2$ is an axiom of CARL, $\Gamma \vdash p_1 \leftrightarrow p_2$. Thus $\overline{p_1} = \overline{p_2}$, whence $\langle \mathcal{F}_\Gamma, v \rangle \models x_1 = x_2$.

Let $(x_1 = x_2) \Rightarrow (x_3 = x_4)$ be a quasi-equation of Ax. Let v be a valuation and $v(x_i) = \overline{p_i}$ for $1 \leq i \leq 4$. Then

$$\frac{p_1 \leftrightarrow p_2}{p_3 \leftrightarrow p_4}$$

is an instance of an inference rule of CARL. Now assume that $\overline{p_1} = \overline{p_2}$ is true in $\mathcal{F}_\Gamma$, whence $\Gamma \vdash p_1 \leftrightarrow p_2$. By the rule above, $\Gamma \vdash p_3 \leftrightarrow p_4$, hence $\overline{p_3} = \overline{p_4}$ in $\mathcal{F}_\Gamma$. Thus $\mathcal{F}_\Gamma \vDash x_1 = x_2 \Rightarrow x_3 = x_4$. □

Proof. (Lemma 5.3) Easy by definition. □

Acknowledgments. The author is grateful to Hajnal Andréka, István Németi and Ildikó Sain. Thanks are also due to Maarten Marx.

References

Andréka, H., and Sz. Mikulás. 1994. Lambek calculus and its relational semantics: completeness and incompleteness. *Journal of Logic, Language and Information* 3(1):1–38.

van Benthem, J. 1991. *Language in Action (Categories, Lambdas and Dynamic Logic)*. Studies in Logic, Vol. 130. North-Holland, Amsterdam.

Benthem, J. van. 1994. A note on Dynamic Arrow Logic. In *Logic and Information Flow*, ed. J. van Eijck and A. Visser. 15–29. Cambridge (Mass.): MIT Press.

Došen, K. 1992. A Brief Survey of Frames for the Lambek Calculus. *Zeitschr. f. math. Logik und Grundlangen d. Math.* 38:179–187.

Jónsson, B., and C. Tsinakis. 1992. Relation Algebras as Residuated Boolean Algebras. Technical report. Vanderbilt University, Nashville. Preprint.

Marx, M. 1995. *Algebraic Relativization and Arrow Logic*. Doctoral dissertation, Institute for Logic, Language and Computation, University of Amsterdam. ILLC Dissertation Series 1995-3.

Németi, I. 1992. A telephone call.

Pratt, V. 1990. Action Logic and Pure Induction. In *Logics in AI*, ed. J. van Eijck, Lecture Notes in Artificial Intelligence, Vol. 478, 97–120. Springer–Verlag.

Roorda, D. 1991. *Resource Logics. Proof-Theoretical Investigations*. Doctoral dissertation, Institute for Logic, Language and Computation, University of Amsterdam.

7
Many-Dimensional Arrow Structures: Arrow Logics II

DIMITER VAKARELOV

ABSTRACT. We introduce the notion of an n-dimensional arrow structure, which for $n = 2$ coincides with the notion of a directed multi-graph. We study several first-order and modal languages connected with arrow structures and compare their expressive power. The second part is devoted to the axiomatization of some arrow logics. Finally, some further perspectives of the "arrow approach" are discussed.

Introduction: a Short History of Arrow Logic

This paper is the second one in a series of papers devoted to Arrow Logic. The term "Arrow Logic" has recently become very fashionable, though used by different people with different meaning. I will start with a short history of the subject to explain my point of view.

The idea to investigate arrow structures, and the modal logics based on them, was suggested to me by Johan van Benthem during the Kleene conference, held in June 1990 in Varna. His idea was that it would be nice to have a simple modal logic, with semantics in two-sorted structures, later to be called "arrow structures", which have points ("states") and arrows ("transitions"). Such arrow structures must have different models: ordered pairs, directed graphs, categories, vectors, states and transitions, and so on. So, two main problems arose: first, the choice of an adequate mathematical structure of arrows, and second, the corresponding choice of an appropriate modal language.

At that time van Benthem (in 1989) had already introduced one notion of arrow structure (called "arrow frame"), originating from the algebra

---

Research partially supported by the Bulgarian Ministry of Science, Education and Technology, contract I-412-94

Arrow Logic and Multi-Modal Logic
M. Marx, L. Pólos and M. Masuch, eds.
Copyright © 1996, CSLI Publications.

of relations. An arrow frame is a relational system of the form $\underline{W} = (W, C, R, I)$, where $W \neq \emptyset$, $C \subseteq W^3$, $R \subseteq W^2$ and $I \subseteq W$, with the following intuition:

- W — a non-empty set, whose elements are called *arrows*,
- $C(x, y, z)$ — x is a *composition* of y and z
- $R(x, y)$ — y is a *reversal* of x
- $I(x)$ — x is an *identity* arrow.

For a picture of these relations, cf. figure 1 in chapter 2.

In this approach it is not supposed that arrows have some explicitly stated internal structure (that they are "ordered pairs", or have "first points" and "last points", etc.). They are treated as abstract objects, equipped with the above three relations, for which initially nothing is supposed. These structures are used as a semantic base of a modal language, extending the language of the classical propositional logic with one binary modality $A \circ B$ (composition), one unary modality $\otimes A$ (reverse) and a propositional constant $\iota\delta$ (identity). The semantics is this:

$x \Vdash A \circ B$ iff $(\exists y, z) C(x, y, z)$ $y \Vdash A$ and $z \Vdash B$

$x \Vdash \otimes A$ iff $(\exists y) R(x, y)$ and $y \Vdash A$

$x \Vdash \iota\delta$ iff $I(x)$.

The resulting logic, and many of its natural extensions, that aims to approach the desirable properties composition, reversal and identity from the algebra of relations was studied in van Benthem 1989. In van Benthem 1991 it is called Arrow Logic. Probably this is the first official appearance of the term "Arrow Logic". In van Benthem 1994 the minimal Arrow Logic (based on frames without any assumptions), here called Dynamic Arrow Logic, is extended with operations like the Kleene star $*$, and various possible extensions and variations are discussed.

Now some words about my approach. As for the mathematical model of arrows, I decided to take less abstract structures than van Benthem's arrow frames. The intended "arrow structures" should have "points", some of them connected by "arrows". In graph theory these are called directed multi-graphs. I have adopted the following, more formal definition; the resulting notion is called "arrow structure", reserving directed multi-graphs as one of its models. Arrow structures are systems of the form $S = (Ar, Po, 1, 2)$, where:

- Ar is a non-empty set, whose elements are called arrows,
- Po is a non-empty set, whose elements are called points,
- 1 and 2 are total functions from Ar to Po, called projections. If $x \in Ar$ then $1(x)$ is called the first end of x and $2(x)$ is called the second end of x;

- Each point is a first end or second end of some arrow.

Graphically: $1(x) \bullet \xrightarrow{\quad x \quad} \bullet 2(x)$

So far so good. The next problem was how to associate an appropriate modal language to arrow structures and how to interpret this language in such structures. As we know, the standard Kripke semantics requires binary relations in one sorted structure. So one solution is to define an appropriate set of binary relations in the set of arrows in such a way, that the new relational system contains the information of the whole arrow structure. The following four relations in the set Ar proved to have such property: for $x, y \in Ar, i, j = 1, 2$,

$$xR_{ij}y \text{ iff } i(x) = j(y).$$

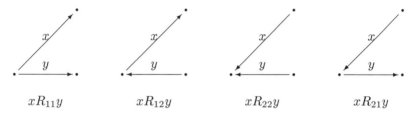

$\qquad xR_{11}y \qquad\qquad xR_{12}y \qquad\qquad xR_{22}y \qquad\qquad xR_{21}y$

The relations R_{ij}, called incidence relations, express the four possible ways for two arrows to have a common end. The relations R_{ij} satisfy the following simple first-order conditions, for $x, y, z \in Ar$ and $i, j, k = 1, 2$:

(ρii) $xR_{ii}x$,
(σij) $xR_{ij}y \to yR_{ji}x$,
(τijk) $xR_{ij}y \ \& \ yR_{jk}z \to xR_{ik}z$.

These conditions are characteristic in the following sense: if in a set W we have four relations R_{ij} satisfying the above conditions, then there exists an arrow structure $S = (Ar_S, Po_S, 1, 2)$ such that $Ar_S = W$ and S determines the same relations R_{ij}. So instead of arrow systems, which are two-sorted systems and therefore not convenient for Kripke interpretations, we can use relational systems of the form $\underline{W} = (W, R_{11}, R_{12}, R_{21}, R_{22})$, satisfying the conditions (ρii), (σij) and (τijk). I have called such systems "arrow frames". Arrow frames in this new sense have two advantages: first, they are in some sense equivalent to arrow structures, so their abstract elements are real arrows; and second, they have a simple first-order relational definition, suitable for modal purposes. Note that the relations C, R and I, used in van Benthem's arrow frames, have natural definitions in terms of the the relations R_{ij} :

$$
\begin{array}{lll}
C(x,y,z) & \text{iff} & xR_{11}y \ \& \ yR_{21}z \ \& \ zR_{22}x \\
R(x,y) & \text{iff} & xR_{21}y \ \& \ xR_{12}y \\
I(x) & \text{iff} & xR_{12}x.
\end{array}
$$

The corresponding modal language for the minimal, or Basic Arrow Logic, *BAL* for short, is now easy to define. It extends the language of propositional logic with four unary modalities $[ij]$ (for $i,j = 1,2$) with standard Kripke interpretation in arrow frames. Fortunately, our axioms of arrow frame are modally definable and canonical, so *BAL* is easy to axiomatize. Moreover it admits filtration; therefore it is decidable, hence it possesses all desirable properties[1]. It is also flexible for extensions with new connectives by definable accessibility relations. For instance an analog of van Benthem's arrow logic will be an extension of *BAL* with modalities $A \circ B$, $\otimes A$ and modal constant $\iota\delta$. An axiomatization of this system is presented by Andrey Arsov in Arsov and Marx 1994.

In 1990 I met the Hungarian logicians Hajnal Andréka and István Németi, who gave a long-term seminar on Algebraic Logic in Warsaw. We had very fruitful discussions, leading us to the conclusion that a good deal of the old Tarskian Algebraic Logic has what one could call an "arrow" nature, where arrows are ordered pairs or n-tuples.

At that time István and Hajnal explained to me the importance of the so called "finitization problem in Algebraic Logic"[2]. In short, the essence of the finitization problem can be explained as follows. Almost all algebraic systems, studied in algebraic logic, such as representable relational algebras, representable cylindric algebras etcetera, possess bad properties: their equational theory is undecidable, they do not have finite axiomatization. So the "finitization problem" consists of finding good enough approximations of these logical algebras, as to have a finite axiomatization or a decidable equational theory. One solution is to take the so-called relativized versions of these algebras; for instance, in the case of relational algebras over some full square $W = U \times U$, to take the appropriate algebra of relations in some subset $W \subseteq U \times U$. In this way, using results of Maddux 1982 and Kramer 1991 we can obtain varieties which are finitely axiomatizable and representable in relativized set algebras; using a result of Németi 1987 we get that their equational theory is decidable.

There is a close analogy between such relativized relational set algebras and the semantics of Basic Arrow Logic. This is because each set $W \subseteq$

---

[1] The first presentation of these results about *BAL* were summarized in the manuscript Vakarelov 1990 that was distributed to some people from Amsterdam and Budapest. The full version will be presented in Vakarelov to appear.

[2] For a discussion of this problem, see the excellent survey of Algebraic Logic given in Németi 1991.

$U \times U$ determines an arrow structure and the interpretation of formulas in this structure are subrelations (subsets) of W and the modal operations on formulas define some relational operations on these subrelations. It is a well-known fact that completeness theorems in Modal Logic correspond to the representability of some logical algebras and that decidability of the logical calculus corresponds to decidability of equational theory of the corresponding algebraic variety. Thus semantic methods from Modal Logic may help to solve some problems in Algebraic Logic, and conversely, some results of Algebraic Logic may have direct implications for some problems of Modal Logic.

The combination of the algebraic methods of Tarskian Algebraic Logic, and semantic methods from Modal Logic proved to be very fruitful. As a result of this combination of methods, Venema 1989, 1991, 1995 axiomatizes many modal logics that correspond to classical systems of Algebraic Logic, using special modal rules originating with Gabbay 1981 and a polymodal generalization of Sahlqvist's theorem on first-order definability Sahlqvist 1975. Maarten Marx 1995, using the so-called graph method, which is in a sense a polymodal and refined version of the Sahlqvist's unraveling construction from modal logic, Sahlqvist 1975, gave a new, very short and elegant proof of Maddux's theorem 1982 and Kramer's theorem 1991 on representation of relativized relational algebras.

The influence of the Budapest school of Algebraic Logic on my work is seen the generalization of arrow structures to n-dimensional arrow structures in Vakarelov 1991b, 1991a, in order to cover the case of n-ary relations. In 1993 I extended the related Basic Arrow Logic of dimension n — BAL^n — with modal operations, corresponding to cylindrifications from cylindric algebras and give the completeness theorem and decidability. Cylindrifications are modal analogs of quantifiers and the first-order fragment corresponding to this logic presents a decidable version of first-order logic.

The present paper includes the results announced in Vakarelov 1991b, 1991a and Vakarelov 1993, and some new results, obtained by the author during his visit in the University of Amsterdam in June–July 1992.

The paper is divided into two parts. Sections 1–3 are devoted to the notion of n-dimensional arrow structures and some first-order and modal languages associated with these. Sections 4–6 discuss the axiomatization of some arrow logics of dimension n.

In section 1 we prove that the relational structure, consisting of the set of arrows of some n-dimensional arrow structure and equipped with the incidence relations R_{ij} $i,j \leq n$, contains in some sense the whole information of the arrow structure. This makes it possible to introduce the notion of n-dimensional arrow frame as one-sorted equivalent of the notion of n-dimensional arrow structure. With the point part of an n-

dimensional arrow structure we associate another relational system, called an n-dimensional point frame, containing one n-place relation ρ, which holds for the sequence of points $A_1, \ldots, A_n$ iff they are points on some arrow.

In section 2 we introduce three first-order languages connected with arrow structures — the language $\mathcal{L}(S)$ of the whole structure, the language $\mathcal{L}(W)$, connected with the arrow frame of the structure, and the language $\mathcal{L}(V)$, connected with the point frame of the structure. The main result here is a theorem stating that all these three languages have equal expressive power. As a consequence we obtain that the first-order theory of one n-place relation can be reduced to the first-order theory of n^2 special relations: the relations R_{ij}.

In section 3 we introduce two modal languages — an "arrow" language corresponding to the n-dimensional arrow frames, extended with the universal modality, and a "point" language, corresponding to the point frames, containing $n-1$-argument modal operations, which is interpreted with the n-place point relation ρ. The main result here is a translation of the "point" language into the "arrow" language which preserves the corresponding semantic validity. This result has several implications. First it is shown that we can associate arrow structures both with "arrow logics", talking about arrows, and with "point logics", talking about points. The translation of "point" language into "arrow" language shows that the second language is at least expressible as the "point" language. In other words, this answers the question why "Arrow Logics". Another implication of this result is that polyadic modalities can be reduced to monadic ones.

In section 4 we axiomatize the Basic Arrow Logic of dimension n — BAL^n — and an extension of BAL^n with the universal modality. We prove several completeness results, including finite model property.

In section 5 we axiomatize an extension of BAL^n with the operations of cylindrification, corresponding to a version of relativized set cylindric algebras without constants δ_{ij} for identity. We prove the corresponding completeness theorem and decidability. The method of proving the completeness theorem, called "copying", consists of transforming non-intended models into intended ones, preserving semantic validity.

In section 6, finally, we discuss some open problems and further perspectives of the "arrow approach". Besides some possible extensions of two-dimensional and many-dimensional basic arrow logics we discuss possible "arrow" versions of Dynamic Logic and Process Logic, and connections of many-dimensional arrow systems with the information systems of Pawlak and the corresponding information logics.

To conclude this introductory section, let us return to the main question: what is Arrow Logic? At least for me it is a branch of Modal Logic that characterizes arrows, which are here considered not only as ordered pairs or n-tuples, and that combines methods from the arsenal of Modal Logic with those from the old Tarskian Algebraic Logic.

I. First-Order and Modal Languages Associated with Arrow Structures of Dimension n

1 Arrow Structures and Arrow Frames of Dimension n

By an arrow structure of dimension n (n-arrow structure for short) we will understand any three-sorted algebraic system $S = (Ar, Po, (n), .)$ where:

- Ar is a non-empty set whose elements are called arrows,
- Po is a non-empty set whose elements are called points,
- $(n) = \{1, \ldots, n\}$ is the set of natural numbers from 1 to n.
- $(.)$ is a total function from $Ar \times (n)$ to Po, called application.

If $x \in Ar$ and $i \in (n)$ then $i.x$ is called the i-th point of x, $1.x$ is also called the beginning of x and $n.x$ is called the end of x, or the head of x.

Graphically: $\quad \underset{\bullet}{1.x} \quad \underset{\bullet}{2.x} \quad \underset{\bullet}{3.x} \quad \cdots \quad \underset{\bullet}{n.x}$

We assume that $Ar \cap Po = \emptyset$ and that the following axiom is satisfied

$(Ax) \quad (\forall A \in Po)(\exists i \in (n))(\exists x \in Ar)(i.x = A).$

The meaning of (Ax) is that each point is a point in some arrow, so that there are no isolated points. It is possible to develop a more general theory without this axiom; we use it only because the formulations of some theorems will be simpler.

Two-dimensional arrow structures are just directed multi-graphs without isolated points. Thus, in some sense n-arrow structures can be considered as a generalization of the notion of directed multi-graph.

Sometimes we will write Ar_S and Po_S to denote that Ar and Po are from a given system S.

An n-arrow structure S is called *normal* if it satisfies the following axiom of normality:

$(Nor) \quad (\forall x, y \in Ar_S)((\forall i \in (n))(i.x = i.y) \to x = y).$

In the two-dimensional case, normal arrow structures correspond to the notion of a directed graph.

The following example of a normal n-arrow structure is very important. Let $\underline{V} = (V, \rho)$, $V \neq \emptyset$, be a relational system such that $\rho \subseteq V^n$ is a non-empty n-ary relation in V satisfying the following condition:

$(\forall A \in V)(\exists i \in (n))(\exists A_1, \ldots, A_n \in V)(\rho(A_1, \ldots, A_n) \,\&\, A = A_i).$

Such a system will be called an n-point frame. We shall construct an n-arrow structure $S = S(V) = (Ar(V), Po(V), (n), .)$ over $\underline{V}$ in the following way. Put $Ar(V) = \rho, Po(V) = V$ and for $i \in (n)$ and $(A_1, \ldots, A_i, \ldots, A_n) \in \rho$, define $i.(A_1, \ldots, A_i, \ldots, A_n) = A_i$. Then obviously the system $S(V)$ is a normal n-arrow structure, called a relational n-arrow structure over the system V.

Now we shall show that each n-arrow structure S determines a normal n-arrow structure in the following way. Let S be given and define in Po_S the following n-ary relation $\rho = \rho_S$: for $A_1, \ldots, A_n \in Po_S$,

$$(A_1, \ldots, A_n) \in \rho_S \text{ iff } (\exists x \in Ar_S)(1.x = A_1 \ \& \ \ldots \ \& \ n.x = A_n).$$

The relational structure $V(S) = (Po_S, \rho_S)$ will be called an n- point frame over S. Now $V(S)$ determines a normal n-arrow structure $S(V(S))$ in a way described above. We shall show that $S(V(S))$ is a homomorphic image of S, and if S is a normal n-arrow structure, then $S(V(S))$ is isomorphic with S in the following sense of homomorphism and isomorphism.

Let S and S' be two n-arrow structures. A pair (f, g) of mappings $f : Ar_S \to Ar_{S'}$ and $g : Po_S \to Po_{S'}$ is called a homomorphism from S into S' if for any $x \in Ar_S$ and $i \in (n)$ we have $g(i.x) = i.f(x)$. If f and g are 1-1-mappings, then (f, g) is called an isomorphism from S onto S'.

Theorem 1.1 *Let S be an n-arrow structure, $V(S) = (Po_S, \rho_S)$ be the n-point frame over S and $S' = S(V(S))$ be the n-arrow structure over $V(S)$. Then S' is a homomorphic image of S and if S is a normal n-arrow structure then S' is an isomorphic image of S.*

Proof. Let S be given. For $x \in Ar_S$ and $A \in Po$ we define $f(x) = (1.x, \ldots, n.x)$ and $g(A) = A$.

Then (f, g) is the required homomorphism from S onto S'. Indeed $f(x) \in \rho_S = Po_{S'}$ so f is correctly defined. The following equalities show that (f, g) is a homomorphism:

$$i.f(x) = i.(1.x, \ldots, n.x) = i.x = g(i.x)$$

Suppose now that S is a normal n-arrow structure, and let $f(x) = f(y)$. Then $(1.x, \ldots, n.x) = (1.y, \ldots, n.y)$, and given the condition of normality this implies $x = y$. So f is an injective function. To show that f is a surjective function, let $(A_1, \ldots, A_n) \in \rho_S$. Then by the definition of ρ_S, there exists an $x \in Ar_S$ such that $(A_1, \ldots, A_n) = (1.x, \ldots, n.x) = f(x)$. This shows that in this case (f, g) is an isomorphism from S onto S'. □

Let S be an n-arrow structure. The following binary relations will play an important role in the theory of n-arrow structures. For $x, y \in Ar_S$ and $i, j \in (n)$ we define a relation $R_{ij} = R_{ij}^S$ as follows:[3]

$$xR_{ij}y \text{ iff } i.x = j.y.$$

---

[3]For $n = 2$ the relation R_{21} is studied by Kuhn 1989 and called "domino relation".

The relation $xR_{ij}y$ says that the i-th point of x coincides with the j-th point of y. Graphically, for arbitrary n:

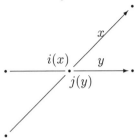

Lemma 1.2 *Let S be an n-arrow structure. Then:*
(i) *The relations R_{ij} satisfy the following conditions for any $x, y, z \in Ar_S$ and $i, j, k \leq n$:*
  (ρii) $xR_{ii}x$,
  (σij) $xR_{ij}y \to yR_{ji}x$,
  (τijk) $xR_{ij}y$ & $yR_{jk}z \to xR_{ik}z$.
(ii) *S is normal n-arrow structure iff the following condition is satisfied for any $x, y \in Ar_S$:*
  (Nor') $xR_{11}y$ & $xR_{22}y$ & .... & $xR_{nn}y \to x = y$.

Proof. By an easy verification. □

Let $W \neq \emptyset$ be a set, $n \geq 2$ be a fixed natural number and $\{R_{ij} \subseteq W^2 \mid i, j \leq n\}$ be a set of n^2 binary relations in W. Then the relational system $\underline{W} = (W, \{R_{ij} \mid i, j \leq n\})$ is called an n-dimensional arrow frame (n-arrow frame for short) if the conditions (ρii), (σij) and (τijk) from lemma 1.2 are satisfied for any $i, j, k \leq n$ and $x, y, z \in W$. If $\underline{W}$ satisfies (Nor'), then it is called a normal n-arrow frame. If $W = Ar_S$ and $R_{ij} = R_{ij}^S$ for some n-arrow structure S, then $\underline{W} = \underline{W}(S)$ is called a standard n-arrow frame (over S).

We now show that each (normal) n-arrow frame is a standard n-arrow frame over some (normal) n-arrow structure S.

Theorem 1.3 *Let $\underline{W} = (W, \{R_{ij} \mid i, j \leq n\})$ be an n-arrow frame. Then there exists an n-arrow structure $S = S(\underline{W})$ such that $Ar_S = W$ and for any $i, j \leq n$ $R_{ij}^S = R_{ij}$, if $\underline{W}$ is normal n-arrow frame, then S is also a normal n-arrow structure.*

Proof. Let $\underline{W}$ be given. We have to construct $S = S(\underline{W})$. We set $Ar_S = Ar(\underline{W}) = W$. To construct Pos we shall introduce the notion of a generalized point in $\underline{W}$.

Let $\alpha_1, \ldots, \alpha_n$ be subsets of W. The n-tuple $(\alpha_1, \ldots, \alpha_n)$ will be called a generalized point in $\underline{W}$, if the following conditions are satisfied for any $x, y \in W$ and $i, j \leq n$:

(1) If $x \in \alpha_i$ and $y \in \alpha_j$ then $xR_{ij}y$,
(2) if $xR_{ij}y$ and $x \in \alpha_i$ then $y \in \alpha_j$,
(3) $\alpha_1 \cup \ldots \cup \alpha_n \neq \emptyset$.

The set of all generalized points in W will be denoted by $Po(W)$. We set $Po_S = Po(W)$.

It remains to define the application function, which here will be denoted by $\circ$. For that purpose we need the following notation: let R be any binary relation in W and $x \in W$, then we set $R(x) = \{y \in W \mid xRy\}$. For any $i \in (n)$ and $x \in W$ we define:

$$i \circ x = (R_{i1}(x), \ldots, R_{in}(x))$$

Now we set $S(\underline{W}) = (Ar(\underline{W}), Po(\underline{W}), (n), \circ)$. The proof that $S(\underline{W})$ is an n-arrow structure follows directly from the next lemma.

Lemma 1.4 *(i) For any $i \in (n)$ and $x \in W$ $i.x$ is a generalized point in $\underline{W}$.*
(ii) $xR_{ij}y$ iff $i.x = j.y$.
(iii) If $A = (\alpha_1, \ldots, \alpha_n)$ is a generalized point in W then there exists $x \in W$ and $i \in (n)$ such that $i \circ x = A$.

Proof. The proof of this technical lemma is long but straightforward and consists of many applications of the axioms (ρii), (σij) and (τijk). Let us for instance prove (iii). Suppose $A = (\alpha_1, \ldots, \alpha_n)$. By condition (3) of the definition of a generalized point $\alpha_1 \cup \ldots \cup \alpha_n \neq \emptyset$, so there exist $i \in (n)$ and $x \in W$ such that $x \in \alpha_i$. We shall show that $i.x = A$, i.e. that for any $j \in (n)$ $R_{ij}(x) = \alpha_j$. Let $y \in R_{ij}(x)$, then $xR_{ij}y$. From $xR_{ij}y$ and $x \in \alpha_i$, by condition (2) of the definition of a generalized point we get $y \in \alpha_j$, hence $R_{ij}(x) \subseteq \alpha_j$. For the converse inclusion suppose that $y \in \alpha_j$. Then by (1) and $x \in \alpha_i$ we get $xR_{ij}y$, so $y \in R_{ij}(x)$ and hence $\alpha_j \subseteq R_{ij}(x)$. Thus $R_{ij}(x) = \alpha_j$. □

Let us turn to the proof of the theorem 1.3. Suppose that $\underline{W}$ is a normal n-arrow frame. We shall show that $S(\underline{W})$ is also a normal n-arrow structure. Let $x, y \in Ar(W) = W$ and suppose that for any $i \in (n)$ we have $i.x = i.y$. Then for any $i \in (n)$ we get $R_{ii}(x) = R_{ii}(y)$. By (ρii) we have $yR_{ii}y$ so $y \in R_{ii}(y)$, hence $y \in R_{ii}(x)$ and consequently $xR_{ii}y$ for any $i \in (n)$. Then by (Nor') we get $x = y$.[4] □

The n-arrow structure $S(\underline{W})$ constructed in theorem 1.3 will be called a canonical n-arrow structure over $\underline{W}$.

Theorem 1.5 *Let S be an n-arrow structure, $\underline{W} = \underline{W}(S)$ be the standard n-arrow frame over S and $S' = S(\underline{W}(S))$ be the canonical n-arrow structure over $\underline{W}(S)$. Then $S(\underline{W}(S))$ is isomorphic with S.*

---
[4] A theorem characterizing domino relations similar to theorem 1.3 is proved by Kuhn in 1989.

Proof. For $i \in (n)$ and $A \in Po_S$, define $i(A) = \{x \in Ar_S \mid i.x = A\}$ and $g(A) = (1(A), \ldots, n(A))$. Let id be the identity function in Ar_S. Then the pair (id, g) is the required isomorphism. This follows from the following lemma. □

Lemma 1.6 (i) $g(A)$ *is a generalized point in* $\underline{W}(S)$,
(ii) g *is a* $1-1$-*function from* Po_S *into* $Po(\underline{W}(S))$
(iii) *for any* $x \in Ar_S$ *and* $i \in (n)$, $g(i.x) = i \circ x = i \circ id(x)$.

Proof. (i) The proof that $g(A)$ is a generalized point in $\underline{W}(S)$ is straightforward.

(ii) Supposing $g(A) = g(B)$, we shall show that $A = B$, which proves that g is an injective function. From $g(A) = g(B)$ we get that for any $i \leq n$ we have $i(A) = i(B)$. By axiom (Ax), we know that for point A there exists $i \leq n$ such that $i.x = A$. From here we get that $x \in i(A)$, and since $i(A) = i(B)$ we obtain $x \in i(B)$, so $B = i.x$; thus we have $A = B$. To prove that g is a function "onto", suppose that $(\alpha_1, \ldots, \alpha_n)$ is a generalized point in $\underline{W}$. We shall show that there exists a point $A \in Po_S$ such that $g(A) = (\alpha_1, \ldots, \alpha_n)$.

By lemma 1.4(*iii*) there exist $x \in Ar_S$ and $i \leq n$ such that $i \circ x = (R_{i1}(x), \ldots, R_{in}(x)) = (\alpha_1, \ldots, \alpha_n)$. Let $i.x = A$. We shall show that for any $j \leq n$, $j(A) = R_{ij}(x)$, which will prove the equality $g(A) = i \circ x$. Suppose that $y \in j(A)$. Then $j.y = A$, so $i.x = j.y$, $xR_{ij}y$ and consequently $y \in R_{ij}(x)$. Hence $j(A) \subseteq R_{ij}(x)$. For the converse inclusion suppose $y \in R_{ij}(x)$, so $xR_{ij}y$, $A = i.x = j.y$, and finally $y \in j(A)$; and therefore $R_{ij}(x) \subseteq j(A)$, and $j(A) = R_{ij}(x)$. From here we get $g(A) = i \circ x = (\alpha_1, \ldots, \alpha_n)$. This proof contains also the proof of (*iii*), because we have shown that $g(i.x) = i \circ x = i \circ id(x)$. □

Theorem 1.5 shows that the whole information of an n-arrow structure is contained in the standard n-arrow frame $\underline{W}(S)$ over S and can be expressed in terms of arrows and the relations R_{ij}. An example of such a correspondence are the conditions (*Nor*) and (*Nor'*). In the next section we shall examine this correspondence in more details. The next lemma, taken from Vakarelov to appear, gives some examples for the case $n = 2$.

Lemma 1.7 *Let* $S = (Ar, Po, \{1, 2\}, .)$ *be a two dimensional arrow structure,* (Po, ρ) *be the point frame over* S *and* $(Ar, \{R_{11}, R_{22}, R_{12}, R_{21}\})$ *be the arrow frame over* S. *Then:*

(i) ρ *is a reflexive relation iff*

$$(\forall x \in Ar)(\exists y \in Ar)(xR_{11}y \ \& \ xR_{12}y) \ \&$$
$$(\forall x \in Ar)(\exists y \in Ar)(xR_{21}y \ \& \ xR_{22}y),$$

(ii) ρ *is a symmetric relation iff* $(\forall x \in Ar)(\exists y \in Ar)(xR_{12}y \ \& \ xR_{21}y)$,

(iii) ρ *is a transitive relation iff*
$$(\forall x, y \in Ar)(xR_{21}y \to (\exists z \in Ar)(xR_{11}z \ \& \ zR_{22}y)).$$

Proof. In Vakarelov to appear. □

2 First-Order Languages Associated with Arrow Structures

Let $S = (Ar_S, Po_S, (n), (.))$ be an n-arrow structure, $V(S) = (Po_S, \rho_S)$ be the n-point frame over S and $W(S) = (Ar_S, \{R_{ij}^S \mid i, j \leq n\})$ be the n-arrow frame over S. We shall construct three first-order languages $\mathcal{L}(S)$, $\mathcal{L}(V)$ and $\mathcal{L}(W)$ corresponding to the systems S, $V(S)$ and $W(S)$ respectively. Since $V(S)$ and $W(S)$ are definable subsystems of S, the corresponding languages $\mathcal{L}(V)$ and $\mathcal{L}(W)$ are in some sense definable sublanguages in the language $\mathcal{L}(S)$ for S. The main result of this section will be the fact that $\mathcal{L}(S)$, $\mathcal{L}(V)$ and $\mathcal{L}(W)$ have equal expressive power. In particular this implies that the first-order theory of one n-place relation can be reduced to the first-order theory of n^2 special binary relations.

The Language $\mathcal{L}(S)$ for n-Dimensional Arrow Structures

The language $\mathcal{L}(S)$ is a three-sorted first-order language. It contains symbols for point variables, arrow variables, the set of integer constants $(n) = \{1, \ldots, n\}$ and integer variables, ranging over (n). The formal definition is the following.

The Alphabet of $\mathcal{L}(S)$.

- $VAR_{Po} = \{A^1, A^2, A^3, \ldots\}$ — a denumerable sequence of different point variables,
- $VAR_{Ar} = \{x^1, x^2, x^3, \ldots\}$ — a denumerable sequence of different arrow variables,
- $(n) = \{1, 2, \ldots, n\}$ integer constants,
- $VAR(n) = \{i^1, i^2, i^3, \ldots\}$ — a denumerable sequence of different integer variables, ranging in (n),
- . — application symbol,
- = — equality,
- $\neg, \wedge, \exists$, — logical symbols,
- (,) — parentheses.

Terms. A^p, $i^p.x^q$, $c.x^q$ for $c \leq n$.

Atomic Formulas.

(i) If A and B are terms then $A = B$ is a formula,
(ii) $x^p = x^q$ is an atomic formula.

Formulas. The set $FOR(\mathcal{L}(S))$ of all formulas of $\mathcal{L}(S)$ is constructed in the usual way from the set of atomic formulas by using the logical symbols and quantifiers over VAR_{Po} and VAR_{Ar}.

Abbreviations. The other logical connectives — $\vee, \Leftrightarrow, \Rightarrow$, the constants $\top, \bot$ and the quantifier $\forall$ are introduced in the standard way. The replacement of all occurrences of an integer variable i^p by a constant c, $c \leq n$ in a formula α will be denoted by $\alpha[i^p/c]$. The quantification over integer variables is defined as follows:

$$(\forall i^p)\alpha \stackrel{def}{=} \alpha[i^p/1] \wedge \alpha[i^p/2] \wedge \ldots \wedge \alpha[i^p/n]$$
$$(\exists i^p)\alpha \stackrel{def}{=} \neg(\forall i^p)\neg\alpha = \alpha[i^p/1] \vee \alpha[i^p/2] \vee \ldots \vee \alpha[i^p/n]$$

We also introduce the following abbreviations:

$$x^p R_{ij} x^q \stackrel{def}{=} i.x^p = j.x^q \text{ where } i,j \in VAR(n) \cup (n)$$
$$\rho(A_1, \ldots, A_n) \stackrel{def}{=} (\exists x^p)(1.x^p = A_1 \wedge 2.x^p = A_2 \wedge \ldots \wedge n.x^p = A_n),$$

where $A_1, A_2, \ldots, A_n$ are terms. As an example of a formula in $\mathcal{L}(S)$ we will give the axiom

(Ax) $\qquad (\forall A^p)(\exists i^q)(\exists x^r)(A^p = i^q.x^r)$

Semantics for $\mathcal{L}(S)$. The standard semantics for the language $\mathcal{L}(S)$ will be given in terms of n-dimensional arrow structures. This is a standard definition but we will formulate it in order to fix the notations we will use.

Let S be an n-arrow structure and let v be a function, called valuation, which assigns to different kinds of variables from the language corresponding objects from S, namely $v(A^p) \in Pos$, $v(x^p) \in Ar_S$ and $v(i^p) \leq n$. Let v be a valuation, x be any variable and a be an object from S of the same type as the type of the variable x; for instance if x is the point variable A^p then $a \in Pos$. Then we define a new valuation v_a^x such that $v(x) = a$ and for any variable y different from x we have $v_a^x(y) = v(y)$. The notation v_{ab}^{xy} means $(v_a^x)_b^y$. If S is an n-arrow structure and v is a valuation in S, then the pair $M = (S, v)$ will be called a model.

The interpretation of a term at a valuation v is the standard:

$$v(i^p.x^q) = v(i^p).v(x^q) \text{ and for } c \leq n, v(c.x^q) = c.v(x^q).$$

The interpretations of the formulas of $\mathcal{L}(S)$ in a model $M = (S, v)$ will be given inductively. We shall use the standard notation $(S, v) \models \alpha$ to be read "α is true in the model (S, v)" or "α is true in S at the valuation v". For the atomic formulas we have:

$(S, v) \models x^p = x^q$ iff $v(x^p) = v(x^q)$
$(S, v) \models A = B$ iff $v(A) = v(B)$ where A and B are terms
$(S, v) \models \neg\alpha$ iff $(S, v) \not\models \alpha$
$(S, v) \models \alpha \wedge \beta$ iff $(S, v) \models \alpha$ and $(S, v) \models \beta$

$(S,v) \models (\exists A^p)\alpha$ iff $(\exists B \in Pos)((S, v_B^{A^p}) \models \alpha)$
$(S,v) \models (\exists x^p)\alpha$ iff $(\exists a \in Ar_S)((S, v_a^{x^p}) \models \alpha)$.

A formula α is true in an n-arrow structure S, in notation $S \models \alpha$, if for every valuation v we have $(S, v) \models \alpha$. The truth of a closed formula in a model (S,v) does not depend on v, so if α is a closed formula, then $S \models \alpha$ if for some v we have $(S,v) \models \alpha$.

The Language $\mathcal{L}(V)$ of Point Frames

The alphabet of $\mathcal{L}(V)$ contains the set of point variables VAR_{Po}, the logical symbols $=$ (equality) and one n-place predicate symbol ρ. The atomic formulas are of the form

$$A^p = A^q, \; \rho(A^{q_1}, \ldots, A^{q_n}).$$

The set of all formulas $FOR(\mathcal{L}(V))$ of $\mathcal{L}(V)$ is constructed from the atomic formulas in a standard way. Obviously $\mathcal{L}(V)$ can be considered as a part of $\mathcal{L}(S)$, namely the part, that talks about points in terms of $=$ and the relation ρ (definable in $\mathcal{L}(S)$).

The standard semantics for $\mathcal{L}(V)$ is given in terms of n-point frames $V(S)$ over n-arrow structures S and can be formulated in the same way as was done for the language $\mathcal{L}(S)$. The relation ρ is interpreted in $V(S)$ by the relation ρ_S.

The Language $\mathcal{L}(W)$ of n-Dimensional Arrow Frames

The alphabet of $\mathcal{L}(W)$ contains the set VAR_{Ar} of arrow variables, $=$ (equality), and for each $i, j \leq n$ a two place predicate symbol R_{ij}. Atomic formulas are

$$x^p R_{ij} x^q, \; x^p = x^q$$

and the set $FOR(\mathcal{L}(W))$ is constructed from the atomic formulas in the standard way.

We shall define a translation of the closed formulas of $\mathcal{L}(S)$ into the closed formulas of $\mathcal{L}(W)$. But since the definition will go inductively for arbitrary formulas, for the translation of open formulas of $\mathcal{L}(S)$ we need a richer version $\mathcal{L}'(W)$ of $\mathcal{L}(W)$, such that the set of closed formulas of $\mathcal{L}'(W)$ will be the same as the set of closed formulas of $\mathcal{L}(W)$.

The definition of $\mathcal{L}'(W)$ is the following. It contains the sets VAR_{Ar}, (n), $VAR(n)$, $=$; and instead of the predicates R_{ij} from $\mathcal{L}(W)$, we now have one four-place predicate symbol R with two places for integer variables or integer constants and two places for arrow variables. The atomic formulas are the following:

$$x^p = x^q, \; x^p R_{ij} x^q, \text{ where } i, j \in VAR(n) \cup (n).$$

When $i, j \leq n$ we will consider R_{ij} as a two-place predicate for arrows and it will be identified with the corresponding predicate R_{ij} from the language $\mathcal{L}(W)$.

The set of all formulas $FOR(\mathcal{L}'(W))$ is defined in the usual way from the set of atomic formulas by means of logical connectives and quantifiers over arrow variables. Quantification over integer variables can be defined as in the language $\mathcal{L}(S)$. Now it is obvious that the sets of closed formulas of the languages $\mathcal{L}(W)$ and $\mathcal{L}'(W)$ coincide.

The standard interpretation of $\mathcal{L}'(W)$ and $\mathcal{L}(W)$ is given in terms of n-arrow frames over n-arrow structures and can be formulated in a way similar to that for the language $\mathcal{L}(S)$.

A Translation τ from $\mathcal{L}(S)$ into $\mathcal{L}'(W)$ and $\mathcal{L}(W)$

We define a translation τ from $\mathcal{L}(S)$ into $\mathcal{L}'(W)$ by induction on the complexity of the formulas in $\mathcal{L}(S)$. For closed formulas in $\mathcal{L}(S)$, τ will be a translation of $\mathcal{L}(S)$ into $\mathcal{L}(W)$:

$$\tau(A^p = A^q) \stackrel{def}{=} x^p R_{i^p i^q} x^q,$$

$$\tau(A^p = j.x^q) \stackrel{def}{=} x^p R_{i^p j} x^q,$$

$$\tau(j.x^q = A^p) \stackrel{def}{=} x^q R_{j i^p} x^p, \text{ where } j \in VAR(n) \cup (n)$$

$$\tau(i.x^p = j.x^q) \stackrel{def}{=} x^p R_{ij} x^q,$$
$$\text{where } i, j \in VAR(n) \cup (n),$$

$$\tau(x^p = x^q) \stackrel{def}{=} x^p = x^q,$$

$$\tau(\neg \alpha) \stackrel{def}{=} \neg \tau(\alpha),$$

$$\tau(\alpha \wedge \beta) \stackrel{def}{=} \tau(\alpha) \wedge \tau(\beta),$$

$$\tau((\exists x^p)\alpha) \stackrel{def}{=} (\exists x^p)\tau(\alpha),$$

$$\tau((\exists A^p)\alpha) \stackrel{def}{=} (\exists i^p)(\exists x^p)\tau(\alpha),$$

$$\tau((\exists i^p)\alpha) \stackrel{def}{=} (\exists i^p)\tau(\alpha).$$

Let S be an n-arrow structure and $W(S) = (Ar_S, \{R_{ij} \mid i,j \leq n\})$ be the n-arrow frame over S. Let v be a valuation of the variables of $\mathcal{L}(S)$ in S and w be a valuation of the variables of $\mathcal{L}'(W)$ in $W(S)$. We say that v and w are connected if the following conditions are satisfied:

(1) $v(A^p) = w(i^p).w(x^p)$,
(2) $v(x^p) = w(x^p)$,
(3) $v(i^p) = w(i^p)$.

If (S, v) is a model for $\mathcal{L}(S)$ in S and $(W(S), w)$ is a model of $\mathcal{L}'(W)$ in $W(S)$, we say that (S, v) and $(W(S), w)$ are connected models if v and w are connected valuations.

Lemma 2.1 *(i) If w is a valuation of $\mathcal{L}'(W)$ in $W(S)$ then there exists a valuation of $\mathcal{L}(S)$ in S such that v and w are connected.*
(ii) If v and w are connected valuations then for any $c \leq n$ and $a \in Ar_S$ the valuations $v_{c.a}^{A^p}$ and $w_{ca}^{i^p x^p}$ are connected.

Proof. (i). Define v by the equations (1)–(3). The proof of (ii) follows immediately from the definitions of connected valuations. □

Lemma 2.2 *Let S be an n-arrow structure and $W(S)$ be the n-arrow frame over S. Then for any formula $\alpha \in FOR\mathcal{L}(S)$ and for any two connected models (S,v) and $(W(S),w)$ the following equivalence is true:*

$$(S,v) \models \alpha \text{ iff } (W(S),w) \models \tau(\alpha).$$

Proof. The proof is by induction on the complexity of α.
$\alpha = (A^p = A^q)$. Then $(S,v) \models \alpha$ iff $(S,v) \models A^p = A^q$ iff $v(A^p) = v(A^q)$ iff $w(i^p).w(x^p) = w(i^q).w(x^q)$ iff $w(x^p)R_{w(i^p)w(i^q)}w(x^q)$ iff $(W(S),w) \models x^p R_{i^p i^q} x^q$ iff $(W(S),w) \models \tau(A^p = A^q)$ iff $(W(S),w) \models \tau(\alpha)$. The proof is similar for the cases where α is another type of atomic formula.

Now suppose as induction hypothesis that the assertion is true for the formulas β and γ. The proof for the cases $\alpha = \neg\beta$, $\alpha = \beta \wedge \gamma$, $\alpha = (\exists x^p)\beta$ and $\alpha = (\exists i^p)\beta$ is straightforward. The remaining case is $\alpha = (\exists A^p)\beta$. We will proceed for the two directions separately.

($\rightarrow$) Suppose $(S,v) \models (\exists A^p)\beta$. Then there exists $B \in Pos$ such that $(S, v_B^{A^p}) \models \beta$. Then by axiom (Ax), there exists $c \leq n$ and $a \in Ar_S$ such that $B = c.a$. By lemma 2.1.(ii) the valuations $v_B^{A^p}$ and $w_{ca}^{i^p x^p}$ are connected and then the induction hypothesis gives us $(W(S), w_{ca}^{i^p x^p}) \models \tau(\beta)$. This shows that $(W(S),w) \models (\exists i^p)(\exists x^p)\tau(\beta)$, so $(W(S),w) \models \tau((\exists A^p)\beta)$.

($\leftarrow$) Let $(W(S),w) \models \tau((\exists A^p)\beta)$. Then $(W(S),w) \models (\exists i^p)(\exists x^p)\tau(\beta)$, so for some $c \leq n$ and $a \in Ar_S$ $(W(S), w_{ca}^{i^p x^p}) \models \tau(\beta)$. Let $B = c.a$. By lemma 2.1.(ii) the valuations $v_B^{A^p}$ and $w_{ca}^{i^p x^p}$ are connected. Then by the induction hypothesis we have $(W(S), v_B^{A^p}) \models \beta$. This shows that $(W(S),v) \models (\exists A^p)\beta$, which completes the proof of the lemma. □

Theorem 2.3 *Let S be an arbitrary n-arrow structure and $W(S)$ be the n-arrow frame over S. Then for any closed formula α of $\mathcal{L}(S)$:*

$$S \models \alpha \text{ iff } W(S) \models \tau(\alpha)$$

Proof. Let w be an arbitrary valuation of $\mathcal{L}'(S)$ in $W(S)$. Then by lemma 2.1.(i) there exists a valuation v of $\mathcal{L}(S)$ in S such that the valuations v and w are connected. Applying lemma 2.2 we have that $(S,v) \models \alpha$ iff $(W(S),w) \models \tau(\alpha)$. But α and $\tau(\alpha)$ are closed formulas, so we have $S \models \alpha$ iff $W(S) \models \tau(\alpha)$. □

A Translation μ of $\mathcal{L}(W)$ into $\mathcal{L}(V)$

We define a translation μ by induction on the complexity of formulas in $\mathcal{L}(W)$. For that purpose we arrange the point variables in $\mathcal{L}(V)$ in the following way: $A_1^1, A_2^1, \ldots, A_n^1, \ldots, A_1^p, A_2^p, \ldots, A_n^p$, and so forth. μ is defined as

$$\begin{aligned}
\mu(x^p R_{ij} x^q) &= A_i^p = A_j^q, i, j \leq n, \\
\mu(x^p = x^q) &= A_1^p = A_1^q \wedge \ldots \wedge A_n^p = A_n^q \text{ when } \mathcal{L}(W) \text{ has } =, \\
\mu(\neg \alpha) &= \neg \mu(\alpha), \\
\mu(\alpha \wedge \beta) &= \mu(\alpha) \wedge \mu(\beta), \\
\mu((\exists x^p) \alpha) &= (\exists A_1^p) \ldots (\exists A_n^p)(\rho(A_1^p, \ldots, A_n^p) \wedge \mu(\alpha)).
\end{aligned}$$

Let S be an n-arrow structure, $V(S) = (Po_S, \rho_S)$ be the n-point frame over S and $W(S) = (Ar_S, \{R_{ij} \mid i, j \leq n\})$ be the n-arrow frame over S. Let v be a valuation of the point variables of $\mathcal{L}(V)$ in $V(S)$ and w be a valuation of arrow variables of $\mathcal{L}(W)$ in $W(S)$. We will say that v and w are connected — in a new sense — if for any $i \leq n$ we have $v(A_i^p) = i.w(x^p)$.

Lemma 2.4 (i) If w is a valuation of $\mathcal{L}(W)$ in $W(S)$ then there exists a valuation v of $\mathcal{L}(V)$ in $V(S)$ such that v and w are connected.

(ii) if v and w are connected and for $a \in Ar_S$ we have $B_1 = 1.a, \ldots, B_n = n.a$, then the valuations

$$v_{B_1, \ldots, B_n}^{A_1^p, \ldots, A_n^p} \text{ and } w_a^{x^p}$$

are connected.

Proof. The proof is similar to the proof of lemma 2.1. □

Lemma 2.5 Let S be an n-arrow structure, $W(S)$ be the n-arrow frame over S and $V(S)$ be the n-point frame over S. Then for any formula α and any two connected models $(W(S), w)$ and $(V(S), v)$ we have the following equivalence:

$$(W(S), w) \models \alpha \text{ iff } (V(S), v) \models \mu(\alpha).$$

Proof. We proceed by induction on the complexity of α.

1. $\alpha = x^p R_{ij} x^q$. Then $(W(S), w) \models x^p R_{ij} x^q$ iff $i.w(x^p) = j.w(x^q)$ iff $v(A_i^p) = v(A_j^q)$ iff $(V(S), v) \models A_i^p = A_j^q$ iff $(V(S), v) \models \mu(x^p R_{ij} x^q)$.

2. $\alpha = (x^p = x^q)$ (In this case we assume that $\mathcal{L}(W)$ has $=$ and that the n-arrow structure S is normal.) We have: $(W(S), w) \models (x^p = x^q)$ iff $w(x^p) = w(x^q)$ iff, by the normality condition, $1.w(x^p) = 1.w(x^q)$ & $\ldots$ & $n.w(x^p) = n.(x^q)$ iff $v(A_1^p) = v(A_1^q)$ & $\ldots$ & $v(A_n^p) = v(A_n^q)$ iff $(V(S), v) \models A_1^p = A_1^q \wedge \ldots \wedge A_n^p = A_n^q$ iff $(V(S), v) \models \mu(x^p = x^q)$.

3. Induction hypothesis: suppose that the assertion is true for β and γ.

4. The proof for the case that $\alpha = \neg \beta$ and $\alpha = \beta \wedge \gamma$ is straightforward.

5. $\alpha = (\exists x^p) \beta$.

($\rightarrow$) Suppose $(W(S), w) \models (\exists x^p)\beta$. Then for some $a \in Ar_S$ we have
$$(W(S), w_a^{x^p}) \models \beta.$$
Let $B_1 = 1.a, \ldots, B_n = n.a$. Then by lemma 2.4.(ii) the valuations
$$w_a^{x^p} \text{ and } v_{B_1 \ldots B_n}^{A_1^p \ldots A_n^p}$$
are connected and by the induction hypothesis we have
$$(V(S), v_{B_1 \ldots B_n}^{A_1^p \ldots A_n^p}) \models \mu(\beta).$$
We also have $\rho_S(B_1, \ldots, B_n)$, so
$$(V(S), v_{B_1 \ldots B_n}^{A_1^p \ldots A_n^p}) \models \rho(A_1^p, \ldots, A_n^p) \wedge \mu(\beta)$$
and hence
$$(S(V), v) \models (\exists A_1^P) \ldots (\exists A_n^p)(\rho(A_1^p, \ldots, A_n^p) \wedge \mu(\beta)).$$
Thus $(V(S), v) \models \mu((\exists x^p)\beta)$.

($\leftarrow$) Suppose now that $(V(S), v) \models \mu((\exists x^p)\beta)$. Then
$$(V(S), v) \models (\exists A_1^P) \ldots (\exists A_n^p)(\rho(A_1^p, \ldots, A_n^p) \wedge \mu(\beta)).$$
This means that for some $B_1, \ldots B_n \in Po_S$ we have
$$(V(S), v_{B_1 \ldots B_n}^{A_1^P \ldots A_n^P}) \models \rho(A_1^p, \ldots, A_n^p) \wedge \mu(\beta).$$
From this we obtain $\rho_S(B_1, \ldots, B_n)$. By the definition of ρ_S, there exists $a \in Ar_S$ such that $B_1 = 1.a, \ldots, B_n = n.a$. Then by lemma 2.4.(ii) the valuations
$$v_{B_1 \ldots B_n}^{A_1^P \ldots A_n^P} \text{ and } w_a^{x^p}$$
are connected and by the induction hypothesis we have $(W(S), w_a^{x^p}) \models \beta$, so $(W(S), v) \models (\exists x^p)\beta$. This ends the proof of the lemma. □

Theorem 2.6 *Let S be arbitrary n-arrow structure, $V(S)$ be the n-point frame over S and $W(S)$ be the n-arrow frame over S. Then:*

(i) *If the language $\mathcal{L}(W)$ is without $=$, then for any closed formula α of $\mathcal{L}(W)$ we have: $W(S) \models \alpha$ iff $V(S) \models \mu(\alpha)$.*

(ii) *if $\mathcal{L}(W)$ has $=$ then the above equivalence is true if S is a normal n-arrow structure.*

Proof. By lemma 2.5. □

Corollary 2.7 *If the language $\mathcal{L}(W)$ does not contain $=$, then the languages $\mathcal{L}(S)$, $\mathcal{L}(W)$ and $\mathcal{L}(V)$ have one and the same expressive power.*

3 Modal Languages Associated with Arrow Structures: Point Logics and Arrow Logics

We have associated with each n-arrow structure S two relational structures: $W(S) = (Ar_S, \{R_{ij}^S \mid i, j \leq n\})$ — the n-arrow frame over S, and $V(S) =$

(Po_S, ρ_S) — the n-point frame over S. Both structures can be used for the interpretation of the corresponding modal languages. The language associated with point frames will be called point language and the language associated with arrow frames will be called arrow language. The main result of this section will be a construction of a translation of a sufficiently rich point language into a certain arrow language which preserves semantic validity. In particular this implies that the modal theory of n-ary modality can be reduced to the theory of special unary modalities.

A Modal Language for Arrow Frames: Arrow Logics

The most natural way to associate a modal language to n- arrow frames of the type $W(S)$, is to extend the language of classical propositional logic with modal connectives $[ij]$ for each relation R_{ij}, where $[ij]$ is interpreted in a Kripke-style manner. We denote this language by $ML^n([ij])$. To be more precise, $ML^n([ij])$ contains:

- $VAR(ML^n([ij])) = \{p^1, p^2, p^3, \ldots\}$ — a denumerable list of distinct propositional variables,
- $\neg, \wedge, \vee$ — Boolean connectives,
- $[ij]$ $i, j \leq n$ — n^2 one argument modal (box) operations, and
- $(,)$ — parentheses.

The set $FOR(ML^n([ij]))$ of formulas is defined in the usual way: all p^i are formulas, if α and β are formulas then so are $\neg \alpha$, $(\alpha \wedge \beta)$, $(\alpha \vee \beta)$ and $[ij]\alpha$ for $i, j \leq n$. We will leave out the parentheses where possible. We abbreviate $\langle ij \rangle \alpha = \neg[ij]\neg \alpha$ and adopt the usual definitions for the other Boolean connectives.

The standard semantics for this language is a Kripke-style semantics over n-arrow frames. The general semantics is over arbitrary relational structures of the type $\underline{W} = (W, \{R_{ij} \mid i, j \leq n\})$, which will be called also frames. Let a frame $\underline{W}$ be given. A function v, which assigns to each propositional variable p^i a subset $v(p^i) \subseteq W$ is called a valuation of propositional variables. The satisfiability relation $x \Vdash_v \alpha$ — the formula α is true in $x \in W$ at the valuation v — is defined inductively in the usual Kripke definition as follows:

$$\begin{aligned}
x \Vdash_v p^i & \quad \text{iff} \quad x \in v(p^i) \\
x \Vdash_v \neg \alpha & \quad \text{iff} \quad x \nVdash_v \alpha \\
x \Vdash_v \alpha \wedge \beta & \quad \text{iff} \quad x \Vdash_v \alpha \text{ and } x \Vdash_v \beta \\
x \Vdash_v \alpha \vee \beta & \quad \text{iff} \quad x \Vdash_v \alpha \text{ or } x \Vdash_v \beta \\
x \Vdash_v [ij]\alpha & \quad \text{iff} \quad (\forall y \in W)(x R_{ij} y \rightarrow y \Vdash_v \alpha).
\end{aligned}$$

The pair $M = (\underline{W}, v)$ is called a model. We say that a formula α is true in a model $M = (\underline{W}, v)$, or that M is a model for α, in symbols $M \models \alpha$, if for any $x \in W$ we have $x \Vdash_v \alpha$. A formula α is true in a frame $\underline{W}$ or that $\underline{W}$ is a frame for α, in symbols $\underline{W} \models \alpha$, if α is true in all models over $\underline{W}$. A formula α is true in a class Σ of frames if α is true in any member of Σ, in symbols $\Sigma \models \alpha$. A class of formulas L is true in a model M if any member of L is true in M. L is true in a class of frames Σ if any member of L is true in Σ. L is called the logic of Σ and denoted by $L(\Sigma)$ if it consists of all formulas true in Σ.

Let $ARROW^n$ denote the class of all n-arrow frames. Then $L(ARROW^n)$ is called the Basic Arrow Logic of dimension n and is denoted by BAL^n. The language $ML^n([ij])$ is called the language of BAL^n. The language of BAL^n can be extended in different ways. In this section we shall study an extension with a new modal symbol $\Box$, called universal modality (see Goranko and Passy 1992). This extension is called $ML^n([ij], \Box)$. The semantics of this modality can be given in any set W as follows:

$$x \Vdash_v \Box \alpha \text{ iff } (\forall y \in W)(y \Vdash_v \alpha).$$

We abbreviate $\neg \Box \neg$ as $\Diamond$.

If Σ is a class of frames then by $L(\Sigma \Box)$ we will denote the class of all formulas from $ML^n([ij], \Box)$ which are true in Σ. Using this notation the logic $L(ARROW^n \Box)$ will be called Basic Arrow Logic of dimension n with universal modality and will be denoted also by $BAL^n \Box$.

A Modal Language for Point Frames: Point Logics

Relational structures of n-point frames are of the form $V(S) = (Pos, \rho_S)$ for certain n-arrow structure S. Here ρ_S is a n-place relation in Pos. Let us consider arbitrary relational systems of the form $\underline{V} = (V, \rho)$ with $V \neq \emptyset$, where ρ is an n-place relation in V. A natural way to generalize the Kripke semantics is to use a generalization of $\Box$ and $\Diamond$ with more than one argument places and to use the relation ρ for interpretation of such operations. Since for one argument modality the Kripke definition requires a two place relation, in our case $\Box$ and $\Diamond$ should have $n-1$ arguments — $\Box(\alpha_2, \ldots, \alpha_n)$ and $\Diamond(\alpha_2, \ldots, \alpha_n)$. The semantic conditions for these operations are the following:

$A \Vdash_v \Box(\alpha_2, \ldots, \alpha_n)$ iff $(\forall A_1, A_2, \ldots, A_n)(\rho(A_1, A_2, \ldots, A_n)$ & $A = A_1$
$\rightarrow A_2 \Vdash_v \alpha_2$ or $\ldots$ or $A_n \Vdash_v \alpha_n)$

$A \Vdash_v \Diamond(A_2, \ldots, A_n)$ iff $(\exists A_1, A_2, \ldots, A_n)(\rho(A_1, A_2, \ldots, A_n)$ &
$A = A_1$ & $A_2 \Vdash_v \alpha_2$ & $\ldots$ & $A_n \Vdash_v \alpha_n)$

Obviously we have that $\Diamond(\alpha_2, \ldots, \alpha_n)$ is equivalent to $\neg\Box(\neg\alpha_2, \ldots, \neg\alpha_n)$ and similarly for $\Diamond$. Let us note that the semantic definition for $\Diamond$ is taken from the Jónsson-Tarski representation theory of Boolean algebras with operators Jónsson and Tarski 1951.

The frames of the form (V, ρ) can be used for interpretation of other

$(n-1)$ argument operations. Let σ be any permutation of the sequence $\langle 1, 2, \ldots, n \rangle$ and let ρ^σ be a new n-place relation in V defined by the equivalence

$$\rho^\sigma(A_1, A_2, \ldots A_n) \text{ iff } \rho(A_{\sigma(1)}, A_{\sigma(2)}, \ldots, A_{\sigma(n)}).$$

Now let for each permutation σ of $\langle 1, 2, \ldots, n \rangle$, $[\sigma]$ be an $(n-1)$ argument modality. Then the interpretation of $[\sigma](\alpha_2, \ldots, \alpha_n)$ is the same as for $\Box(\alpha_2, \ldots, \alpha_n)$, where instead of ρ we use ρ^σ. It is obvious now that the operation $\Box$ corresponds to $[id]$, where id is the identity permutation. In (V, ρ) we can also interpret the universal modality $\Box$.

Hence, our point language, called $ML^n([\sigma], \Box)$, contains:

- $VAR(ML^n([\sigma], \Box)) = \{p^1, p^2, \ldots\}$ — infinite sequence of different propositional variables,
- $\neg, \wedge, \vee$ — Boolean connectives,
- $[\sigma]$ — $(n-1)$ argument box operators for each permutation σ of $\langle 1, 2, \ldots n \rangle$,
- $\Box$ — the universal modality,
- $(,)$ — parentheses.

The set of the formulas $FOR(ML^n([\sigma], \Box))$ is defined now in an obvious way. The standard semantics of this language is a Kripke-like semantics, as described above, in n-point frames of the form $V(S) = (Po_S, \rho_S)$ under some n-arrow structure S. The general semantics is in arbitrary relational systems of the form (V, ρ), where $\rho \subseteq V^n$. The notions of a model, validity and so on in this semantics can be defined in the same way as for the arrow case.

Note that in this language we have $n!$ modal operators of the form $[\sigma]$ since we have $n!$ different permutations in $\langle 1, 2, \ldots, n \rangle$. But not all of these modal operators are independent. For instance let σ be a permutation such that $\sigma(1) = 1$ and let σ^{-1} be the converse of σ. Then $[\sigma](\alpha_2, \ldots, \alpha_n)$ is semantically equivalent to the formula $[id](\alpha_{\sigma^{-1}(2)}, \ldots, \alpha_{\sigma^{-1}(n)})$. Of course, we can take some basis to define the others, but for the sake of symmetry, we take all operations $[\sigma]$. Note that in the case $n = 2$ we have only two permutations and hence two modalities, which in tense logic would correspond to "future" and "past".

A Translation of the Point Language $ML^n([\sigma], \Box)$ into the Arrow Language $ML^n([ij], \Box)$

We shall define a translation τ of the language $L_1 = ML^n([\sigma], \Box)$ into the language $L_2 = ML^n([ij], \Box)$, that preserves semantic validity. It will be composed of several translations.

Let the propositional variables of L_1 and L_2 be arranged as follows:

$$VAR(L_1) = \langle p^1, p^2, \ldots, p^k, \ldots \rangle$$

$$VAR(L_2) = \langle p_1^1, p_2^1, \ldots, p_n^1, \ldots, p_1^k, p_2^k, \ldots, p_n^k, \ldots \rangle$$

Now we define inductively n different translations of L_1 into L_2. For $1 \leq i \leq n$ set:

$$\begin{aligned} i(p^k) &= p_i^k, \\ i(\neg\alpha) &= \neg i(\alpha), \\ i(\alpha \wedge \beta) &= i(\alpha) \wedge i(\beta), \\ i(\langle\sigma\rangle(\alpha_2, \ldots, \alpha_n)) &= \langle i\sigma^{-1}(1)\rangle(\sigma^{-1}(2)(\alpha_2) \wedge \ldots \wedge \sigma^{-1}(n)(\alpha_n)) \\ i(\Diamond\alpha) &= \Diamond(1(\alpha) \vee \ldots \vee n(\alpha)) \\ \nu(\alpha) &= 1(\alpha) \wedge \ldots \wedge n(\alpha) \end{aligned}$$

Let $VAR(\alpha)$ be the set of the propositional variables occurring in α. Then for each $p^k \in VAR(\alpha)$ and $i, j \leq n$, define the formula $\langle ij\rangle j(p^k) \Rightarrow i(p^k)$ and let $\mu(\alpha)$ be the conjunction of all these formulas, i.e.

$$\mu(\alpha) = \wedge\{\langle ij\rangle j(p^k) \Rightarrow i(p^k) \mid i, j \leq n\} \text{ and } p^k \in VAR(\alpha)\}$$

Now the definition of the translation τ is

$$\tau(\alpha) = \Box\mu(\alpha) \Rightarrow \nu(\alpha).$$

Let S be a given n-arrow structure and let $V(S) = (Pos, \rho_S)$ and $W(S) = (Ar_S, \{R_{ij}^S \mid i, j \leq n\}$ be the corresponding n-point frame and n-arrow frame over S and let $(V(S), v)$ and $(W(S), w)$ be two models over $V(S)$ and $W(S)$, respectively. Let $\Phi \subseteq VAR(L_1)$ be a set of propositional variables of the language L_1 and let $FOR(\Phi)$ be the set of all formulas of L_1 with variables taken from Φ. We say that the valuations v and w are Φ-connected if for any $x \in Ar_S, i \leq n$ and $p^k \in \Phi$ we have

$$i.x \in v(p^k) \text{ iff } x \in w(p_i^k).$$

Lemma 3.1 *If the valuations v and w are Φ-connected then for any $x \in Ar_S, i \leq n$ and $\alpha \in FOR(\Phi)$ we have $i.x \Vdash_v \alpha$ iff $x \Vdash_w i(\alpha)$.*

Proof. We shall proceed by induction on α. If α is a propositional variable, then the assertion is true by the definition of Φ-connected valuations. Since the translations i commute with Boolean connectives these cases do not present difficulties.

Now suppose that the assertion is true for the formulas $\alpha_2, \ldots, \alpha_n$ from $FOR(\Phi)$ and let $\alpha = \langle\sigma\rangle(\alpha_2, \ldots, \alpha_n)$. Then

$(\rightarrow)$ $i.x \Vdash_v \langle\sigma\rangle(\alpha_2, \ldots, \alpha_n)$ iff $(\exists A_1, A_2, \ldots A_n \in Pos)(\rho_S^\sigma(A_1, A_2, \ldots, A_n)$ & $A_1 = i.x$ & $A_2 \Vdash_v \alpha_2$ & $\ldots$ & $A_n \Vdash_v \alpha_n)$.

From this we get $\rho_S(A_{\sigma(1)}, A_{\sigma(2)}, \ldots, A_{\sigma(n)})$. By the definition of ρ_S we have: $(\exists u \in Ar_S)(\forall j \leq n)(j.u = A_{\sigma(j)})$. Since σ has a converse σ^{-1} we get that for any $j \leq n$ $\sigma^{-1}(j).u = A_j$. So we have:

$$\sigma^{-1}(1).u = A_1 = i.x, \sigma^{-1}(2).u = A_2 \Vdash_v \alpha_2, \ldots, \sigma^{-1}(n).u = A_n \Vdash_v \alpha_n.$$

Then by the induction hypothesis we get: $xR_{i\sigma^{-1}(1)}^S u$, $u \Vdash_w \sigma^{-1}(2)(\alpha_2)$,

..., $u \Vdash_w \sigma^{-1}(n)(\alpha_n)$. Hence
$$x \Vdash_w \langle i.\sigma^{-1}(1)\rangle(\sigma^{-1}(2)(\alpha_2) \wedge \ldots \wedge \sigma^{-1}(n)(\alpha_n))$$
and
$$x \Vdash_w i(\langle \sigma \rangle(\alpha_2, \ldots, \alpha_n)).$$
($\leftarrow$) Suppose that $x \Vdash_w i(\langle \sigma \rangle(\alpha_2, \ldots, \alpha_n))$. Then there exists $u \in Ar_S$ such that
$$xR^S_{i\sigma^{-1}(1)}u, u \Vdash_w \sigma^{-1}(2)(\alpha_2), \ldots, u \Vdash_w \sigma^{-1}(n)(\alpha_n).$$
Using the induction hypothesis we get: $i.x = \sigma^{-1}(1).u$ and $\sigma^{-1}(2).u \Vdash_v \alpha_2$, ..., $\sigma^{-1}(n).u \Vdash_v \alpha_n$.

Let for $j = 1, \ldots, n$ denote $A_j = \sigma^{-1}(j).u$. From this we obtain $A_{\sigma(j)} = j.u$ for $j = 1, \ldots n$. This gives $\rho_S(A_{\sigma(1)}, A_{\sigma(2)}, \ldots, A_{\sigma(n)})$ and consequently $\rho_S^\sigma(A_1, A_2, \ldots, A_n)$. Since we have $A_1 = i.x$, $A_2 \Vdash_v \alpha_2$, ..., $A_n \Vdash_v \alpha_n$, we get that $i.x \Vdash_v \langle \sigma \rangle(\alpha_2, \ldots, \alpha_n)$. This completes the proof of the case.

The case $\alpha = \Diamond \beta$ is easy and can be dealt with in a similar way. □

Lemma 3.2 *Let $(V(S), v)$ and $(W(S), w)$ be two models and let $\Phi \subseteq VAR(L_1)$.*

(i) if v and w are Φ-connected then w satisfies the following condition for any $p^k \in \Phi, x, y \in Ar_S$ and $i, j \leq n$,

(∗) $\quad xR^S_{ij}y \ \& \ y \in w(j(p^k)) \to x \in w(i(p^k)).$

(ii) Let $\alpha \in FOR(L_1)$ and $\Phi = VAR(\alpha)$. Then the condition (∗) is fulfilled for w iff $(W(S), w) \models \mu(\alpha)$.

(iii) Suppose for any $p^k \in \Phi, x, y \in Ar_S$ and $i, j \leq n$, we have (∗) and define v for $p^k \in \Phi$ as follows:

$$v(p^k) = \{A \in Pos \mid \exists j \leq n \exists y \in Ar_S \ A = j.y \ and \ y \in w(j(p^k))\}.$$

Then the valuations v and w are Φ-connected.

(iv) Let for any $p^k \in \Phi$ and $i \leq n$ the valuation v be defined as follows:

$$w(i(p^k)) = \{x \in Ar_S / i.x \in v(p^k)\}.$$

Then the valuations v and w are Φ-connected.

Proof. Straightforward. □

Theorem 3.3 *Let S be a n-arrow structure and $V(S)$ be an n-point frame, and $W(S)$ be an n-arrow frame over S. Then for any formula α we have:*

$$V(S) \models \alpha \ \text{iff} \ W(S) \models \tau(\alpha).$$

Proof. ($\rightarrow$) Suppose that $V(S) \models \alpha$ and with a view to reaching a contradiction, assume that $W(S) \not\models \tau(\alpha)$, so $W(S) \not\models \Box\mu(\alpha) \Rightarrow \nu(\alpha)$. Then there exists a valuation w and $x \in Ar_S$ such that $x \Vdash_w \Box\mu(\alpha)$ and $x \not\Vdash_w \nu(\alpha)$.

From $x \Vdash_w \Box\mu(\alpha)$ we have that for any $y \in Ar_S$ we have $y \Vdash_v \mu(\alpha)$, so $(W(S), w) \models \mu(\alpha)$. Then by lemma 3.2.$(ii)$ we obtain that the valuation w satisfies the condition $(*)$ for $\Phi = VAR(\alpha)$. Then by lemma 3.2.(iii) there exists a valuation v of $VAR(L_1)$ into $V(S)$ such that v and w are Φ-connected.

From $x \nVdash_w \nu(\alpha)$, i.e. from $x \nVdash_w 1(\alpha) \wedge \ldots \wedge n(\alpha)$ we conclude that for some $i, 1 \leq i \leq n$, we have $x \nVdash_w i(\alpha)$. Since v and w are Φ-connected, we obtain by lemma 3.1 that $i.x \nVdash_v \alpha$. But $i.x \in Pos$, so $V(S) \not\models \alpha$, contrary to the assumption.

($\leftarrow$) Suppose that $W(S) \models \tau(\alpha)$ and with a view to reaching a contradiction, assume that $V(S) \not\models \alpha$. Then there exists $A \in Pos$ and a valuation v such that $A \nVdash_v \alpha$. Then by lemma 3.2.(iv) there exists a valuation w of $VAR(L_2)$ into Ar_S such that v and w are Φ-connected for $\Phi = VAR(\alpha)$. Then by lemma 3.2.(i) condition $(*)$ is fulfilled and by lemma 3.2.(ii) we have $(W(S), w) \models \mu(\alpha)$, so for any $y \in Ar_S$ we have $y \Vdash_w \mu(\alpha)$. Thus $x \Vdash_w \Box\mu(\alpha)$.

Let $A = i.x$, so $i.x \nVdash_v \alpha$. Then by lemma 3.1 $x \nVdash_w i(\alpha)$, so $x \nVdash_w \nu(\alpha)$. We obtain from here that $x \nVdash_w \Box\mu(\alpha) \Rightarrow \nu(\alpha)$, so $x \nVdash_w \tau(\alpha)$. Thus $W(S) \not\models \tau(\alpha)$, which contradicts the assumption. This completes the proof of the theorem. □

Theorem 3.4 *The translation τ preserves first-order definability in a sense that a formula $\alpha \in L_1$ is first-order definable in the standard semantics of L_1 if and only if its translation $\tau(\alpha)$ is first-order definable in the standard semantics of L_2.*

Proof. ($\rightarrow$) Let α be a formula, first-order definable in L_1 by a first-order sentence F. Of course this sentence is in the first-order language $\mathcal{L}(V)$ of point frames. We have to show that the translation $\tau(\alpha)$ is first-order definable by some sentence in the first-order language $\mathcal{L}(W)$ of arrow frames. By theorem 2.3 we can conclude that there exists a translation $*$ — in the theorem this translation is denoted by τ — from the language $\mathcal{L}(V)$ to the language $\mathcal{L}(W)$ such that for any n-arrow structure S and sentence G in the language $\mathcal{L}(V)$ we have: $V(S) \models G$ iff $W(S) \models G*$. We shall show that $\tau(\alpha)$ is first-order definable by the sentence $F*$. Suppose that this is not true. Then there exists a n-arrow structure S such that the n-arrow frame $W(S)$ satisfies $F*$ but $W(S) \not\models \tau(\alpha)$. By the above remark we have that $V(S)$ satisfies F. From $W(S) \not\models \tau(\alpha)$ by theorem 3.3 we get that $V(S) \not\models \alpha$. But α is modally definable by sentence F. From here we obtain that F is not true in $V(S)$ — a contradiction.

($\leftarrow$) Suppose now that $\tau(\alpha)$ is first-order definable by a sentence F in the language $\mathcal{L}(W)$. By theorem 2.6 there exists a translation $+$ — in the theorem this translation is denoted by ν — of the language $\mathcal{L}(W)$ into the language $\mathcal{L}(V)$ such that for any sentence G, and n-arrow structure S

we have $V(S) \models G^+$ iff $W(S) \models G$. We shall show that α is first-order definable by F^+. Suppose that this is not true. Then, as in the above case, by applying theorem 3.3 we obtain a contradiction. This completes the proof of the theorem. □

Problem 3.5 *Is there a translation from the language $ML^n([\sigma])$ into the language $ML^n([ij])$ — □ is dropped — with properties similar to those of τ?*

Problem 3.6 *Is there a translation from the language $ML^n([ij])$ to the language $ML^n([\sigma], □)$ or some extension of $ML^n([\sigma], □)$ with properties similar to that of τ?*

II. Axiomatizations of Some n-Dimensional Arrow Logics

4   Axiomatization of BAL^n and $BAL^n□$

We have defined the logic BAL^n — the Basic Arrow Logic of dimension n — as the set $L(ARROW^n)$ of all formulas in the language $ML^n([ij])$ true in the class $ARROW^n$ of all n-arrow frames. This is a semantic definition of BAL^n. Let us introduce the following classes of arrow frames:

$ARROW^nFIN$ the class of finite arrow frames of dimension n,

$ARROW^nNOR$ the class of normal arrow frames of dimension n,

$ARROW^nFINNOR$ the class of finite normal arrow frames of dimension n.

In this section we shall propose an axiomatization of BAL^n and prove that it is sound and complete in each of the above classes of arrow frames. We propose the following set of axioms and rules for BAL^n:

Axiom Schemes for BAL^n.

(*Bool*)   All or enough Boolean tautologies,
($K[ij]$) $[ij](A \Rightarrow B) \Rightarrow ([ij]A \Rightarrow [ij]B)$, $i,j \leq n$,
$Ax(\rho ii)$ $[ii]A \Rightarrow A$, $i \leq n$,
$Ax(\sigma ij)$ $A \vee [ij]\neg[ji]A$, $i,j \leq n$,
$Ax(\tau ijk)$ $[ik]A \Rightarrow [ij][jk]A$, $i,j,k \leq n$.

Rules of Inference.

Modus Ponens   (MP) $A, A \Rightarrow B/B$,
Necessitation   ($N[ij]$) $A/[ij]A$, $i,j \leq n$.

We will denote the set of theorems of this formal system also by BAL^n.

Note that the axioms $Ax(\rho ii)$, $Ax(\sigma ij)$ and $Ax(\tau ijk)$ are just modal translations in the sense of modal definability theory van Benthem 1983 of the conditions (ρii), (σij) and (τijk), which immediately shows that BAL^n is sound in its intended semantics.

Theorem 4.1 (Completeness theorem for BAL^n) *The following are equivalent for any formula A of BAL^n:*

(i) *A is a theorem of BAL^n,*

(ii) *A is true in the class $ARROW^n$.*

Proof. (i) $\to$ (ii) This follows from the observation that BAL^n is sound with respect to its standard semantics — $ARROW^n$.

(ii) $\to$ (i) This implication can be proved by a standard application of the canonical construction known from Modal Logic. For details, the reader can consult for instance Segerberg 1971 or Hughes and Creswell 1984. $\square$

Corollary 4.2 $BAL^n = L(ARROW^n)$.

Now, using the method of filtration, known from mono-modal logic we shall show that BAL^n is complete in the class $ARROW^n FIN$ of all finite arrow frames of dimension n. For that purpose we shall give the relevant definition of filtration in the form of Segerberg 1971, adapted for the language of BAL^n.

Let Γ be a finite set of formulas, closed under subformulas and let $\underline{W} = (W, \{R_{ij} \mid i,j \leq n\})$ be any frame and $M = (\underline{W}, v)$ be any model over $\underline{W}$. By means of Γ and M we define an equivalence relation $\sim$ in W in the following way: for $x, y \in W$, we define

$$x \sim y \text{ iff } (\forall A \in \Gamma)(x \Vdash_v A \leftrightarrow y \Vdash_v A).$$

Let $|x| = \{y \in W / x \sim y\}$, $|W| = \{|x|/x \in W\}$ and for $A \in VAR$ $v'(A) = \{|x| \in |W|/x \in v(A)\}$.

We say that the model $M' = (|W|, \{R'_{ij} \mid i,j \leq n\}, v')$ is a filtration of M through Γ if the relations R'_{ij} satisfy the following two conditions for any $x, y \in W$:

$(FilR_{ij}1)$ if $xR_{ij}y$ then $|x|R'_{ij}|y|$,

$(FilR_{ij}2)$ if $|x|R'_{ij}|y|$ then $(\forall [ij]A \in \Gamma)(x \Vdash_v [ij]A \to y \Vdash_v A)$.

In the next section we shall use filtration for another kind of frames. The general definition is that, for each relation R, we suppose two conditions $(FilR1)$ and $(FilR2)$ like the above.

The importance of filtration is seen in the following:

Lemma 4.3 (Filtration lemma) (i) *For any $A \in \Gamma$ and $x \in W$ the following equivalence is true: $x \Vdash_v A$ iff $|x| \Vdash_{v'} A$,*

(ii) $Card|W| \leq 2^m$, *where $m = Card\Gamma$.*

Proof. As in the mono-modal case Segerberg 1971, this is by induction on the construction of A. Note that the clauses $(FilR_{ij}1)$ and $(FilR_{ij}2)$ from the definition are used in the case when A is in the form $[ij]B$. $\square$

Lemma 4.4 (Filtration for BAL^n) Let $\underline{W} = (W, \{R_{ij} \mid i,j \leq n\})$ be an n-arrow frame, $M = (\underline{W}, v)$ be a model over $\underline{W}$ and A be a formula of BAL^n. Then there exist a finite set Γ of formulas, containing A and closed under subformulas, and a filtration $M' = (|W|, \{R'_{ij} \mid ij \leq n\}, v')$ through Γ such that $\underline{W}' = (|W|, \{R'_{ij} \mid ij \leq n\})$ is a finite arrow frame of dimension n such that
$$Card|W| \leq 2^{n^2 \cdot m},$$
where m is the number of subformulas of A.

Proof. Let Γ be the smallest set of formulas closed under subformulas and containing A, satisfying the following closure condition:

(γ) If for some $i, j \leq n$ $[ij]B \in \Gamma$ then for any $i, j \leq n$ $[ij]B \in \Gamma$.

Obviously Γ is a finite set of formulas, containing no more than $n^2 \cdot m$ elements, where m is the number of the subformulas of A. Define $|W|$ and v' as in the definition of filtration. By lemma 4.3.(ii) we have
$$Card\Gamma \leq 2^{n^2 \cdot m}.$$

For the relations R'_{ij} we set:

$|x|R'_{ij}|y|$ iff $(\forall [ij]B \in \Gamma)(\forall k \leq n)(x \Vdash_v [ik]B \leftrightarrow y \Vdash_v [jk]B)$.

It can be easily proved that the definition of R'_{ij} is correct in the sense that

if $|x| = |x'|$ and $|y| = |y'|$ then $|x|R'_{ij}|y|$ iff $|x'|R'_{ij}|y'|$.

First we shall show that the frame $\underline{W}' = (|W|, \{R'_{ij}/ij \leq n\})$ is an n-arrow frame. The conditions (ρii) and (σij) follow directly from the definition of R'_{ij}. For the condition (τijk) suppose $|x|R'_{ij}|y|$ and $|y|R'_{jk}|z|$. To prove $|x|R'_{ik}|z|$, suppose $[ik]A \in \Gamma$, $l \leq n$ and for the direction ($\to$) suppose $x \Vdash_v [il]A$ and proceed to show that $z \Vdash_v [kl]A$. From $[ik]A \in \Gamma$ we get $[ij]A, [jl]A \in \Gamma$. Then $|x|R'_{ij}|y|$, $[ij]A \in \Gamma$ and $x \Vdash_v [il]A$ imply $y \Vdash_v [jl]A$. Together with $[jl]A \in \Gamma$ and $|y|R'_{jk}|z|$, this implies $z \Vdash_v [kl]A$.

The converse direction ($\leftarrow$) can be proved in a similar way.

It remains to show that the conditions of filtration $(FilR_{ij}1)$ and $(FilR_{ij}2)$ are satisfied.

For the condition $(FilR_{ij}1)$, suppose $xR_{ij}y$, $[ij]A \in \Gamma$, $k \leq n$; for the direction ($\to$) suppose $x \Vdash_v [ik]A$, $yR_{jk}z$ and proceed to show that $z \Vdash_v A$. From $xR_{ij}y$ and $yR_{jk}z$ we get $xR_{ik}z$, and since $x \Vdash_v [ik]A$ we get $z \Vdash_v A$. For the direction ($\leftarrow$) suppose $y \Vdash_v [jk]A$, $xR_{ik}z$ and proceed to show that $z \Vdash_v A$. From $xR_{ij}y$, we get $yR_{ji}x$ and by $xR_{ik}z$ we get $yR_{jk}z$. From this and $y \Vdash_v [jk]A$, we get $z \Vdash_v A$. This ends the proof of $(FilR_{ij}1)$.

For the condition $(FilR_{ij}2)$ suppose $|x|R'_{ij}|y|$, $[ij]A \in \Gamma$ and $x \Vdash_v [ij]A$. From here we obtain $y \Vdash_v [jj]A$ and since $yR_{jj}y$ we get $y \Vdash_v A$. This completes the proof of the lemma. □

Theorem 4.5 (Finite completeness theorem for BAL^n) *The following are equivalent for any formula A of BAL^n:*

(i) A is a theorem of BAL^n,

(ii) A is true in all finite n-arrow structure with cardinality $\leq 2^{n^2 \cdot m}$ where m is the number of the subformulas of A.

Proof. By theorem 4.1 and lemma 4.4. □

Corollary 4.6 (i) $BAL^n = L(ARROW^n FIN)$

(ii) BAL^n possesses finite model property and is decidable.

Now we shall show that $BAL^n = L(ARROW^n NOR)$. For that purpose we shall use a construction, called *copying*, which can transform each model over an arrow frame into an equivalent model over a normal arrow frame. First we shall give the definition of copying adapted for frames like arrow frames.

Let $\underline{W} = (W, \{R_{ij} \mid ij \leq n\})$ and $\underline{W}' = (W', \{R'_{ij} \mid ij \leq n\})$ be any two frames — not necessary arrow — and let $M = (\underline{W}, v)$, $M' = (\underline{W}', v')$ be models over $\underline{W}$ and $\underline{W}'$ respectively. Let I be a non- empty set of mappings from W into W'. We say that I is a copying from $\underline{W}$ to $\underline{W}'$ if the following conditions are satisfied for any $i, j \leq n$, $x, y \in W$ and $f, g \in I$:

(I1) $(\forall y' \in W')(\exists y \in W)(\exists g \in I) g(y) = y'$

(I2) If $f(x) = g(y)$, then $x = y$,

($R_{ij}1$) If $xR_{ij}y$, then $(\forall f \in I)(\exists g \in I) f(x) R'_{ij} g(y)$,

($R_{ij}2$) If $f(x) R'_{ij} g(y)$, then $xR_{ij}y$.

We say that I is a copying from M to M', if in addition the following condition is satisfied for any $p \in VAR$, $x \in W$ and $f \in I$:

(V) $\qquad x \in v(p)$ iff $f(x) \in v'(p)$.

For $x \in W$ and $f \in I$, $f(x)$ is called f-th copy of x and $f(W) = \{f(x) \mid x \in W\}$ is called f-th copy of W. By (I1) we obtain that $W' = \bigcup \{f(W) \mid f \in I\}$, so W' is a sum of copies of W. If I is one element set $\{f\}$ then f is an isomorphism from $\underline{W}$ onto $\underline{W}'$.

In the next section we shall use copying for other relational structures. The obvious adaptation of the definition is that for each binary relation R in the structure we have two conditions $(R1)$ and $(R2)$ like $(R_{ij}1)$ and $(R_{ij}2)$.

The importance of the copying construction is in the following lemma.

Lemma 4.7 (Copying Lemma) (i) *Let I be a copying from the model M to the model M'. Then for any formula $A, x \in W$ and $f \in I$ the following equivalence holds:*

$$x \Vdash_v A \text{ iff } f(x) \Vdash_{v'} A,$$

(ii) If I is a copying from the frame $\underline{W}$ to the frame $\underline{W}'$ and v is a valuation, then there exists a valuation v' such that I is a copying from the model $M = (\underline{W}, v)$ to the model $M' = (\underline{W}', v')$.

Proof. (i) The proof is by induction on the complexity of the formula A. For $A \in VAR$ the assertion holds by the condition (V) of copying. If A is a Boolean combination of formulas, the proof is straightforward. Let $A = [ij]B$ and by an induction hypothesis, suppose that the assertion for B holds.

($\rightarrow$) Suppose $x \Vdash_v [ij]B$ and $f \in I$. To show that $f(x) \Vdash_{v'} [ij]B$ suppose $f(x)R'_{ij}y'$ and proceed to show that $y' \Vdash_v B$. By $(I1)$ $(\exists y \in W)(\exists g \in I)g(y) = y'$, so $f(x)R'_{ij}g(y)$ and by $(R_{ij}2)$ we get $xR_{ij}y$. From $xR_{ij}y$ and $x \Vdash_v [ij]B$ we get $y \Vdash_v B$. Then by the induction hypothesis we get $g(y) \Vdash_{v'} B$, so $y' \Vdash_{v'} B$.

($\leftarrow$) Suppose $f(x) \Vdash_{v'} [ij]B$. To show that $x \Vdash_v [ij]B$ suppose $xR_{ij}y$ and proceed to show that $y \Vdash_v B$. From $xR_{ij}y$ we obtain by $(R_{ij}1)$ that there exists $g \in W$ such that $f(x)R'_{ij}g(y)$. Then, since $f(x) \Vdash_{v'} [ij]B$, we get $g(y) \Vdash_{v'} B$ and by the induction hypothesis $y \Vdash_v B$.

(ii) Define for $p \in VAR$:

$$v'(p) = \{x' \in W'/(\exists x \in W)(\exists f \in I)f(x) = x' \text{ and } x \in v(p)\}$$

We shall show that the condition (V) of copying is fulfilled. Let $x \in W$ and $f \in I$ and suppose $x \in v(p)$. Then by the definition of v', we have $f(x) \in v'(p)$. For the converse implication suppose $f(x) \in v'(p)$. Then there exist $y \in W$ and $g \in I$ such that $f(x) = g(y)$ and $y \in v(p)$. By $(I2)$ we get $x = y$, so $x \in v(p)$. □

Lemma 4.8 *Let $\underline{W} = (W, \{R_{ij} \mid ij \leq n\})$ be an n-arrow frame. Then*

(i) *There exists a normal arrow frame $\underline{W}' = (W', \{R'_{ij} \mid ij \leq n\})$ and a copying I from $\underline{W}$ to $\underline{W}'$.*

(ii) *If $\underline{W}$ is a finite n-arrow frame then the frame $\underline{W}'$ from (i) is also a finite frame.*

Proof. Let $\underline{B}(W) = (B(W), 0, 1, +, .)$ be the Boolean ring over the set W, namely $B(W)$ is the set of all subsets of W, $0 = \emptyset$, $1 = W$, $A + B = (A \backslash B) \cup (B \backslash A)$ and $A.B = AB = A \cap B$. Note that in Boolean rings $a - b = a + b$. Let $I = B(W)^n = B(W) \times \ldots \times B(W)$ n times. For $f \in I$ we denote by f_i the i-th coordinate of f, so $f = (f_1, \ldots, f_n)$.

We set $W' = W \times B(W)^n$. The elements of $I = B(W)^n$ can be treated as functions from W into W' as follows: for $f \in I$ and $x \in W$ we set $f(x) = (f, x)$. Obviously the conditions $(I1)$ and $(I2)$ from the definition of copying are fulfilled, and each element of W' is in the form of $f(x)$ for some $f \in I$ and $x \in W$. For the relations R'_{ij} we have the following definition: $f(x)R'_{ij}g(y)$ iff

$$xR_{ij}y \ \& \ (f_1 + \ldots + f_n + \{x\} = g_1 + \ldots + g_n + \{y\}) \ \& \ f_i = g_j.$$

To verify the condition $(R_{ij}1)$, suppose $xR_{ij}y$ and $f \in I$. Define g as follows. Set $g_j = f_i$ and choose $g_1, \ldots, g_{j-1}, g_{j+1}, \ldots, g_n$ in such a way as to satisfy the equation $f_1 + \ldots + f_n + \{x\} = g_1 + \ldots + g_n + \{y\}$. This is possible because $B(W)$ is a ring and $n \geq 2$.

Condition $(R_{ij}2)$ follows directly from the definition of R'_{ij}. So I is a copying.

The proof that W' with the relations R'_{ij} is an arrow frame is straightforward. For the condition of normality, suppose $f(x)R'_{ii}g(y)$ for $i = 1, \ldots, n$ and proceed to show that $f(x) = g(y)$. By the definition of R'_{ii} we have $f_1 + \ldots + f_n + \{x\} = g_1 + \ldots + g_n + \{y\}$ and $f_i = g_i$ for $i = 1, \ldots, n$, so $f = g$. Since $B(W)$ is a ring we obtain $\{x\} = \{y\}$, so $x = y$ and hence $f(x) = g(y)$. Thus W' is a normal n-arrow frame.

Suppose now that $\underline{W}$ is a finite n-arrow frame. Then the Boolean ring over W is finite too and hence $\underline{W}'$ is a finite n-arrow frame. □

Theorem 4.9 (Completeness theorem for BAL^n with normal frames) *The following are equivalent for any formula A of BAL^n:*

(i) A is a theorem of BAL^n,
(ii) A is true in the class $ARROW^n NOR$,
(iii) A is true in the class $ARROW^n FINNOR$.

Proof. The implication $(i) \to (ii)$ follows from the soundness part of theorem 4.1. For the converse implication $(ii) \to (i)$, suppose that A is true in $ARROW^n NOR$ and in order to reach a contradiction, assume that A is not a theorem of BAL^n. Then, by theorem 4.1, A is falsified in some n-arrow frame

$$\underline{W} = (W, \{R_{ij} \mid ij \leq n\}),$$

so there exists a valuation v and $x \in W$ such that $x \not\Vdash_v A$. By theorem 4.8 there exists a normal n-arrow structure

$$\underline{W}' = (W', \{R'_{ij} \mid ij \leq n\})$$

and a copying I from $\underline{W}$ to $\underline{W}'$. Then by lemma 4.7.(i) there exists a valuation v' such that I is a copying from the model $(\underline{W}, v)$ to the model $(\underline{W}', v')$. By lemma 4.7.(ii) we obtain that for any $f \in I$ $f(x) \not\Vdash_v A$, so A is not true in the normal n-arrow structure $\underline{W}'$ — a contradiction.

The equivalence $(i) \leftrightarrow (iii)$ can be proved in a similar way, using theorem 4.5 and theorem 4.8.(ii). □

Corollary 4.10 $BAL^n = L(ARROW^n NOR) = L(ARROW^n FINNOR)$.

Corollary 4.11 *The normality condition is not modally definable.*

Proof. Suppose the contrary. Then there exists a formula A such that A is true in an n-arrow frame $\underline{W}$ iff $\underline{W}$ is normal. Let $\underline{W}_0$ be a non-normal n-arrow frame. Then A is not true in $\underline{W}_0$. Then by lemma 4.8 there exist a normal n-arrow frame $\underline{W}'$ and a copying I from $\underline{W}_0$ to $\underline{W}'$. By the copying lemma A is not true in $\underline{W}'$, which, by the property of A, implies that W' is not normal — a contradiction. □

The following theorem summarizes the completeness theorems for BAL^n.

Theorem 4.12 (Completeness theorem for BAL^n) *The following conditions are equivalent for any formula A of BAL^n:*

(i) A is a theorem of BAL^n,
(ii) A is true in the class $ARROW^n$,
(iii) A is true in the class $ARROW^n FIN$,
(iv) A is true in the class $ARROW^n NOR$,
(v) A is true in the class $ARROW^n FINNOR$.

Now we turn to the axiomatization of $BAL^n\square$. The language of $BAL^n\square$ — $ML^n([ij],\square)$ — is an extension of the language $ML^n([ij])$ of BAL^n with the universal modality $\square$. We have introduced $L(ARROW^n\square)$ as the the semantic definition of the logic $BAL^n\square$. We propose the following axiomatization of $BAL^n\square$, which is an extension of the axiomatizations of BAL^n.

Axiom Schemes for $BAL^n\square$.

$(Bool)$ All or enough Boolean tautologies,
$(K[ij])$ $[ij](A \Rightarrow B) \Rightarrow ([ij]A \Rightarrow [ij]B), i,j \leq n$,
$Ax(\rho ii)$ $[ii]A \Rightarrow A, i \leq n$,
$Ax(\sigma ij)$ $A \vee [ij]\neg[ji]A, i,j \leq n$,
$Ax(\tau ijk)$ $[ik]A \Rightarrow [ij][jk]A, i,j,k \leq n$,
$(K\square)$ $\square(A \Rightarrow B) \Rightarrow (\square A \Rightarrow \square B)$,
$Ax(S5\square)$ $\square A \Rightarrow A, A \vee \square\neg\square A, \square A \Rightarrow \square\square A$,
$Ax(incl)$ $\square A \Rightarrow [ij]A$.

Rules of Inference.

Modus ponens $(MP)\ A, A \Rightarrow B/B$,
Necessitation $(N\square)\ A/\square A$.

Note that the rule $(N[ij])\ A/[ij]A$ can be obtained by $(N\square)$ and $Ax(incl)$. The axiom $Ax(incl)$ expresses the fact that the relations R_{ij} are contained in the universal relation.

Theorem 4.13 (Completeness theorem for $BAL^n\square$) *The following conditions are equivalent for any formula A of $BAL^n\square$:*

(i) A is a theorem of $BAL^n\square$,
(ii) A is true in the class $ARROW^n$,
(iii) A is true in the class $ARROW^n FIN$,

(iv) A is true in the class $ARROW^n NOR$,

(v) A is true in the class $ARROW^n FINNOR$.

Proof. This proof is similar to the proof of the completeness theorems for BAL^n and therefore it is left to the reader. □

We end this section by noticing that the reader can find in Vakarelov to appear completeness theorems for several extensions of BAL^2, called there BAL.

5 An Extension of BAL^n with Cylindric Operators: the Logic CAL^n

In this section we shall study an extension of the logic BAL^n with new modal formulas $[i]A$ and $\langle i \rangle A$, $i \leq n$, interpreted as $\forall x_i$ and $\exists x_i$ respectively. The modalities $\langle i \rangle$ are called cylindric operators and the obtained polymodal logic, denoted by CAL^n, is called cylindric arrow logic of dimension n. The name "cylindric" comes from the theory of cylindric algebras in Algebraic Logic, in which the cylindric operations are algebraic analogs of the existential quantifiers.

The formal definition of CAL^n is the following. The language, denoted by $ML^n([ij],[i])$, is an extension of the language $ML^n([ij])$ of BAL^n with n box modalities $[i]$ $i = 1, \ldots, n$. Diamonds are defined as usual: $\langle i \rangle = \neg[i]\neg$.

The standard semantics of $ML^n([ij],[i])$ is defined as follows. A relational system of the form $\underline{W} = (W, \{R_{ij} \mid ij \leq n\}, \{R_i \mid i \leq n\})$ is called a standard frame for CAL^n if the reduct $(W, \{R_{ij} \mid ij \leq n\})$ is a normal n-arrow frame and the relations R_i, $i \leq n$, are defined by the equations

$$xR_iy \text{ iff } (\forall j \neq i)xR_{jj}y, (*)$$

or in another form

$$R_i = R_{11} \cap \ldots \cap R_{i-1i-1} \cap R_{i+1i+1} \cap \ldots \cap R_{nn}, i \leq n.$$

The relations R_{ij} are used for the interpretation of modalities $[ij]$ as before and the relations R_i are used for the interpretation of the new modalities $[i]$ and $\langle i \rangle$:

$$x \Vdash_v [i]A \text{ iff } (\forall y \in W)(xR_iy \to y \Vdash_v A),$$
$$x \Vdash_v \langle i \rangle A \text{ iff } (\exists y \in W)(xR_iy \ \& \ y \Vdash_v A).$$

By theorem 1.1, each normal arrow structure can be represented as a normal arrow structure over some relational system (U, ρ) where ρ is a nonempty n-place relation in U. Then by theorem 1.3. each normal n-arrow frame $\underline{W} = (W, \{R_{ij} \mid ij \leq n\})$ can be identified with the frames of the following form: there exists $U \neq \emptyset$ and $W \subseteq U^n$, and for $(x_1, \ldots, x_n)$, $(y_1, \ldots, y_n) \in W$ and $i, j \leq n$ the relation R_{ij} is defined by the equivalence

$$(x_1, \ldots, x_n)R_{ij}(y_1, \ldots, y_n) \text{ iff } x_i = y_j.$$

Then for the relations R_i we obtain:
$$(x_1,\ldots,x_n)R_i(y_1,\ldots,y_n) \text{ iff } (\forall j \neq i)x_j = y_j.$$
Now the semantics of $ML^n([ij],[i])$ can be reformulated as follows. For $(x_1,\ldots,x_n) \in W$,

$(x_1,\ldots,x_n) \Vdash_v A$ iff $(x_1,\ldots,x_n) \in v(A)$, for $A \in VAR$

$(x_1,\ldots,x_n) \Vdash_v \neg A$ iff $(x_1,\ldots,x_n) \not\Vdash_v A$,

$(x_1,\ldots,x_n) \Vdash_v A \wedge B$ iff $(x_1,\ldots,x_n) \Vdash_v A$ and $(x_1,\ldots,x_n) \Vdash_v B$,

$(x_1,\ldots,x_n) \Vdash_v A \vee B$ iff $(x_1,\ldots,x_n) \Vdash_v A$ or $(x_1,\ldots,x_n) \Vdash_v B$,

$(x_1,\ldots,x_n) \Vdash_v \langle ij \rangle A$ iff $(\exists y_1 \ldots \exists y_{j-1} \exists y_{j+1} \ldots \exists y_n \in U)(y_1,\ldots,y_{j-1},x_i,y_{j+1},\ldots,y_n) \in W$ & $(y_1,\ldots,y_{j-1},x_i,y_{j+1},\ldots,y_n) \Vdash_v A$,

$(x_1,\ldots,x_n) \Vdash_v \langle i \rangle A$ iff $(\exists y_i \in U)(x_1,\ldots,x_{i-1},y_i,x_{i+1},\ldots,x_n) \in W$ & $(x_1,\ldots,x_{i-1},y_i,x_{i+1},\ldots,x_n) \Vdash_v A$,

and dually for $[ij]A$ and $[i]A$. This form of the semantics shows that the interpretation of $\langle i \rangle$ is indeed like the interpretation of the existential quantifier $\exists y_i$. The meaning of $\langle jj \rangle$ is more complex and can be represented by a prefix of $n-1$ existential quantifiers $\exists y_1 \ldots \exists y_{j-1} \exists y_{j+1} \ldots \exists y_n$. The meaning of $\langle ij \rangle$ is the same quantification, followed by a substitution, changing the value of j-th variable with the value of x_i.

A modal study of operations of the form $\langle i \rangle$ is given by Venema in 1991. The standard semantics of the modality $\langle i \rangle$ in Venema's paper is the same as above, with the difference that Venema considers the more classical case $W = U^n$ — the full n-cube of some set U. Venema's logic contains also propositional constants δ_{ij}, formalizing equality with the following semantics:
$$(x_1,\ldots,x_n) \Vdash_v \delta_{ij} \text{ iff } x_i = x_j.$$
The n-dimensional modal logic obtained in this way appears to be undecidable and non-finitely axiomatizable by normal modal rules — see Venema 1991, 1995. In contrast with this result, we show in this section that CAL^n, with the semantics given above, is finitely axiomatizable by the orthodox rules of modus ponens and necessitation, and that it is decidable. Note that the price for this result is that we take our semantics to be based not on full n-cubes but on subsets of full cubes. An implication of this is that CAL^n determines a natural decidable version of first-order logic. The first announcement of this result is in Vakarelov 1993 and can be considered as a contribution to the so-called "finitization problem in Algebraic Logic", described in Németi 1991. A similar result was obtained by Németi 1995 for a version of first order logic.

The axiomatization of CAL^n presents some difficulties because the equalities (∗) defining R_i are not modally definable, in the sense of modal definability theory van Benthem 1983. For that purpose we introduce a non-standard semantics for CAL^n, for which it is easy to prove the corresponding completeness theorem. Then, using the copying construction introduced in section 4, we prove the completeness of CAL^n with respect to its standard semantics.

The non-standard semantics for CAL^n is defined as follows. A relational structure $\underline{W} = (W, \{R_{ij} \mid ij \leq n\}, \{R_i \mid i \leq n\})$ is called a non-standard frame for CAL^n if $(W, \{R_{ij} \mid ij \leq n\})$ is an n-arrow frame and the relations R_i satisfy the following conditions for any $x, y, z \in W$ and $i, j \leq n$:

(ρi) $xR_i x$,
(σi) $xR_i y \to yR_i x$,
(τi) $xR_i y$ & $yR_i z \to xR_i z$,
($incl, i \neq j$) $i \neq j$ & $xR_i y \to xR_{jj} y$.

Note that each standard frame for CAL^n is also a non-standard frame for CAL^n and that a non-standard frame $\underline{W}$ is a standard one iff it is normal n-arrow frame and satisfies the following condition for any $x, y \in W$ and $i \leq n$:

(Sti) $(\forall j \neq i)(xR_{jj} y) \to xR_i y$, $i \leq n$

Note also that all conditions of standard frames are modally definable, except normality and the condition (Sti); this will be proved later.

We propose the following axiomatization of CAL^n.

Axiom Schemes for CAL^n ($i, j \leq n$).

($Bool$)	All or enough Boolean tautologies,
($K[ij]$)	$[ij](A \Rightarrow B) \Rightarrow ([ij]A \Rightarrow [ij]B)$, $i, j \leq n$
$Ax(\rho ii)$	$[ii]A \Rightarrow A$, $i \leq n$
$Ax(\sigma ij)$	$A \vee [ij]\neg[ji]A$, $i, j \leq n$,
$Ax(\tau ijk)$	$[ik]A \Rightarrow [ij][jk]A$, $i, j, k \leq n$,
($K[i]$)	$[i](A \Rightarrow B) \Rightarrow ([i]A \Rightarrow [i]B)$, $i \leq n$,
$Ax(\rho i)$	$[i]A \Rightarrow A$, $i \leq n$,
$Ax(\sigma i)$	$\langle i \rangle [i]A \Rightarrow A$, $i \leq n$,
$Ax(\tau i)$	$[i]A \Rightarrow [i][i]A$, $i \leq n$,
$Ax(incl, i \neq j)$	$[jj]A \Rightarrow [i]A$, $i \neq j, i, j \leq n$.

Rules of Inference.

Modus ponens (MP) $A, A \Rightarrow B/B$,
Necessitation ($N[ij]$) $A/[ij]A$ $i, j \leq n$, ($N[i]$)$A/[i]A$, $i \leq n$

Note that axioms ($Ax(\rho i)$) — ($Ax(incl, i \neq j)$) are modal translations of the corresponding conditions of non-standard frames.

Theorem 5.1 (Completeness for CAL^n with non-standard semantics) *The following conditions are equivalent for any formula A of CAL^n:*

(i) A is a theorem of CAL^n,
(ii) A is true in all non-standard frames for CAL^n.

Proof. (i) $\to$ (ii) — in a standard way by proving that the axioms are true in each non-standard frame and that the rules of inference preserve validity.

(ii) $\to$ (i). Use the canonical construction and the fact that all axioms are canonical in a sense that they determine the corresponding semantic condition in the canonical model. $\square$

To prove the completeness of CAL^n with respect to its standard semantics we shall use the copying construction with an obvious modification of the notion of copying for frames for CAL^n.

Lemma 5.2 (Copying for non-standard frames for CAL^n)
Let $\underline{W} = (W, \{R_{ij} \mid ij \leq n\}, \{R_i / i \leq n\})$ be a non-standard frame for CAL^n. Then:
(i) There exists a standard frame $\underline{W}' = (W', \{R_{ij} \mid ij \leq n\}, \{R_i \mid i \leq n\})$ and a copying I from $\underline{W}$ to $\underline{W}'$.
(ii) If $\underline{W}$ is a finite frame then $\underline{W}'$ from (i) is a finite frame too.

Proof. Let $\underline{B}(W) = (B(W), 0, 1, +, .)$ be the Boolean ring over the set W, namely let $B(W)$ be the set of all subsets of W, $0 = \emptyset$, $1 = W$, $A + B = (A \backslash B) \cup (B \backslash A)$ and $A.B = AB = A \cap B$. Let I be the set of all matrices over $B(W)$ in the following form

$$f = \begin{pmatrix} f_1^0 & f_2^0 & \cdots & f_n^0 \\ f_1^1 & f_2^1 & \cdots & f_n^1 \\ \cdots\cdots\cdots\cdots\cdots\cdots \\ f_1^n & f_2^n & \cdots & f_n^n \end{pmatrix}$$

For $f \in I$ and $x \in W$ define

$$f_i = \begin{pmatrix} f_i^0 \\ f_i^1 \\ \vdots \\ f_i^n \end{pmatrix}, \quad f^i[x] = \sum_{k=1, k \neq i}^{n} f_k^i + R_i(x), \quad f[x] = \begin{pmatrix} f^0[x] \\ f^1[x] \\ \vdots \\ f^n[x] \end{pmatrix},$$

where $R_0(x) = \{x\}$ and $R_i(x) = \{y \in W \mid xR_i y\}$. Since R_i is an equivalence relation, then $R_i(x)$ is just the equivalence class of R_i determined by x. Now define $W' = W \times I$, and for $x \in W$ and $f \in I$ let $f(x) = (f, x)$. Obviously the conditions $(I1)$ and $(I2)$ from the definition of copying — given in section 4 — are fulfilled. Then each element of W' has the form $f(x)$ for some $f \in I$ and $x \in W$. Now we define the relations R'_{ij} and R'_i, $i, j \leq n$, as follows. Let $f, g \in I$ and $x, y \in W$, then:

(R'_{ij}) $f(x)R'_{ij}g(y)$ iff $xR_{ij}y$ and $f[x] = g[y]$ and $f_i = g_j$,
(R'_i) $f(x)R'_i g(y)$ iff $xR_i y$ and $f[x] = g[y]$ and $(\forall k \neq i) f_i = g_i$.
The remaining part of the lemma follows from the next lemmas.

Lemma 5.3 *The relations R'_{ij} and R'_i satisfy the conditions of copying.*

Proof. For the conditions of the type $(R2)$ the assertion is obvious. Now we will proceed to prove the condition $(R_i 2)$:

$$(R_i 2) \quad xR_i y \rightarrow (\forall f \in I)(\exists g \in I) f(x) R'_i g(y).$$

For that purpose it is enough to prove the implication:

$$xR_i y \rightarrow (\forall f \in I)(\exists g \in I) f[x] = g[y] \text{ and } (\forall k \neq i)(f_k = g_k)$$

Suppose $xR_i y$. Then we have $R_i(x) = R_i(y)$. Let $f \in I$ be given. Set for $k \neq i$ $g_k = f_k$. In this way we have defined all columns of g except g_i. The equality $f[x] = g[y]$ can be considered as a system of n linear equations for the unknown elements of g_i. The j-th equation is $f^j[x] = g^j[y]$, or in more details

$$f_1^j + \cdots + f_{j-1}^j + f_{j+1}^j + \cdots + f_i^j + \cdots + f_n^j + R_j(x) =$$
$$g_1^j + \cdots + g_{j-1}^j + g_{j+1}^j + \cdots + g_i^j + \cdots + g_n^j + R_j(y).$$

In this equation the only unknown element is g_i^j. Since $B(W)$ is a ring we can solve this equation with respect to g_i^j. The same can be done with the other equations, except the i-th one:

$$f_1^i + \cdots + f_{i-1}^i + f_{i+1}^i + \cdots + f_n^i + R_i(x) = g_1^i + \cdots + g_{i-1}^i + g_{i+1}^i + \cdots + g_n^i + R_i(y).$$

This equation does not contain unknown elements, but nevertheless it is true, because for all $k \neq i$ we have $f_k^i = g_k^i$ and $R_i(x) = R_i(y)$.

The proof of the condition $(R_{ij} 1)$ is simple and can be done in the same way. □

Lemma 5.4 *The structure $\underline{W}'$ is a standard frame for CAL^n.*

Proof. The proof that $\underline{W}'$ is a non-standard frame for CAL^n is straightforward and follows directly from the definitions of R'_{ij} and R'_i. We shall show the condition of normality and the condition (Sti) for $i \leq n$.

For the condition of normality let $f, g \in I$, $x, y \in W$, and suppose that for any $i \leq n$ we have $f(x) R'_{ii} g(y)$. Then we have $f[x] = g[y]$ and $\forall i \leq n$ $f_i = g_i$. From this we obtain $f = g$, and from the first equation of $f[x] = g[y]$ we obtain $\{x\} = \{y\}$. Consequently $x = y$ and $f(x) = f(y)$ — the normality is proven.

For the condition (Sti) suppose that, for any $j \neq i$, $f(x) R'_{jj} g(y)$ and proceed to show that $f(x) R'_i g(y)$. From this assumption we get that for any $j \neq i$: $f[x] = g[y]$ and $f_j = g_j$. From here we obtain that for any $j \neq i$ $f_j^i = g_j^i$. Then from the i-th equality of $f[x] = g[x]$ we obtain

$R_i(x) = R_i(y)$, so xR_iy. Then by the definition of R'_i we obtain $f(x)R'_ig(y)$, which ends the proof of the lemma. □

Finally, then, the proof of theorem 5.2 follows from lemma 5.3 and lemma 5.4. □

Theorem 5.5 (Completeness of CAL^n with standard semantics) *The following are equivalent for any formula A of CAL^n:*

(i) *A is a theorem of CAL^n,*
(ii) *A is true in all standard frames for CAL^n.*

Proof. The implication $(i) \to (ii)$ is trivial. To prove the converse implication $(ii) \to (i)$ suppose (ii) and with a view to reaching a contradiction, assume that A is not a theorem of CAL^n. Then by theorem 5.1 A is falsified in some non-standard frame $\underline{W}$ for CAL^n, that is, there exists valuation v in W and $x \in W$ such that $x \not\Vdash_v A$. Then by theorem 5.2 there exist a standard frame $\underline{W}'$ and a copying I from $\underline{W}$ to $\underline{W}'$. By the copying lemma (a lemma like lemma 4.7, which is true for any copying) there exists a valuation v' in $\underline{W}'$ such that for any $f \in I$ $f(x) \not\Vdash_{v'} A$. This contradicts the assumption that A is true in all standard frames for CAL^n. □

Corollary 5.6 *The condition (Sti) $(\forall j \neq i)(xR_{jj}y) \to xR_iy$ is not modally definable.*

Proof. See the proof of corollary 4.11. □

Now we shall show the decidability of CAL^n, showing that it possesses the finite model property. For that purpose we shall use the method of filtration, applying to the non-standard models of CAL^n.

Lemma 5.7 (Filtration for CAL^n) *Let $\underline{W} = (W, \{R_{ij}|i,j \leq n\}, \{R_i|i \leq n\})$ be a non-standard frame for CAL^n, $M = (\underline{W}, v)$ be a model over $\underline{W}$ and A be a formula of CAL^n. Then there exist a finite set Γ of formulas, containing A and closed under subformulas, and a filtration $M' = ((|W|, \{R'_{ij} \mid ij \leq n\}, \{R'_i|i \leq n\}), v')$ through Γ, such that $\underline{W}' = (|W|, \{R'_{ij} \mid ij \leq n\}, \{R'_i|i \leq n\}))$ is a finite non-standard frame for CAL^n, such that $Card|W| \leq 2^{(n^2+n) \cdot m}$, where m is the number of the subformulas of A.*

Proof. Let the formula A be given and let Γ be the smallest set of formulas containing A, closed under subformulas and satisfying the following condition:

(γ) if $[ij]A \in \Gamma$, then for any $ij \leq n$ $[ij]A \in \Gamma$.

Obviously Γ is a finite set of formulas, containing no more than $(n^2+n).m$ elements, where m is the number of subformulas of A. Define $|W|$ and the valuation v' as in the definition of filtration given in section 4 and R'_{ij} as in lemma 4.4. The relations R'_i are defined as follows:

$|x|R'_i|y|$ iff $(\forall [i]B \in \Gamma)(x \Vdash_v [i]B \leftrightarrow y \Vdash_v [i]B)$ & $(\forall k \neq i)|x|R'_{kk}|y|$.

The proof that the conditions of the filtration are fulfilled and that

$$(|W|, \{R'_{ij} \mid ij \leq n\}, \{R'_i \mid i \leq n\})$$

is a non-standard frame for CAL^n is similar to the proof of lemma 4.4, so we leave it to the reader. □

Theorem 5.8 (Finite completeness theorem for CAL^n) *The following are equivalent for any formula A of CAL^n:*

(i) *A is a theorem of CAL^n,*

(ii) *A is true in all finite non-standard frames $\underline{W}$ of CAL^n with $CardW \leq 2^{(n^2+n) \cdot m}$, where m is the number of the subformulas of A,*

(iii) *A is true in all finite standard frames for CAL^n.*

Proof. The implication $(i) \to (ii)$ is standard, and $(ii) \to (i)$ follows from lemma 5.7. $(i) \to (iii)$ is also standard, so we have $(ii) \to (iii)$. The implication $(iii) \to (ii)$ follows from lemma 5.2.(ii). □

Corollary 5.9 (i) *CAL^n possesses the finite model property,*

(ii) *CAL^n is decidable.*

Theorem 5.10 *The following conditions are equivalent for any formula A of CAL^n:*

(i) *A is a theorem of CAL^n,*

(ii) *A is true in all non-standard frames for CAL^n,*

(iii) *A is true in all standard frames for CAL^n,*

(iv) *A is true in all finite non-standard frames $\underline{W}$ for CAL^n such that $Card\underline{W} \leq 2^{(n^2+n) \cdot m}$, where m is the number of subformulas of A,*

(v) *A is true in all finite standard frames for CAL^n.*

6 Open Problems and Further Perspectives

Some open problems have been put forward in the main text. Now we shall discuss some new problems and further perspectives of the "arrow" approach.

Extensions of BAL^2 with Additional Connectives

In Vakarelov to appear, we studied some extensions of BAL^2 with operators having their standard semantics in terms of two-dimensional arrow frames. There are many possibilities of such extensions, depending of what kind of relations between arrows we want to describe in a modal setting. The main scheme is the following: for each $n+1$-ary relation $R(x_0, x_1, \ldots, x_n)$

we introduce an n-place modal box operation $[R](A_1,\ldots,A_n)$ with the following semantics:
$x_0 \Vdash_v [R](A_1,\ldots,A_n)$ iff
$(\forall x_1,\ldots,x_n \in W)(R(x_0,x_1,\ldots,x_n) \to x_1 \Vdash_v A_1$ or $\ldots$ or $x_n \Vdash_v A_n)$
The dual operator $\langle R \rangle(A_1,\ldots,A_n)$ is defined by $\neg[R](\neg A_1,\ldots,\neg A_n)$. In the following we list some natural relations between arrows, which are candidates for a modal study (see also Vakarelov to appear):
$Path_n(x_1,\ldots,x_n)$ iff $x_1 R_{21} x_2$ & $x_2 R_{21} x_3$ & $\ldots$ & $x_{n-1} R_{21} x_n$, $n \geq 2$

•───►•───►• $\cdots$ •───►•
   x_1 x_2 x_n

$Path_\infty(x_1,x_2,x_3,\ldots)$ iff $(\forall n) Path_n(x_1,\ldots,x_n)$.
$Loop_n(x_1,\ldots,x_n)$ iff $Path_{n+1}(x_1,\ldots,x_n,x_1)$, $n \geq 1$;
$Loop(x) = Loop_1(x)$ iff $x R_{21} x$, viz.

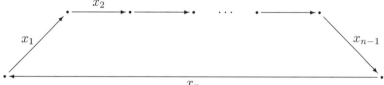

Converse: xSy iff $Loop_2(x,y)$, viz.

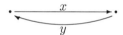

$Trapezium_n(x_1,\ldots,x_n,y)$ iff $Path_n(x_1,\ldots,x_n)$ & $x_1 R_{11} y$ & $x_n R_{22} y$, viz.

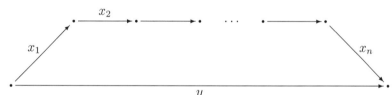

$Triangle(x,y,z)$ iff $Trapezium_2(x,y,z)$, viz.

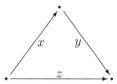

Connectedness:
$Con(x,y)$ iff $\exists n \geq 2 \exists x_1 \ldots x_n : x = x_1$ & $x_n = y$ & $Path_n(x_1,\ldots,x_n)$, viz.

Double side connectedness: $Dcon(x,y)$ iff $Con(x,y)$ & $Con(y,x)$, viz

The relations $Path_n$, $Path_\infty$, $Loop_n$ can also be used to define the semantics for suitable propositional constants:
Path$_n$, Path$_\infty$, Loop$_n$, Loop as follows.
$x_1 \Vdash_v$ **Path$_n$** iff $(\exists x_2, \ldots x_n \in W) Path_n(x_1, x_2, \ldots, x_n)$,
$x_1 \Vdash_v$ **Path$_\infty$** iff $(\exists x_2, x_3, \ldots) Path_\infty(x_1, x_2, x_3, \ldots)$,
$x_1 \Vdash_v$ **Loop$_n$** iff $(\exists x_2, \ldots, x_n) Loop_n(x_1, x_2, \ldots, x_n)$
$x \Vdash_v$ **Loop** iff $x \Vdash_v$ **Loop$_1$**.

These considerations motivate the following general problem: develop a modal theory — axiomatization, definability, (un)decidability — of some extensions of BAL^2 with modal operations corresponding to the above defined relations in arrow structures.

For example, the extension of BAL^2 with the modal operations

$$A \bullet B = \langle Triangle \rangle (A, B),$$

$A^{-1} = [Converse]A$ and the propositional constant $Id = $ **Loop** is a natural generalization of the modal logic of binary relations (van Benthem 1989, 1991, 1994, Venema 1989, 1991) and has a close connection with various versions of representable relativized relational algebras (Kramer 1991, Maddux 1982, Németi 1987, 1991). Extensions of BAL^2 with $\langle Triangle \rangle$ and $\langle Converse \rangle$, taken alone, were axiomatized by Andrey Arsov in his Master thesis Arsov 1993. Axiomatization of the extension of BAL^2 with all tree operations — $\langle Triangle \rangle$, $\langle Converse \rangle$ and **Loop** is given by Arsov in Arsov and Marx 1994. Versions of this logic with semantics, where arrows are considered as ordered pairs — arrows in normal 2-arrow structures in our terminology — have been studied in Marx 1995.

Arrow Semantics of Lambek Calculus and its Generalizations

Let S be a 2-arrow structure, and let A/B and $A\backslash B$ be "duals" of $A \bullet B$ with the following semantics:
$x \Vdash_v B/A$ iff $(\forall yz \in Ar_S)(Triangle(x, y, z)$ & $y \Vdash_v A \to z \Vdash_v B)$,
$y \Vdash_v A\backslash B$ iff $(\forall xz \in Ar_S)(Triangle(x, y, z)$ & $x \Vdash_v A \to z \Vdash vB)$.

The modal connectives $A \bullet B$, $A\backslash B$, and A/B can be considered as the operations in the Lambek Calculus. Andréka and Mikulás 1994 proves a completeness theorem for the Lambek Calculus with respect to a relational semantics of the above type over transitive normal arrow frames — this is an equivalent reformulation of Mikulás' result in our "arrow" terminology.

Roorda 1991 studies a modal version of Lambek Calculus extended with classical Boolean operations. Similar is the logic studied by Mikulás in chapter 6. So it is natural to study extensions of BAL^2 with the above diadic modal operations, which will give another intuition for the operations in the Lambek Calculus.

Connections between Monomodal Logics and Arrow Logics of Dimension 2

Every usual Kripke frame $\underline{W} = (W, \rho)$, with $\rho \subseteq W^2$ and $\rho \neq \emptyset$ determines a normal arrow structure of dimension 2 and the corresponding arrow frame in an obvious way. Each class Σ of usual Kripke frames with a non-empty relation determines a corresponding class $Ar(\Sigma)$ of arrow frames. A general question is the comparative study of the modal logic $L(\Sigma)$ and the corresponding arrow logic $L(Ar(\Sigma))$ for a given class Σ. For instance, BAL^2 corresponds to the modal logic K. In Vakarelov to appear we axiomatize the arrow analogs of KT (ρ reflexive), $K4$ (ρ transitive), KB (ρ symmetric), $S4$ (ρ reflexive and transitive), $S5$ (ρ an equivalence relation). One kind of connection between $L(\Sigma)$ and $L(Ar(\Sigma))$ is given in part I of this paper.

Extensions of BAL^n and CAL^n

1. Extend and axiomatize the logic BAL^n with propositional constants δ_{ij}, $i, j \leq n$ with the following semantics:

$$x \Vdash_v \delta_{ij} \text{ iff } xR_{ij}x$$

2. Extend and axiomatize the logic CAL^n with the constants δ_{ij}. The obtained logic will correspond to the cylindric modal logic of Venema 1995. The standard meaning of δ_{ij} is the equality of i-th and j-th coordinates of a normal n-arrow $x = (x_1, \ldots, x_n)$.
3. Extend and axiomatize the logic BAL^n with modalities corresponding to arbitrary intersections of the relations R_{ij}. For instance the $\langle R \rangle$ modality corresponding to the intersection $R = R_{i_1 i_1} \cap \ldots \cap R_{i_k i_k}$ gets the interpretation of a quantification prefix which can be obtained from the prefix $\exists x_1 \ldots \exists x_n$ by omitting all quantifiers of the form $\exists x_{i_j}$, $j = 1, \ldots, k$. If $R = R_{ij} \cap R_{ji} \cap \bigcap_{k=1, k \neq i,j}^{n} R_{kk}$, then the modality $\langle R \rangle$ will correspond to the permutation of the i-th and j-th coordinates of a normal n-arrow x.

Dynamic Arrow Logics

There are various possibilities to give arrow analogs of PDL, Propositional Dynamic Logic. Here we shall describe one that is based on 2-arrow structures, mentioned in Vakarelov 1991a. Other versions are given by van Benthem in van Benthem 1994 and Marx in chapter 5.

In the standard interpretation of PDL, formulas are interpreted by subsets of a given set W, and programs by subsets of $W \times W$ (relations in

W). But the set $W \times W$ consists of just arrows of the full directed graph (i.e., a normal arrow structure) with the set of points W. Generalizing, we can reformulate the semantics of PDL in arbitrary 2-arrow structures $S = (Ar, Po, \{1, 2\}, (.))$, interpreting the formulas as subsets of Po and programs as subsets of Ar. Each subset $\alpha \subseteq Ar$ determines a binary relation, denoted also by α, in Po in the following obvious way: for $A, B \in Po$

$$A\alpha B \text{ iff } (\exists x \in \alpha)(1.x = A \text{ and } 2.x = B).$$

Analogs of the Boolean operations on relations are just the Boolean operations of the subsets of Ar. The operation of composition of two subsets of α, β of Ar is defined as follows:

$$\alpha \circ \beta = \{z \in Ar \mid (\exists x \in \alpha, y \in \beta) Triangle(x, y, z)\}$$

Then the relation corresponding to $\alpha \circ \beta$ is the following: for $A, B \in Po$:

$$A\alpha \circ \beta B \quad \text{iff} \quad (\exists z \in Ar)(\exists x \in \alpha)(\exists y \in \beta) Triangle(x, y, z) \ \& $$
$$1.z = A \ \& \ 2.z = B.$$

This definition obviously simulates the definition of composition of two relations; when S is a full directed graph over some set W, it becomes the standard definition of composition. In general the composition $\alpha \circ \beta$ is not associative and cannot be used to define composition of more than two members. So, it is natural to introduce composition of n subsets of Ar for each $n \geq 2$. The following definition generalizes the associative case. For $n \geq 2$ and $\alpha_1, \ldots \alpha_n \subseteq Po$ (the definition of Trapezium is given above):

$Com_n(\alpha_1, \ldots, \alpha_n) =$
$\{y \in Ar \mid (\exists x_1 \in \alpha_1 \ldots \exists x_n \in \alpha_n) Trapezium(x_1, \ldots, x_n, y)\}$

Using this we define:

$$\alpha^0 = Loop, \ \alpha^1 = \alpha \text{ and for } n \geq 2 \ \alpha^n = Com_n(\alpha, \ldots, \alpha), \ \alpha^* = \bigcup_{i=0}^{n} \alpha^i.$$

The operation α^* corresponds to the star operation in PDL. Our definition is different from the one proposed by van Benthem in van Benthem 1994.

The test operation ? can be defined by generalizing the standard definition in full squares. For each subset $a \subseteq Po$ we define a set of arrows $a?$ as follows

$$a? = \{x \in Ar \mid 1.x = 2.x \ \& \ 1.x \in a\}.$$

For the interpretation of modal formulas in PDL, we need the following operation on subsets of Po. For each $a \subseteq Po$ and $\alpha \subseteq Ar$ we define

$$[\alpha]a = \{A \in Po \mid (\forall B \in Po)(A\alpha B \to B \in a)\}.$$

These considerations propose the following definition of the "arrow" PDL, $ArPDL$ for short. The language of $ArPDL$ consists of

Φ_0 — infinite set of propositional variables (atomic formulas),
$\neg, \land, \lor$ — Boolean connectives,
Π_0 — non-empty set of program variables (atomic programs),
$-, \cap, \cup,$ — Boolean operations on programs,
Com_n for $n = 2, 3, \ldots, *, ?,$
— n-composition, iteration and test program operations,
$(,)[,]$ — round and square brackets.

The set Φ of formulas and the set Π of programs are defined by simultaneous induction as follows:

(i) $\Phi_0 \subseteq \Phi$,
(ii) If $a, b \in \Phi$ then $\neg a, (a \land b)$ and $(a \lor b) \in \Phi$,
if $\alpha \in \Pi$ and $a \in \Phi$ then $[\alpha]a \in \Phi$,
(i') $\Pi_0 \subseteq \Pi$,
(ii') If $\alpha, \beta, \alpha_1, \ldots, \alpha_n \in \Pi$ then $-\alpha, (\alpha \cap \beta), (\alpha \cup \beta)$,
$Com_n(\alpha_1, \ldots, \alpha_n),$ and α^* are in Π,
if $a \in \Phi$ then $a? \in \Pi$.

Note that in the original language of PDL, instead of Com_n for each $n \geq 2$, we have only binary composition and no intersection "$\cap$" and complement "$-$" of programs. But in some versions of PDL intersection and complement are also considered.

The semantics of $ArPDL$ can be given in two-dimensional arrow systems as follows. Let S be a 2-arrow structure. By a dynamic model over S we mean any triple $M = (S, V, R)$ where V is a valuation of propositional variables and R is a valuation of program variables. For each $A \in \Phi_0$, $V(A)$ is a subset of Po_S, and for each $\alpha \in \Pi_0$, $R(\alpha)$ is a subset of Ar_S. We extend V and R for the elements of Φ and Π respectively, by simultaneous induction on the construction of formulas and programs.

(i) For $a \in \Phi_0$ and $\alpha \in \Pi_0$ $V(a)$ and $R(\alpha)$ are given.
(ii) Let for $a, b \in \Phi$ $V(a)$ and $V(b)$ be defined, and for $\alpha, \beta, \alpha_1, \ldots, \alpha_n \in \Pi$ $R(\alpha)$, let $R(\beta), R(\alpha_1), \ldots, R(\alpha_n)$ be defined. Then:
$V(\neg a) = Po_S \setminus V(a)$, $V(a \land b) = V(a) \cap V(b)$, $V(a \lor b) = V(a) \cup v(B)$,
$V([\alpha]a) = [R(\alpha)]V(a)$, $R(\alpha \cup \beta) = R(\alpha) \cup R(\beta)$, $R(\alpha^*) = R(\alpha)^*$,
$R(Com_n(\alpha_1, \ldots, \alpha_n)) = Com_n(R(\alpha_1), \ldots, R(\alpha_n))$, $R(a?) = V(a)?$

The satisfiability relation $A \Vdash_v a$ ("a is true in the point A at the valuation v") is defined as follows: $A \Vdash_v a$ iff $A \in V(a)$.

A formula a of $ArPDL$ is true in a model $M = (S, V, R)$ if $V(a) = Po$; a is true in a 2-arrow structure S if a is true in every model over S; a is semantically true if it is true in each 2-arrow structure S. The main problem is to axiomatize the proposed notion of semantic validity for formulas of $ArPDL$. Another problem is whether $ArPDL$ is decidable?

Arrow Structures of Dimension n as a Semantic Base for Some Process Connectives

In $ArPDL$ we interpret programs as sets of arrows, considering each arrow as a concrete process of the program. Since in Dynamic Logic of Programs we are interested mainly in the input and output of the programs, two-dimensional arrows are sufficient. But if we are interested in some properties of the data during the process of the program, we may model this by arrows with more dimensions. Then the intermediate points of an arrow will denote changes of the input during the action of the program. In this way we may give semantics of some process connectives. We will briefly consider some examples.

Let S be an n-arrow structure and let (S, R, V) be a model, i.e. as in $ArPDL$, R maps programs into subsets of Ar_S and V maps formulas into subsets of Po_S. Intuitively a subset of Ar will denote some action (program) and their members will denote some concrete processes of the action. Besides the standard PDL modalities $[\alpha]$ and $\langle \alpha \rangle$, we will consider some "process" modalities. Examples:

- "Sometimes at the beginning of α a", in symbols $\langle 1.\alpha \rangle a$,
  $A \Vdash_v \langle 1.\alpha \rangle a$ iff $(\exists x \in R(\alpha))(1.x = A \,\&\, 1.x \Vdash_v a)$.
- "Always at the beginning of α a", in symbols $[1.\alpha]a$,
  $A \Vdash_v [1.\alpha]a$ iff $(\forall x \in R(\alpha))(1.x = A \to 1.x \Vdash_v a)$.

Obviously $[1.\alpha]a$ is equivalent to $\neg \langle 1.\alpha \rangle \neg a$. Later on we will give only the "sometimes" versions of the connectives.

- "Sometimes just after the beginning a", in symbols $\langle 2.\alpha \rangle a$,
  $A \Vdash_v \langle 2.\alpha \rangle a$ iff $(\exists x \in R(\alpha))(1.x = A \,\&\, 2.x \Vdash_v a)$.
- "Sometimes, at some moment during α a", in symbols $\langle \exists i\alpha \rangle a$,
- $A \Vdash_v \langle \exists i\alpha \rangle a$ iff $(\exists x \in R(\alpha))(1.x = A \,\&\, (\exists i \leq n)(i.x \Vdash_v a))$.
- "Sometimes at all moments during α a", in symbols $\langle \forall i\alpha \rangle a$,
  $A \Vdash_v \langle \forall i\alpha \rangle a$ iff $(\exists x \in R(\alpha))(1.x = A \,\&\, (\forall i \leq n)(i.x \Vdash_v a))$.
- "Sometimes, at the end of α a", in symbols $\langle END\alpha \rangle a$,
  $A \Vdash_v \langle END\alpha \rangle a$ iff $(\exists x \in R(\alpha))(1.x = A \,\&\, n.x \Vdash_v a)$.

The problem of axiomatization of the above modalities remains open.

Arrow Structures and Pawlak Information Systems

By an information system — in the sense of Pawlak 1981 — we will mean any system $S = (OB, AT, \{VAL(a) \mid a \in AT\}, f)$ where

- $OB \neq \emptyset$ is a set, whose elements are called objects,
- $AT \neq \emptyset$ is a set, whose elements are called attributes (like "color")
- for each element $a \in AT$, $VAL(a)$ is a set, whose elements are called values of the attribute a (like "green"),

- f is a total function, called information function, which assigns to each object x and attribute a a value $f(x,a) \in VAL(a)$ (for instance $f(x, color) = green$).

If f assigns to the pair (x, a) not a single value but a subset $f(x,a) \subseteq VAL(a)$, then the system S is called a non-deterministic information system (for this notion see Orlowska and Pawlak 1984, Vakarelov 1991d, Orlowska 1985).

There is a close connection between Pawlak's information systems and arrow structures of dimension n. Each n-arrow structure S determines an information system as follows: put $OB = Ar_S$, $AT = (n) = \{1, 2, \ldots, n\}$, $VAL(i) = \{A \in Po_S \mid (\exists x \in Ar_S)i.x = A\}$ and $f(x, i) = i.x$. In this information system the objects are the arrows of S, the attributes are the numbers $1, 2, \ldots, n$ and the value of each attribute is a point (which may stay at the i-th place of some arrow).

Conversely, each information system $S = (OB, AT, \{VAL(a) \mid a \in AT\}, f)$ with a finite set of attributes $AT = \{a_1, \ldots, a_n\}$ determines an n-arrow structure S' as follows: $Ar_{S'} = OB$, $Po_{S'} = \bigcup VAL(a) \mid a \in AT\}$ and for $x \in OB$ and $i \leq n$, $i.x = f_S(x, a_i)$.

In this way some results from the theory of arrow systems may be used in the theory of information systems.

The analogy with information systems suggests some generalizations of arrow systems. First, instead of the set (n) we may take an arbitrary set, or even an ordinal if we want to have some ordering on the points on the arrows. Second, as in non-deterministic information systems, we may assume $i.x$ to be not a unique point, but a set of points, i.e. $i.x \subseteq Po$. In this way we obtain a more general notion where we have a full duality between points and arrows: each arrow x determines n sets of points $i.x, i \leq n$; and each point A determines n sets of arrows $i.A = \{x \in Ar \mid A \in i.x\}, i \leq n$. In these generalized arrow structures, besides the relations R_{ij}, we may define many others. Examples:

$x \leq_{ij} y$ iff $i.x \subseteq j.y$
Obviously we have
$xR_{ij}y$ iff $x \leq_{ij} y$ and $y \leq_{ji} x$
$x \leq y$ iff $(\forall i \leq n)x \leq_{ii} y$
$xI_{ij}y$ iff $i.x \cap j.y \neq \emptyset$,
$x\sigma y$ iff $(\forall i \leq n)xI_{ii}y$,
$x\Sigma y$ iff $(\exists i \leq n)xI_{ii}y$.

In non-deterministic information systems (cf. Orlowska 1985, Vakarelov 1991d) the relations $\leq$, σ and Σ are called "informational inclusion", "strong similarity" and "weak similarity" respectively.

Acknowledgments. Thanks are due to Johan van Benthem, for pointing out to me the importance of modal analysis of arrows; and to Hajnal

Andréka and István Németi for many fruitful discussions about "arrow" interpretations of some classical problems of Algebraic Logic.

References

Andréka, H., and Sz. Mikulás. 1994. Lambek calculus and its relational semantics: completeness and incompleteness. *Journal of Logic, Language and Information* 3(1):1–38.

Arsov, A. 1993. Completeness Theorems for Some Extensions of Arrow Logic. Master's thesis, Sofia University.

Arsov, A., and M. Marx. 1994. Basic Arrow Logic with Relation Algebraic Operators. In *Proceedings of the 9th Amsterdam Colloquium*, ed. P. Dekker and M. Stokhof, 93–112. Universiteit van Amsterdam. Institute for Logic, Language and Computation.

van Benthem, J. 1983. *Modal Logic and Classical Logic*. Naples: Bibliopolis.

van Benthem, J. 1989. Modal Logic and Relational Algebra. Manuscript, May.

van Benthem, J. 1991. *Language in Action (Categories, Lambdas and Dynamic Logic)*. Studies in Logic, Vol. 130. North-Holland, Amsterdam.

van Benthem, J. 1994. A Note on Dynamic Arrow Logics. In *Logic and Information Flow*, ed. J. van Eijck and A. Visser. 15–29. MIT Press, Cambridge (Mass.).

Gabbay, D.M. 1981. An irreflexivity lemma with applications to axiomatizations of conditions on linear frames. In *Aspects of Philosophical Logic*, ed. U. Mönnich. 67–89. D. Reidel Publishing Company.

Goranko, V., and S. Passy. 1992. Using the universal modality: gains and questions. *Journal of Logic and Computation* 2:5–30.

Hughes, G., and M. Creswell. 1984. *A Companion to Modal Logic*. Methuen.

Jónsson, B., and A. Tarski. 1951. Boolean algebras with operators. Parts I and II. *American Journal of Mathematics* 73:891–939. and 74:127–162, 1952.

Kramer, R. 1991. Relativized relation algebras. In *Algebraic logic*, ed. H. Andreka, D. Monk, and I. Nemeti, 293–349. Amsterdam. Colloq. Math. Soc. J. Bolyai, North-Holland, Amsterdam.

Kuhn, S. 1989. The domino relation: flattening a two-dimensional modal logic. *Journal of Philosophical Logic* 18:173–195.

Maddux, R. D. 1982. Some varieties containing relation algebras. *Transactions of the American Mathematical Society* 272:501–526.

Marx, M. 1995. *Algebraic Relativization and Arrow Logic*. Doctoral dissertation, Institute for Logic, Language and Computation, University of Amsterdam. ILLC Dissertation Series 1995-3.

Németi, I. 1987. Decidability of relation algebras with weakened associativity. *Proc. Amer. Math. Soc.* 100, 2:340–344.

Németi, I. 1991. Algebraizations of quantifier logics: an overview. *Studia Logica* L(3/4):485–569.

Németi, I. 1995. Decidability of weakened versions of first-order logic. In *Logic Colloquium '92*, ed. L. Csirmaz, D. Gabbay, and M. de Rijke, 177–242. Studies in Logic, Language and Information, No. 1. Stanford. CSLI Publications.

Orlowska, E. 1985. Logic of Nondeterministic Information. *Studia Logica* 91–100.

Orlowska, E., and Z. Pawlak. 1984. Representation of Nondeterministic Information. *Theoretical Computer Science* 29:27–39.

Pawlak, Z. 1981. Information Systems — Theoretical Aspects. *Information Systems* 6:205–218.

Roorda, D. 1991. *Resource Logics. Proof-Theoretical Investigations*. Doctoral dissertation, Institute for Logic, Language and Computation, University of Amsterdam.

Sahlqvist, H. 1975. Completeness and Correspondence in the First and Second Order Semantics for Modal Logic. In *Proc. of the Third Scandinavian Logic Symposium Uppsala 1973*, ed. S. Kanger. Amsterdam. North-Holland.

Segerberg, K. 1971. An Essay in Classical Modal Logic. *Filosofika Studier* 13.

Vakarelov, D. 1990. Arrow Logics. Manuscript, September.

Vakarelov, D. 1991a. Modal Logic for Reasoning about Arrows. Lecture notes of a talk at the 9th International Congress of Logic Methodology and Philosophy of Sciences, Uppsala and at Banach Semester 1991, Warsaw, August.

Vakarelov, D. 1991b. Modal Logics for Reasoning about Arrows: Arrow Logics. In *Proceedings of the 9th International Congress of Logic Methodology and Philosophy of Sciences,*. Uppsala, August.

Vakarelov, D. 1991c. Rough Polyadic Modal Logics. *Journal of Applied Non-Classical Logics* 1(1):9–35.

Vakarelov, D. 1991d. Modal Logics for Knowledge Representation Systems. *Theoretical Computer Science* 90:433–456.

Vakarelov, D. 1993. Arrow Logics with Cylindric Operators, abstract of a paper presented at the 1992 European Summer Meeting of the ASL. *The Journal of Symbolic Logic* 58(3):1135–1136.

Vakarelov, D. to appear. A Modal Theory of Arrows. Arrow Logic I. In *Advances in Intensional Logic*, ed. M. de Rijke. Kluwer. Available as ILLC Prepublication ML-92-04.

Venema, Y. 1989. Two-dimensional modal logics for relational algebras and temporal logic of intervals. ITLI-prepublication series LP-89-03. University of Amsterdam.

Venema, Y. 1991. *Many-Dimensional Modal Logic*. Doctoral dissertation, Institute for Logic, Language and Computation, Universiteit van Amsterdam.

Venema, Y. 1995. Cylindric Modal Logic. *Journal of Symbolic Logic* 60(2):591–623.

Part II
Multi-Modal Logic

8
What is Modal Logic?

MAARTEN DE RIJKE

ABSTRACT. This paper is concerned with methodological issues in *extended modal logic*, a rapidly expanding and active field that comprises of modal formalisms that differ in important aspects from the traditional format by extending or restricting it in a variety of ways. The paper surveys the parameters along which extensions of the standard modal format have been carried out, it proposes a unifying framework for modal logic, and identifies several research topics that arise naturally in this setting.

1 Introduction

It has been a long time since modal logic (ML) dealt with just two operators $\Diamond$ and $\Box$. Virtually every possible way of deviating from the syntactic, semantic and algebraic notions pertaining to this familiar duo is being or has been explored. One very rich source for such extended modal logics has been the increased interest in *dynamic* aspects of Logic, Language and Information of the past two decades; indeed, a whole landscape of so-called *modal transition logics* is emerging (see Venema, Chapter 1).

The creation of extended modal logics is largely application driven, with little or no concern for methodological issues. In particular, it is difficult to determine what makes a logic a modal logic.[1] The purpose of this paper is three-fold. First, to deal with the current diversity of the field, it offers a unifying framework for modal logic, which will help avoid future 'explosions' of techniques serving essentially similar goals. Second, our systematic view of modal logic will help identify new possibilities and questions; these include classifying the universe of modal logics and languages, generalizing known results, and combining modal logics. Third, we advocate an ab-

---

[1] One general approach is given by the theory of boolean algebras with operators, but it is too restricted for our purposes. For instance, the temporal logic with Since and Until is not captured by this approach.

Arrow Logic and Multi-Modal Logic
M. Marx, L. Pólos and M. Masuch, eds.
Copyright © 1996, CSLI Publications.

stract approach to modal logic; just as Abstract Model Theory is a large industry of interesting results at a high level of abstraction, so may an abstract approach to modal logic reveal general properties of, and connections between modal logics.

We proceed as follows. Section 2 introduces the framework, and Section 3 contains examples of systems of ML that fit inside this framework. Section 4 discusses the framework and research topics it naturally gives rise to, and the fifth and last section has some concluding remarks.

2 A Framework for Modal Logic

To have an example at hand, recall that the standard modal language $\mathcal{ML}(\diamond)$ has its formulas built up from proposition letters p taken from a set Φ, according to the rule $\phi ::= p \mid \bot \mid \neg\phi \mid \phi_1 \wedge \phi_2 \mid \diamond\phi \mid \Box\phi$. A model for $\mathcal{ML}(\diamond)$ is a tuple $M = (W, R, V)$ with $W \neq \emptyset$, $R \subseteq W^2$, and V a valuation, that is: a function $\Phi \longrightarrow \mathcal{P}(W)$, the powerset of W. Truth is defined by $M, x \models p$ iff $x \in V(p)$, and the usual clauses for the Booleans, while $M, x \models \diamond\phi$ iff for some y, xRy and $M, y \models \phi$, and $M, x \models \Box\phi$ iff $M, x \models \neg\diamond\neg\phi$.

$\mathcal{ML}(\diamond)$ expresses (first-order) properties of relational structures via its truth conditions. These properties don't exhaust the full first-order spectrum — only simple graphical *patterns* of the form "there is an R-successor with a property P" (and combinations thereof) are considered.

To present a system of ML we need to give its syntax, its semantic structures, and an interpretation linking the two. We start with the syntax.

Definition 2.1 The *syntax* of a system of modal logic is given by a vocabulary $(\mathcal{S}, \mathcal{V}, \mathcal{C}, \mathcal{O}, \mathcal{F})$ where

- $\mathcal{S}$ is a non-empty set of sort symbols;
- $\mathcal{V}_s$ $(s \in \mathcal{S})$ is a set of (propositional) variables;
- $\mathcal{C}_s$ $(s \in \mathcal{S})$ is a set of constants;
- $\mathcal{O}_s$ $(s \in \mathcal{S})$ is a set of connectives;
- $\mathcal{F}$ is a set of function symbols.

The elements of $\mathcal{F}$ are *modal operators*; via the semantics these will describe simple patterns in the structures in which the modal language is interpreted. Each (propositional) variable and each constant is assumed to be equipped with a sort symbol as are the argument places of the modal operators and connectives. Connectives of sort s take formulas of sort s, and return formulas of sort s, and modal operators are assumed to be marked with the sort they return.

The formulas $Form_s$ of sort s $(s \in \mathcal{S})$ are built up as follows.

Connectives:	•	Atomic formulas:	$p_s \in \mathcal{V}_s \cup \mathcal{C}_s$,
Modal operators:	#,	Formulas:	$\phi \in Form_s$,

$$\phi ::= p_s \mid \bullet(\phi_{1,s},\ldots,\phi_{n,s}) \mid \#(\phi_{1,s_1},\ldots,\phi_{n,s_n}),$$

where it is assumed that • and # return values of sort s. (A side remark: according to the above set-up there is little difference between connectives and modal operators: connectives operate within only one sort, whereas for modal operators the input and output formulas can all have different sorts. But this is what we actually encounter in many systems of ML; Section 4 contains further remarks on this issue.)

Example 2.2 In the standard modal language we have two sorts: p (propositions) and r (relations), the usual set of propositional variables ($p_0, p_1, \ldots$) and constants ($\bot, \top$), and only one constant but no variables of the relational sort (R); the connectives are the usual Booleans, while there is only one modal operator, $\langle \cdot \rangle \cdot$, taking a first argument of the relational sort, a second argument of the propositional sort, and returning a formula of the propositional sort.

A system of modal logic not only specifies the syntax of legal formulas, it also provides a semantics to interpret these formulas. The semantic desiderata include multiple domains, a uniform approach to dealing with the modal operators, and a flexible way of incorporating side-conditions on the interpretations of the symbols in our language.

Generalizing our intuitions from the standard modal format, modal operators will be interpreted as describing simple patterns in the relational structures underlying our languages. Such patterns are given as formulas of a classical language $\mathcal{L}$. Here we take a classical logic to be any logic in the sense of Barwise and Feferman 1985, Chapter 2, but one can simply think of first-order logic to have a concrete example in mind.

Definition 2.3 Let $\mathcal{F}$ be as in Definition 2.1, and let $\# \in \mathcal{F}$. A *pattern* or *$\mathcal{L}$-pattern* $\delta_\#$ for $\#$ is a formula in a classical logic $\mathcal{L}$ that specifies the semantics of $\#$. A pattern has the form

$$\lambda x_{s_1} \ldots \lambda x_{s_n}. \phi(x_{s_1}, \ldots, x_{s_n}; x_{s_{n+1}}, \ldots, x_{s_m}),$$

where $x_{s_1}, \ldots, x_{s_n}, x_{s_{n+1}}, \ldots, x_{s_m}$ are variables of sort $s_1, \ldots, s_n, s_{n+1}, \ldots, s_m$, respectively, the variables $x_{s_{n+1}}, \ldots, x_{s_m}$ are free variables, and all non-logical symbols occurring in ϕ are either among these variables or constants from $\mathcal{C}$.

The semantic structures for a system of ML consist of a number of domains W_s (for s a sort symbol in $\mathcal{S}$) with the modal operators describing configurations of elements of these domains. Moreover, there is an interpretation function I assigning a subset of W_s to every formula in $Form_s$ as follows. First, there is a valuation V that assigns subsets of W_s to atomic symbols, that is, to elements of $\mathcal{V}_s \cup \mathcal{C}_s$. I will tacitly assume that the function I 'knows' how to deal with the elements in $\mathcal{O}_s$. Truly modal aspects pop up with formulas of the form $\#(\phi_1, \ldots, \phi_n)$, for $\# \in \mathcal{F}$: to compute the value of such a formula at a tuple $\vec{x}$ one checks whether the pattern for $\#$ applied to $\phi_1, \ldots, \phi_n$, with $\vec{x}$ assigned to its free variables, is satisfied.

But this is not quite good enough. We may have special interpretations in mind for some of the sorts.

Example 2.4 Consider the standard modal language with its syntax as given in Example 2.2. The intended interpretation of the symbol R is not just $V(R) \subseteq W_r$, but rather $V(R) \subseteq W_r = (W_p \times W_p)$: R is to be interpreted as a binary relation on W_p. With the standard modal format as in Example 2.2, and the constraint as above, the pattern for $\langle R \rangle \cdot$ is $\lambda x_p. \exists y \, ((x, y) \in V(R) \land y \in V(x_p))$.

Example 2.4 motivates the following. A *constraint on an interpretation* I is a formula expressing some meta-level condition on I and on the way I interacts with the domains W_s. Familiar constraints include 'non-empty domain,' and 'persistence of proposition letters' in intuitionistic logic.

We arrive at the following definition of the semantics for an ML.

Definition 2.5 Let $\tau = (\mathcal{S}, \mathcal{V}, \mathcal{C}, \mathcal{O}, \mathcal{F})$ be a modal vocabulary. Let I be an interpretation defined on top of a valuation V and a set of patterns $\delta_\#$ ($\# \in \mathcal{F}$). A τ-*structure* is a tuple $M = (\{W_s\}_{s \in \mathcal{S}}, I, \Gamma)$, where W_s is a non-empty domain corresponding to the sort s, and Γ is a set of constraints on I and satisfied by I.

The value in M of a formula ϕ (of sort s) at an element $\vec{x} \in W_s$ is computed as follows:

$$M, \vec{x} \models p_s \quad \text{iff} \quad \vec{x} \in V(p_s),$$
$$M, \vec{x} \models \bullet(\phi_1, \ldots, \phi_n) \quad \text{iff} \quad \vec{x} \in I(\bullet)(I(\phi_1), \ldots, I(\phi_n)),$$
$$M, \vec{x} \models \#(\phi_1, \ldots, \phi_n) \quad \text{iff} \quad M, \vec{x} \models \delta_\#(I(\phi_1), \ldots, I(\phi_n)).$$

The upshot of Definitions 2.1 and 2.5 is that a modal logic is a formalism whose expressions involve multiple sorts, and whose distinguishing feature is the presence of functions between sorts, and the use of restricted relational patterns as truth definitions to describe simple geometric aspects of the underlying structures. In addition, there may be constraints governing both the sorts and the interaction between the sorts.

In what sense is a modal language a description language? And what does it mean to take a modal perspective on phenomena to be described? To answer the first question, let $\mathcal{ML}$ be a modal language. If all connectives and patterns of $\mathcal{ML}$ live in a classical language $\mathcal{L}$, then $\mathcal{ML}$ may be seen as a fragment of $\mathcal{L}$. This connection is given by a *standard translation* that takes $\mathcal{ML}$-formulas to $\mathcal{L}$-formulas by following the truth definition of $\mathcal{ML}$. For each sort s we fix sufficiently many individual variables x_s, and assume that there are predicate symbols corresponding to the propositional variables p_s; we will simply write p_s for the corresponding predicate letter. Then $ST(p_s) = P_s \vec{x}_s$, ST commutes with connectives •, and $ST(\#(\phi_1, \ldots, \phi_n)) = \delta_\#(ST(\phi_1), \ldots, ST(\phi_n))$, where some provisos have to be made to perform substitutions in $\delta_\#$. For instance, the modal case for $\mathcal{ML}(\Diamond)$ reads $ST(\Diamond \phi) = \exists y \, (xRy \wedge [y/x] ST(\phi))$, where $[y/x]\chi$ is χ with y substituted for x. This means that every modal language 'is' a restricted fragment of a classical language that precisely contains all descriptions of (combinations of) the geometric patterns in relational structures that underlie its modal operators. At another level modal languages are also used as description languages. A *frame* F is simply a τ-structure in which we 'forget' the valuation V; modal formulas are interpreted on frames by quantifying over all valuations: $F, x \models \phi$ iff for all models $M = (F, V)$ on F: $M, x \models \phi$. If $\mathcal{ML}$ and $\mathcal{L}$ are such that the standard translation maps $\mathcal{ML}$ into $\mathcal{L}$, then, on frames, $\mathcal{ML}$-formulas are equivalent to Π_1^1-conditions over $\mathcal{L}$, having the form $\forall \vec{p} ST(\phi)$. For example, on frames the standard modal formula $p \longrightarrow \Diamond p$ corresponds to $\forall p \, (p(x) \longrightarrow \exists y \, (xRy \wedge p(y)))$. However, in many cases a reduction is possible from such second-order conditions to first-order ones; for instance, the above formula is equivalent to Rxx. The mathematics of such reductions is explored in de Rijke 1993.

The framework suggests a 'modal perspective' on using logical description languages. Given certain phenomena to be described one *modalizes* them by assigning a sort to each of the objects under investigation, and introducing relations to capture the important configurations in which these objects occur, and modal operators to explore these relations; the idea is that initially one should impose as few constraints as possible. Of course this may allow for very unrealistic models, so the next step will be to gradually impose further constraints while trying to strike a balance between realistic modeling and having well-behaved modal description languages (i.e., decidable, admitting a finite axiomatization, ...).[2] This is not a novel idea: PDL, propositional dynamic logic, may be viewed as an example of this strategy, and two recent examples of this modalizing strategy are Roorda 1991 and Venema 1995. See also van Benthem, Chapter 9 for the broader methodological picture.

3 Examples

This section contains examples of modal logics fitting the framework of Section 2. Our examples differ from the standard format in a variety of

---

[2]The analogy to keep in mind here is with second-order logic, where a recursive axiomatization isn't possible on standard interpretations of the language, while it is possible if one allows non-standard models in which the variables ranging over predicates are simply interpreted as a separate sort, not necessarily coinciding with all possible subsets of the universe.

ways, but they all fit the framework of Section 2 rather easily. We first describe the standard modal format once more.

The Standard Modal Format. The standard modal format has two sort symbols p (propositions) and r (relations). The variables of sort p are $\mathcal{V}_p = \{p_0, p_1, \ldots\}$; the constants in $\mathcal{C}_p$ are $\bot$, $\top$; the connectives $\mathcal{O}_p$ of sort p are $\neg$, $\wedge$. For the relational sort we have $\mathcal{V}_r = \emptyset$, $\mathcal{C}_r = \{R\}$, $\mathcal{O}_r = \emptyset$. $\mathcal{F} = \{\langle \cdot \rangle \cdot\}$, where the first argument place is marked for symbols of sort r, the second for symbols of sort p, and the result sort is p. The sole constraint is that $W_r \subseteq (W_p)^2$. The pattern for $\langle \cdot \rangle \cdot$ is $\lambda x_r.\, \lambda x_p.\, \exists y ((x, y) \in V(x_r) \wedge y \in V(x_p))$. As there is only one possible input relation for this pattern we often use the 'old' notation $\Diamond \cdot$.

Arrow Logic. As a second example we look at arrow logic. There is one propositional sort p, and three relational sorts r_1, r_2, and r_3. The variables of sort p are $\mathcal{V}_p = \{p_0, p_1, \ldots, \}$; the constants in $\mathcal{C}_p$ are $\bot$, $\top$; the connectives are the Booleans. There are no variables or connectives of the relational sorts; $\mathcal{C}_{r_1} = \{C\}$, $\mathcal{C}_{r_2} = \{R\}$, $\mathcal{C}_{r_3} = \{I\}$. $\mathcal{F} = \{\circ^3, \otimes^2, \delta^1\}$; each of the modal operators returns a value of sort p; $\circ^3$ has its first argument place marked for symbols of sort r_1, $\otimes^2$ for symbols of sort r_2, and δ^1 for symbols of sort r_3; the remaining argument places (if any) are marked for symbols of sort p. There are multiple constraints: $W_{r_1} \subseteq (W_p)^3$, $W_{r_2} \subseteq (W_p)^2$, and $W_{r_3} \subseteq W_p$; various further constraints may be put on the structure of the domain W_p, such as $W_p \subseteq (U \times U)$, for some base set U. The pattern for $\circ^3$ is $\lambda x_{r_1}.\, \lambda x_p \lambda x'_p.\, \exists yz\, ((x, y, z) \in V(x_{r_1}) \wedge y \in V(x_p) \wedge z \in V(z_p))$; the patterns for the remaining modalities are similar.

More Operators. A common motivation for extending a modal formalism is the need to increase its expressive power. The most obvious way to go about things is to add operators. One can simply add an operator to the existing stock and give its pattern in terms of the material already present. One powerful example to this effect is the *D-operator* whose pattern reads $\lambda x_p.\, \exists y\, (y \neq x \wedge y \in V(x_p))$. Adding it to $\mathcal{ML}(\Diamond)$ overcomes important deficiencies in expressive power (de Rijke 1992a). But new operators and their patterns can also be defined in terms of new relations. Provability logic, for example, where the dual $[\cdot] \cdot$ of $\langle \cdot \rangle \cdot$ simulates provability in an arithmetical theory, has been expanded with modal operators simulating interpretability whose pattern is based on an additional ternary relation, (Berarducci 1990).

More Complex Modes of Evaluation. In the standard modal language one evaluates formulas at a single state in a model. As is brought out clearly by the contributions on Arrow Logic to this volume, one may need to switch to more complex sites of evaluation, such as arrow or transitions, ordered pairs of points, intervals (van Benthem 1991b), or even computation paths (Manna and Pnueli 1992).

More Sorts. To fully exploit the presence of multiple sorts, one may have to enrich the modal language and allow for evaluation at objects of all sorts. For example, one particular modal transition logic called *dynamic modal logic* (DML, van Benthem 1989, de Rijke 1992b) has both states and transitions, although it does not allow explicit reference to the latter. Peirce algebras extend DML to allow for such explicit references; they add all of relation algebra (or relativized versions thereof) to DML, as well as modalities linking the boolean and relational sort (de Rijke 1995c); see also Marx, Chapter 5 for a further example of a two-sorted system, called Dynamic Arrow Logic (DAL).

More Structure. A further mode of extending modal logic concerns the need to add structure to a sort already present, rather than add a new one. Marx' Chapter 5 provides an example: in DAL one obtains further means to structure transitions beyond mere compositions and converses by using finite iterations. But instead of adding more structure *amongst* elements of a sort, it has also been proposed to add *internal* structure to elements of a sort. Again, several contributions to this volume provide examples to this effect. In particular, in Venema's contribution one finds a landscape of logics in between abstract arrow logic and the logical counterpart of full relation algebra, and large parts of this landscape may be ordered by the extent to which their arrows are more similar to real ordered pairs.

Restricting the Language. An adaptation of the standard modal format that also fits the framework of Section 2 easily is restricting the language. Examples of modal languages in which the *relational* repertoire is restricted for 'meta-purposes' are abundant. For computational reasons van der Hoek and de Rijke 1995 restrict the propositional part of modal languages arising in AI. Likewise, the Lambek Calculus is a decidable reduct of arrow logic.

Adding and Changing Variables and Constants. The final mode of altering the standard modal format that I mention here is restricting the admissible valuations on (some of) the variables. To increase the expressive power of $\mathcal{ML}(\Diamond)$, Blackburn 1993 discusses the addition of special variables whose value in a model should always be a singleton. In interval logic, Venema 1990 restricts valuations to make his languages *less* expressive, as a result of which some meta-properties are regained (like admitting a finite axiomatization).

To prevent this section from becoming purely taxonomical, further ways of altering the standard modal format will be left out here. It should be clear, however, that according to the framework of Section 2 a modal logic is a many-sorted formalism that is concerned with the fine-structure of models, one that has multiple components each of which may be varied, and which has functions going from sorts to sorts that describe simple relational patterns of the underlying structures.

4 Questions and Comments

With the examples behind us it is time to address broader issues our framework for modal logic raises.

We first consider how the framework of Section 2 relates to other general approaches to logic, like *Abstract Model-Theoretic Logic*. Recall that the two important aspects of abstract model-theoretic logic are:

1. the isolation and study of specific logics for the analysis of various (mathematical) properties, and
2. the investigation into the properties of and the relations between such logics.

Within a general framework for ML similar research issues become clearly visible. As a tool for reasoning about relational structures ML has long been occupied with the first aspect above — but at a more fine-structural level than abstract model theory: most modal systems have just a few simple patterns with low quantifier rank for their operators; there is an interest in finite variable fragments, and in the finite model property or low complexity results rather than in Löwenheim-Skolem numbers or definability in the analytic hierarchy.

The second aspect is only now gaining attention in modal logic; general analyses of properties of modal systems are currently being developed, resulting, among others, in modal versions of theorems from abstract model theory: Andréka et al. 1995 show how model-theoretic results for basic modal logic can be stretched to cover far richer fragments of first-order logic, and de Rijke 1995a has a Lindström-style characterization of so-called basic modal languages.

ML broadly conceived is a part of abstract model-theoretic logic with a very special status; being many-sorted by nature and paying special attention to simple relational truth-definitions, it is concerned with the finestructure of models.

Another important point is this: what research topics does the framework of Section 2 give rise to? There are many; here we will only mention four fairly broad ones, most of them overlapping with the above items 1 and 2: *exploring* the modal universe, *classifying*, *generalizing*, and *combining*.[3]

Exploring. Just as a large part of abstract model theory is devoted to studying relations between logics, so should modal logic. For instance, there is a wide gap in expressive power between ordinary tense logic with F and P as its modal operators, and *Until, Since*-logic. For one thing, the operators

---

[3] A side remark: of course one can pose the 'old questions' for every system of ML, such as questions concerning completeness, expressive power, definability, decidability (or its refinement complexity) and truth preserving relations. But these old questions are not my prime concern here.

Until, Since are known to be irreducibly binary — are there elegant *unary* extensions of F, P-logic approximating *Until, Since*-logic in expressive power? Put more generally, how can such gaps be filled? Another example is this. According to general results in abstract model theory, a classical logic enjoys interpolation only if it has the Beth definability property — can similar general implications be found in modal logic?

Classifying. In a large universe of objects insight is often gained through classification. In the present setting patterns underlying modal operators need classifying. Which are the 'nice' ones? Which yield 'nice' modal logics? Patterns may be organized according to their quantificational structure. Let a *basic modal pattern* be one of the form $\lambda R.\, \lambda \vec{p}.\, \exists \vec{y}\, (Rx\vec{y} \wedge \bigwedge_i p_i(y_i))$.[4] The minimal modal logics of such basic patterns (i.e., counterparts of the standard modal logic **K**) admit a decent sequent-style axiomatization, are decidable, and enjoy interpolation (van Benthem 1993); by using bisimulations their model theory may be developed in parallel with basic first-order model theory (cf. de Rijke 1995b). These results may fail for patterns with a more complex structure, but Andréka et al. 1995 show that results for the basic modal logic can be stretched considerably. Modal patterns may also be classified according to their behavior with respect to relations between models. This is how modal operators and connectives can be distinguished: connectives (and their patterns) are not sensitive to the relational and sortal structure of models, while modal operators and their patterns are. More generally, broad criteria for classifying modal operators and their patterns have yet to be invented, although certain case studies have been carried out.[5] One desideratum here is an explanation of how patterns determine the axiomatic and complexity theoretic properties of the resulting ML.

Generalizing. How do the 'old' results from standard modal logic generalize to novel systems of ML? And if generalizations aren't possible, can natural counterexamples be given? Here are two examples. De Rijke 1993 presents a generalization of Sahlqvist's Correspondence Theorem. In its original form the result describes a class of $\mathcal{ML}(\Diamond)$-formulas that reduce to first-order formulas when interpreted on frames; this may be generalized to arbitrary modal languages; and de Rijke 1995b generalizes the model theory of the standard modal language with $\Diamond$, $\Box$, based on the use of bisimulations as its central tool, to a large class of modal languages. The benefit of striving towards such generalizations may not just be achieving greater generality, but also gaining a better understanding of what made the 'old' results work in the first place.

---

[4] These correspond exactly to normal, additive boolean algebras with operators.

[5] An example: van Benthem 1991a classifies functions between certain sorts with respect to their being a boolean homomorphism or not.

Combining. This has to do with the general architecture of ML, and arises naturally in a setting with multiple components and various links between them. How do those aspects *combine*? How do the components influence each other? What kind of communication is there between them? How do properties of the parts *transfer* to the larger system? Some instances of this question are studied by Kracht and Wolter 1991, Fine and Schurz 1991, Goranko and Passy 1992; they all consider the case of 'independently axiomatized' polymodal logics, and show how properties like completeness and the finite model property do or do not transfer. The results of Finger and Gabbay 1992 provide a further example, as does the work of Spaan 1993 on complexity of modal logics.

One fair complaint about the framework presented here is that it does not mention the proof theory of modal logic at all. Indeed, although the position of this paper has its roots in the Amsterdam school of modal logic which has emphasized the semantic aspects of the enterprise, I feel that a framework for modal logic should also address the issue of proof theory. The proof theory of ML has not kept pace with its model theory, mainly due to the fact that innovating ideas in ML often arise from its semantic use as a description language, where proof theory is not the most obvious research topic.

There may be conflicting requirements here. Should one demand that an ML has a complete proof procedure? In order to answer this, one has to keep in mind the role ML is supposed to play. As many systems of ML are designed with applications in mind, a complete proof theory seems desirable. But completeness is not just another property a system might have or not have. It may be that completeness is too stringent a requirement; even when using ML as a deductive machine, completeness might be sacrificed for other advantages, such as greater expressive power. And even when one does strive for and obtain completeness results in ML a lot still remains to be done. Sequent calculi seem to be needed for most practical purposes, and in this respect the approach advocated by Wansing 1994 seems promising, as the sequent style proof theory presented there seems to allow generalizations to arbitrary modal languages.

Finally, it is clear that many modal logics do indeed count as modal logics according to our framework. But isn't the framework so general that any formal system counts as a modal logic? My reply to this objection is 'Yes, and that's how it should be.' First of all, this attitude pays off, as is witnessed, for instance, by Venema's work on cylindric modal logic in which the modal perspective leads to new results on first-order logic. Second, what is important is that the framework captures and clarifies our intuitions about ML, that it offers a unifying approach to ML, and that it pays off in terms of new insights and questions — I think it does. Third, taking a modal perspective and 'modalizing' a given logic may be profitable by revealing new ways of improving the meta-logical properties of the calculus; see Andréka et al. 1995 for further discussion.

5 Concluding Remarks

Several influential presentations of modal logic have had a rather 'metaphysical' flavour, obscuring the real use and power of modal languages by paying (too) much attention to notions like 'possible world' and 'transworld identity.' At the risk of overdoing it, let me repeat the different picture of modal logic presented here. A modal language is a restricted description language for relational structures; it is a many-sorted formalism in which the modal operators emerge as functions from sorts to sorts that describe simple, restricted patterns in the underlying relational structures. Thus, the main concern of modal logic is best described as 'the fine-structure of many-sorted model theory.'

Although the examples of 'truly' many-sorted systems of modal logic given in this note are still quite traditional, I think that we will see the development of lots of many-sorted modal formalisms in the near future, especially with the growing interest in richer, mixed calculi that allow for more subtle notions of transition than just sets of ordered pairs, and in which multiple cognitive activities can be adequately represented alongside programs or actions. Further examples of multi-sorted modal formalisms are bound to arise in areas where one works with different kinds of information, some of which may influence others.

Finally, it may be that the global perspective outlined here is not the best approach to actually *work* in or with systems of modal logic. But my ideology is meant only as a unifying approach in which important research lines become visible — not as a tool.

Acknowledgments. I wish to thank Johan van Benthem, Patrick Blackburn, Wiebe van der Hoek, and especially Maarten Marx for good advice.

References

Andréka, H., J. van Benthem, and I. Németi. 1995. Back and Forth between Modal Logic and Classical Logic. *Journal of the IGPL* 3(3):685–720.

Barwise, J. and S. Feferman (ed.). 1985. *Model-Theoretic Logics*. New York: Springer-Verlag.

Benthem, J. van. 1989. Modal Logic as a Theory of Information. Report LP-89-05. ILLC, University of Amsterdam.

Benthem, J. van. 1991a. *Language in Action*. Studies in Logic, No. 131. Amsterdam: North-Holland.

Benthem, J. van. 1991b. *The Logic of Time*. Dordrecht: Kluwer. Second edition.

Benthem, J. van. 1993. What Makes Modal Logic Tick? Manuscript, ILLC, University of Amsterdam.

Berarducci, A. 1990. The Interpretability Logic of Peano Arithmetic. *Journal of Symbolic Logic* 55:1059–1089.

Blackburn, P. 1993. Nominal Tense Logic. *Notre Dame Journal of Formal Logic* 34:56–83.

Fine, K., and G. Schurz. 1991. Transfer Theorems for Multimodal Logics. In *Proceedings Arthur Prior Memorial Conference, Christchurch, New Zealand*. To appear.

Finger, M., and D.M. Gabbay. 1992. Adding a Temporal Dimension to a Logic System. *Journal of Logic, Language and Information* 1:203–233.

Goranko, V., and S. Passy. 1992. Using the Universal Modality: Gains and Questions. *Journal of Logic and Computation* 2:5–30.

Hoek, W. van der, and M. de Rijke. 1995. Counting Objects. *Journal of Logic and Computation* 5:325–345.

Kracht, M., and F. Wolter. 1991. Properties of Independently Axiomatizable Bimodal Logics. *Journal of Symbolic Logic* 56:1469–1485.

Manna, Z., and A. Pnueli. 1992. *The Temporal Logic of Reactive and Concurrent Systems. Vol. 1 Specification*. Springer-Verlag, New York.

Rijke, M. de. 1992a. The Modal Logic of Inequality. *Journal of Symbolic Logic* 57:566–584.

Rijke, M. de. 1992b. A System of Dynamic Modal Logic. Report # CSLI-92-170. Stanford University. To appear in *Journal of Philosophical Logic*.

Rijke, M. de. 1993. Correspondence Theory for Extended Modal Logics. Report ML-93-16. ILLC, University of Amsterdam. To appear in *Math. Logic Quarterly*.

Rijke, M. 1995a. A Lindström Theorem for Modal Logic. In *Modal Logic and Process Algebra*, ed. A. Ponse, M. de Rijke, and Y. Venema. 217–230. CSLI Publications.

Rijke, M. de. 1995b. Modal Model Theory. Report CS-R9517. CWI, Amsterdam. To appear in *Annals of Pure and Applied Logic*.

de Rijke, M. 1995c. The Logic of Peirce Algebras. *Journal of Logic, Language and Information* 4:227–250..

Roorda, D. 1991. *Resource Logics: Proof-theoretical Investigations*. Doctoral dissertation, ILLC, University of Amsterdam.

Spaan, E. 1993. *Complexity of Modal Logics*. Doctoral dissertation, ILLC, University of Amsterdam.

Venema, Y. 1990. Expressiveness and Completeness of an Interval Tense Logic. *Notre Dame Journal of Formal Logic* 31:529–547.

Venema, Y. 1995. A modal logic of Quantification and Substitution. In *Logic Colloquium '92*, ed. L. Csirmaz, D.M. Gabbay, and M. de Rijke, 293–309. Studies in Logic, Language and Information. CSLI Publications.

Wansing, H. 1994. Sequent Calculi for Normal Propositional Modal Logics. *Journal of Logic and Computation* 4(2):125–142.

9
Content versus Wrapping: an Essay in Semantic Complexity
JOHAN VAN BENTHEM

1 Contents and Wrappings
The Weight of Logical Tools
Any description of a subject carries its own price in terms of complexity. To understand what is being described, one has to understand the mechanism of the language or logic employed, adding the complexity of the encoder to the subject matter being encoded. Put more succinctly, "complexity is a package of subject matter plus analytic tools". This price is inevitable, and scientific or common sense insight does result all the same. Nevertheless, there is also a persistent feeling that one should never pay more than is necessary. Aristotle already formulated the necessary intellectual 'lightness' as follows. "It is the hall-mark of a scientific mind to give a subject no more formal structure than it can support" (as paraphrased in Kneale and Kneale 1961). Critics of formal logic would certainly agree with this dictum, and they have pointed at many cases in philosophy, linguistics, computer science and even the foundations of mathematics where general mathematical sophistication or just essayistic common sense is the more appropriate road towards insight than elaborate logical formal systems. But also inside formal logic, this question seems a legitimate concern. Are our standard modelings really appropriate for certain phenomena of reasoning, and are the received conclusions that we draw about complexity of phenomena in our field (using qualifications like 'undecidable' or 'higher-order') warranted, or rather an artefact of those modelings? More disturbingly, could it be that much of the respectable literature in our journals, which aims at proving difficult theorems as a sign of the academic 'worthiness' of a topic, mostly derives its continuity from the fact that one encounters the same issues over and over again, precisely because they derive from the

formalisms employed, not from the subject matter at hand? I think these are serious questions, that deserve constant attention. Of course, many logicians do care for these, either implicitly or explicitly, but it will do no harm to keep them high on the agenda by means of some occasional extra advertisement. Now, of course, the question arises how to distinguish the two sources of complexity in specific cases. Often, one has a suspicion that received views as to what is difficult in some form of reasoning or computation and what is not, might be challenged, but how to separate the two components? There may not be any systematic method of separation here — but there is certainly a roundabout answer in the practice of our field, which tends to generate scores of alternative formal modelings. Thus, the role of wrappings becomes visible indirectly, by comparison with alternatives (e.g., think of set-theoretic versus algebraic or category-theoretical formulations of the same problem). In other words, although formal logical modelings may be part of the complexity problem themselves, logicians mitigate this drawback by producing so many of them!

Thresholds of Complexity

There is room for more detailed analysis here, by considering more concrete examples. Received views on complexity in formal semantics often come in the form of warnings concerning certain dangerous 'thresholds' where complexity is generated or increased. For instance, one well-known danger zone is the transition from finite to infinite structures, and another that from first-order to higher-order objects. These moves often produce undecidability or even non-axiomatizability in the description of computational or linguistic phenomena. For instance, in the semantics of programming languages, undecidability may arise through the introduction of infinitary structures for iteration or recursion (Harel 1984, Goldblatt 1987). Likewise, temporal semantics for concurrency usually employ higher-order logics of temporal branches or histories, which involve quantification over sets (Burgess 1984, Stirling 1990). In the semantics of natural languages, similar thresholds arise. For instance, many quantificational phenomena are generally considered to be essentially second-order–such as linguistic 'branching quantifiers' which involve parallel rather than serial processing, with corresponding non-linear quantifier prefixes going beyond first-order logic (cf. Barwise 1979). Other sources of higher-orderness are traditionally located in the semantics of plural quantification (cf. van der Does 1992). Undecidability has also been claimed for variable binding phenomena in natural language anaphora (cf. Hintikka 1979 on the so-called "Any-Thesis"). Finally, new thresholds of complexity have emerged in Artificial Intelligence, with corresponding forms of 'received wisdom'. Notably, non-monotonic reasoning mechanisms, such as 'circumscription', are usually considered to be essentially higher-order and highly complex in general (McCarthy 1980) — and indeed, the same seems to hold, perhaps disturbingly, for any meaningful human cognitive task (cf. Kugel 1986). Do these views square with our intuitive expectations on the basis of a plausible initial estimate of the intrinsic complexity of a subject, prior to its formalization? Not always. In general, it is hard to make uniform predictions. Sometimes, formal analysis does confirm our intuitive hopes or suspicions concerning complexity of a phenomenon, witness the complexity theory of logical formalisms (cf. Spaan 1993). But at other times, plausible expectations turn out wrong, and we stand corrected by the formal analysis. A well-known historical instance is the undecidability of predicate logic, understandable eventually via Gödel's and Church's arguments, which clashed with the tradition of searching for decision procedures in the logical literature. Moreover, there are no infallible rules for keeping complexity down by simply avoiding the above danger zones. For instance, a blanket restriction to finite models has been advocated in natural language semantics (cf. van Benthem 1986). But this same restriction may also make many results harder to obtain (if available at all), witness the additional complexities of 'finite model theory' over general model theory found in Gurevich 1985. Likewise, the use of infinitary logics does increase complexity in some ways, but it also decreases complexity in other ways, witness the discovery of smooth infinitary axiomatizations for various programming logics (Goldblatt 1983), which can be more perspicuous than their first-order counterparts. Let us now turn to some specific cases where influential styles of modeling have supported well-known received views on complexity. These will provide more specific points for our subsequent discussion.

2 From Higher-Order to First-Order
Standard Set-Theoretic Models

There are many cases in semantics where higher-order modeling is thought appropriate and inevitable, even though it brings the cost of employing a formal system whose logical validities are non-axiomatizable (and indeed non-arithmetically definable). Pointing at these dangers is indeed one of the well-known social bonding procedures in our community, where collective shudders run through lecture halls at the mention of this Evil One. Nevertheless, upon closer inspection, in all these cases, an important distinction should be made. It is one thing to employ a higher-order *language*, referring to non-first-order individuals such as sets, choice functions or branches — but quite another to insist that this language should have full set-theoretic *standard models*, making it behave in the above-mentioned fashion. For, insisting on the latter expresses an additional commitment: namely that we want to use one particular mathematical implementation of our formalism, whose complexities will tend to 'pollute', naturally, the validities of our logic which was designed to mirror the core phenomenon. An insidious term in the practice of our field confuses the issue, namely the oft-praised "concreteness" of set-theoretic models. This amounts to insisting on one particular mathematical structure for our wrappings. But, is a set-theoretic model really 'more concrete' than an algebraic structure or a geometrical picture? Intuitively, in the latter case, the opposite would seem true. Moreover, the specificity of a mathematical structure may be the very source of complexity indicated above. Thus, there are reasons for preferring a more neutral stance. Another insidious groove of thought is the 'separation of concerns' favored by authors of semantic innovations. They prefer not to raise too many issues at once, and therefore choose "standard set-theoretic modeling" as their working theory. (Cf. the dominant style of presentation in current 'dynamic semantics', Kamp 1984, Groenendijk and Stokhof 1991, Veltman to appear). And of course, this is an excellent research strategy. But in fact, as we shall see below, working with this standard background may not always be an orthogonal decision. It can even be detrimental to the new proposals, as it may involve them in the hereditary sins, complexity-wise, of the old paradigms — precisely when dynamic semantics is partly inspired by the need to provide some cognitive relief from these.

Many-Sorted General Models

In higher-order logic, a more neutral perspective has been around for a long time. 'General models' for higher-order languages allow restricted ranges for set quantification (Henkin 1950). Standard models are the limit where all mathematically possible sets or predicates must be present — whether needed or not for the phenomenon under study. This broadening of the model class reinstates the usual properties of first-order logic, but it would be misleading to view it as just an opportunistic tactic. More importantly, on this view, higher-order logic becomes a *many-sorted first-order* logic treating, amongst others, individuals, sets and predicates on a par (cf. Enderton 1972). This move has a good deal of independent philosophical justification, witness the 'property theories' advocated in Bealer 1982, Turner 1989. Indeed, we achieve a moral rarity, being a combination of philosophical virtue with computational advantage, in trading set-theoretic complexity for new sorts of individuals. Even so, this move has always carried a stigma of ad-hoc-ness, and one common complaint is its lack of canonicity. No unique behaviour is specified for the sort of sets, so that the resulting 'logic' is subject to intensive manipulation. In other words, this solution is 'too easy' — again, a harmless shibboleth of the field around which we celebrate our professional consensus. But is it really? From another angle, general models do just the right thing. They radically block importation of extraneous set-theoretic truths, letting the subject described stand out. And they replace Platonic complacency by honest work. If we want to add an explicit new sort of objects ('sets', 'predicates' or whatever), then it should indeed be our task to analyze those principles about these objects that are germane to our subject, and formulate them explicitly. Just what about the behaviour of 'sets' is relevant to programming semantics, or to natural language?

A Geometrical Perspective

Answering questions like this suggests a compromise. The usual semantic restriction to standard set-theoretic models is too complex, and unilluminating — while liberalization to all general models is too weak, and unilluminating for the opposite reason. But after some hard work, we may find some appropriate model class in between. Examples of this style of analysis abound, once one perceives matters in this light. For instance, second-order semantics with quantification over branches of some sort may always be replaced by many-sorted theories of suitable 'individuals' (points in time, states, etc.) and 'branches' or 'paths', where one now has to find the key principles concerning branches plus their interaction with the individuals on them. E.g., Stirling 1990 identifies interesting modal-temporal second-order principles — such as 'fusion closure' stating that, for any state occurring in two histories, its past in the one and its future in the other may be glued together so as to form a new history. Likewise, in branching temporal logic, second-order axioms of choice support 'confluence principles' stating when two points on different histories may come together in some common future by suitable further histories (de Bakker et al. 1989). Such principles are best viewed as geometrical conditions on the availability of branches in two-sorted first-order models, on a par with standard geometrical axioms concerning points and lines. This is at least as good mathematics as set theory. After all, standard axiomatizations of geometry use 'points', 'lines', 'planes' on a par, rather than construing the latter outlandishly as point sets. (Notice the historical inversion. Geometry has been reduced to set theory: while we are advocating the opposite route.) For a final example, consider two closely related semantics for intuitionistic logic. 'Kripke models' are first-order, with truth conditions referring only to possible worlds and accessibility, whereas 'Beth models' are second-order, involving also branches (Troelstra and van Dalen 1988). For instance, Beth's truth clause for a disjunction A-or-B says that there exists some barrier of states across all possible future histories such that each state on the barrier verifies either A or B. But again, the latter models can be reformulated by viewing worlds and branches on a par, and analyzing what (little) explicit theory about branches — rather than some Platonic oracle about sets — is needed to explain validity for intuitionistic logic. (The Appendix to Rodenburg 1986 contains a first exploration of the resulting two-sorted model theory, including the interaction of states and branches.)

Correspondences for Geometrical Modeling

Localizing key principles on many-sorted models may be made more concrete and systematic with the tools of *Correspondence Theory* (cf. van Benthem 1984 & 1985). Consider the case of temporal logic. There are two complementary ways of arriving at our desired 'branch theory'. The first involves general reflection on 'branches', sorting out general logical principles governing these from more extravagant mathematical existence claims. The other approach takes the kind of temporal reasoning that is to be analyzed as a guide-line, looking for 'correspondences' between its intended principles and assumptions about branches. This interplay is well-documented for pure temporal logic (cf. van Benthem 1983). Here is a simple example in the branching semantics. Some forms of temporal reasoning allow a quantifier shift between 'future possibility' and 'possible futurity'. This will correspond to a simply computable condition on the pattern of states and branches: namely that time-travel along my current history and then switching to another future history may also be performed by first switching to another history and then traveling into its future. Correspondence Theory has only been investigated very systematically in Modal Logic, but in principle, this style of analysis is available everywhere — witness the powerful extensions presented in Venema 1991, de Rijke 1993. (There is still an issue of methodological consistency. Correspondence analysis in its usual format itself involves computation in higher-order logic! One line to take here says that we are merely using this technique as a semantic heuristic, and that 'defeating second-order logic from within' is an elegant philosophical stratagem. But also, many relevant insights would be forthcoming too on suitable general frames.)

Not every principle that comes up in the above way should be taken for granted. Some correspondences for putative principles of temporal reasoning turn up clear desiderata on two-sorted point-branch models, such as linearity for branches, or unicity of their initial points. Others may turn up interesting but negotiable options. For instance, more complex modal-temporal quantifier shifts in temporal reasoning correspond to axioms of choice, reflecting 'fullness' of state-branch models. No very clear demarcation line exists between the first case and the second — although there have been some systematic proposals. For instance, if genuine logical principles should be entirely free from existential import (cf. Etchemendy 1990), one might admit just the purely universal first-order requirements arising in a correspondence analysis as genuine core conditions on our semantics, relegating all conditions with existential import to some negotiable mathematical part. (Van Benthem 1983 even defends the further restriction to mere universal Horn clauses, viewing propositional disjunctions as existential too.) The methodology advocated here carries no general presumption that all reasonable new semantics will be simple, or even first-order. E.g., in computational applications, temporal branches might well have to be *finite* (a non-first-order condition). Also, in an opposite direction, it has been argued with some force that the essential control structures for programming are *infinitary*, and hence non-first-order (Goldblatt 1983). Either way, such discussions do not count against our broader framework: it rather speaks for its fruitfulness. Whether computation hinges essentially on finiteness for its 'traces' or its control structures is a substantive issue, which should be on the agenda explicitly, and not prejudged in the use of 'concrete standard modeling'.

Further Illustrations

All these points may also be demonstrated for the case of natural language. Consider the following linguistic phenomenon mentioned earlier on. From the start (cf. Hintikka 1979), 'branching quantification' has been considered a typical example of a second-order semantic mechanism, requiring quantification over Skolem functions not represented by ordinary first-order linear quantifier prefixes. To be sure, there is still some opposition to this linguistic claim, and first-order guerrilleros are still active. But, even granting the move toward using non-linearly orderable Skolem functions here, all that is shown by its proponents is that we need to consider a certain family of *relevant* 'choice functions' explicitly among our semantic individuals — not necessarily the whole mathematical space (containing lots of linguistically irrelevant items). Moreover, we can even defend this move on the basis of independent linguistic evidence. For instance, so-called 'functional answers' to questions ("Whom does every man love? His mother.") suggest that we need more abstract 'functional objects' as citizens in our semantic world. But even then, the real issue remains how many of these choice functions are required to explain natural language branching patterns, satisfying which combinatorial constraints. More generally, this point applies to more general linguistic 'polyadic quantification' (van Benthem 1989, Keenan and Westerståhl to appear), whose higher-orderness has been taken for granted so far. The complexity of the set-theoretic denotations arising here might also point at constraints on some family of 'available' predicates and functions. Likewise, considerations of this kind apply to several current discussions of complexity in Artificial Intelligence. For instance, the above-mentioned method of 'circumscription' might just employ minimization in general models, performed with respect to some family of 'relevant predicates' — say, those that are explicitly represented in our computational environment. (Morreau 1985 investigates the resulting two-sorted first-order model theory, and shows how it can deal with many of the non-monotonic reasoning patterns that originally motivated circumscription.)

Other Remodeling Strategies

It should be emphasized that not all broader semantic spaces must arise from a 'general model' strategy. The latter provides one systematic way of stepping back and rethinking the semantic issues. Without independent semantic evidence for its outcomes, however (as in the preceding linguistic example), the resulting many-sorted model might remain an ad-hoc theoretical curiosity. And in fact, eventually, one may come to prefer some alternative modeling altogether. A general illustration of this freedom is the ubiquitous switch found in mathematics between algebraic and geometric viewpoints — and a more specific one, the emergence of some recent analyses of branching quantification in natural language that employ a different circle of ideas concerning 'groups' and collective predication (cf. Landman 1989, Hoeksema 1983, van der Does 1992).

Summing up, changing to broader first-order model classes is not an ad-hoc move of desperation or laziness in taming complexity. It rather represents a more finely-tuned style of analysis, forcing us to do our conceptual homework, and eventually — another potential benefit — suggesting new applications beyond the original field, precisely because of the available new semantic models. Moreover, this is not a risky new-fangled approach of uncertain prospects. For, when all is said and done, this move is precisely what abstract mathematical analysis has always been about.

3 From First-Order to Decidable

Standard First-Order Models

The preceding move from higher-order to first-order logic brings a clear gain in complexity. But how satisfactory is the latter system as a universal semantic medium? Although effectively axiomatizable, predicate logic is undecidable — and again, this feature may import external complexity into the description of subjects whose 'natural complexity' would be decidable. One factor responsible for this situation, as was observed above, is the lure of 'concrete set-theoretic models'. We all think of standard Tarskian models as the essence of concreteness and simplicity (although there has been some underground opposition from the earlier-mentioned 'property theorists', cf. also Zalta 1993). But here again, are we perhaps still importing extraneous set theory, which might account for the undecidability of our logic? For instance, working logicians in linguistics or computer science often have a gut feeling that the styles of reasoning they are analyzing are largely decidable (cf. the percentual estimate given in Bacon 1985, or the analysis of 'natural logic' in Sanchez Valencia 1991), but it is hard to give any mathematical underpinning to these working intuitions.

Algebraic Semantics

Can there be well-motivated decidable versions of predicate logic, arising from giving up certain standard semantic prejudices? As it happens, there are even several roads towards decidability. A traditional one employs only restricted fragments of predicate logic (monadic, universal or otherwise, as is treated in many standard text books), while a very modern route is provided by recent linear logics of occurrences (Girard 1987, van Benthem 1991). But indeed, here too, a very general standard strategy exists for broadening our model class. There is a whole mathematical spectrum running from concrete set-theoretic models for predicate logic to abstract algebraic ones. This is the domain of *algebraic logic*, which has produced a good deal of information concerning these very issues: cf. Andréka 1991, Németi 1991, 1995, Venema 1991. In principle, this method will work whenever some modest minimal requirements are met by the underlying base logic. Of course, as with the above strategy of general models, the interesting possibilities will lie in between. Now, just like the general model move, the algebraization strategy has been criticized in the literature for its ad-hoc-ness and lack of clear constraints. As an old saying goes, 'algebraic semantics is syntax in disguise'. But this is only true for the bottom level (where syntactic Lindenbaum algebras would do the job), whereas usually, algebras of independent interest emerge in abundance through further semantic considerations. Thus, a key concern in our present setting should be the search for interesting independently motivated 'semantic parameters' that can be set differently from the specific choices made in the standard Tarskian paradigm. But then, this search does not seem hopeless. For instance, on the more traditional ontological side, we have already seen that property theorists want to treat individuals and properties on a par as intensional entities, viewing the usual Tarskian 'set-tuple style' of treating predicate denotations as just one, extensional, option out of many.

Dynamic Remodeling

Algebraic remodeling may be motivated concretely through a broad shift in current attitudes concerning logical semantics. From a more 'procedural' or computational viewpoint, various intriguing new ideas have been put forward recently concerning first-order interpretation. In particular, one may view the 'cylindric modal algebra' of Venema 1991, and indeed cylindric algebra in general (Henkin et al. 1971 & 1985) as a more fine-structured account of regimented access to successive variable assignments when interpreting quantified formulas, which may lead to natural decidable predicate logics (witness the contribution by Németi to this Volume). More general models will carry some accessibility pattern on assignments, or abstract states for predicate logic: $(S, \{R_x \mid x \in VAR\})$. Existentially quantified statements $\exists x \phi$ then become existential modalities $\langle x \rangle \phi$ referring to some new state verifying ϕ which must be R_x-accessible from the current state. Correspondence theory will now identify the procedural import of various predicate-logical principles over these abstract models, such as S5-laws for x-accessibility or more delicate 'path principles' (cf. van Benthem to appear). Judicious combinations of such conditions will then produce a whole landscape of attractive decidable predicate logics. Independent motivations for such transitions between assignments exist already. One example is the 'dynamic predicate logic' of Groenendijk and Stokhof 1991, which views first-order formulas as explicit programs for effecting transitions between assignments when interpreting predicate logic. But there are further aspects to dynamics. In some recent semantics of generalized quantification, individual domains are allowed to carry a 'dependence' structure, determining how individuals may become available in the course of interpreting an existence statement. Concrete motivations may be found in the theory of 'arbitrary objects' in Fine 1985, or the work on probabilistic independence relations in van Lambalgen 1991, as well as its 'modal first-order' version proposed in Alechina and van Benthem 1993. This gives a much broader space of models for the language of first-order predicate logic, with the original Tarskian ones becoming the special case where one has 'random access' to the full Cartesian space of all mathematically possible assignments of objects to variables. (An up-to-date systematic discussion may be found in Alechina 1995.)

As before, this style of analysis makes predicate logic into a more finely-tuned tool, with a decidable 'core' and a 'periphery' of more demanding principles. This is of purely logical interest, as it allows us to see new distinctions among what used to be one uniform batch of logical truths, that need not correspond to traditional classifications of propositions (say, using prenex forms or predicate arities). It is also of applied interest, as the richer picture may better fit the natural structure of a subject. 'Accessibility constraints' arise from many viewpoints, mathematical, philosophical and linguistic. Moreover, the decidable subsystems of first-order logic that emerge in the resulting landscape of logical systems may fit with proposed mechanisms of 'natural logic' in human reasoning (cf. Sommers 1982, Sanchez Valencia 1991). Taking this viewpoint seriously is not a minor philosophical move either — as it brings radical repercussions. For instance, from our new stance, what remains the point of Gödel's celebrated Completeness Theorem, stating a natural fit between Tarskian semantics and Fregean axiomatization? One might say that Gödel's result ties one particular natural choice of predicate-logical validity, which may be motivated independently from various angles (model-theoretic, proof-theoretic, game-theoretic) to 'standard set-theoretic modeling'. Kreisel 1967 and Etchemendy 1990 have stressed the surprise in this outcome, noting how it manages to ensnare natural intuitive validity by means of exact mathematical notions. We would have to disagree here. From equally natural semantic points of view, other logical equilibria may arise, this time having a decidable set of validities. What still remains is Gödel's lasting methodological achievement of having put completeness issues on the map, and made them amenable to mathematical statement and proof.

Logical Architecture

Actually, we could go one step further here, as well as earlier. The very multiplicity of modelings for predicate logic suggests that there is no one unique preferred choice — and that one should rather view all options as living together. Having many modelings around in the literature for a phenomenon is not a nuisance or disease, to be eliminated by natural competition, in favor of one canonical solution. It may be a definite virtue to be prized. It is quite natural to view semantic modeling as an open-ended process, which allows ever finer levels of 'grain size'. Examples are recent proposals by Zalta 1993, following Etchemendy, to split first-order semantics into two stages. The first goes from syntactic forms to abstract propositions (this is the domain of 'linguistic change'), and the second goes from these propositions to standard models (this is the domain of 'real change'). The burden of 'predicate logic' could then be divided between these two stages, with a possibly decidable component for the first stage, and an undecidable one for the second. Another example would be the grounding of decidable linear predicate logics in Lorenzen-style or Hintikka-style game-theoretical semantics for first-order logic (Mey 1992), or in intermediate Tarskian evaluation algorithms serving as Fregean senses (Moschovakis 1991). Of course, this does raise new issues of what may be called 'logical architecture' (van Benthem and ter Meulen to appear), as the interplay between the various components now becomes the explicit business of logic, too. Thus, traditional semantics and its rivals might coexist in one broader logical framework, whose task it would be to indicate precisely at which fault lines complexity emerges.

4 Digression: Illustrations in Arrow Logic

Arrow Logic Proper

A more compact testing site for the above general considerations may be found in the 'Arrow Logic' which is so prominent in this volume. Here are a few illustrations of the above general themes at work in this more limited domain. One guiding motivation in the arrow logic research programme has been to challenge another well-known 'complexity threshold' in logics of computation, namely that undecidability is inevitable once a full relational algebra of programming operations is allowed. For, how much of this undecidability is due to the essential complexity of computation, and how much is merely a by-product of the mathematics of ordered pairs that lies underneath standard modeling in this area? As it happens, the resulting theory nicely demonstrates the above general points (cf. chapter 1). First, a new sort of individual objects has been put on the map, namely transitions or 'arrows', which occur in most computational intuitions, and which arguably should be first-class citizens in any analysis of computation. Moreover, we are forced to reflect on the essential structure among these arrows, such as their composition, conversion or possibly other modes of combination (including parallel constructions of 'sheafs of arrows'). At least in the simplest similarity type, here, various decidable core calculi have been discovered, which can do quite a bit of the basic combinatorics of program combination. In particular, these calculi support all Boolean operators on top of relational composition, without tripping over any purported undecidability threshold. On top of that, correspondence analysis reveals the additional semantic content of further programming principles, distinguishing between purely universal ones (whose totality might form the largest desirable 'core theory') and more demanding existence principles. Outcomes have not been totally routine here, in that, e.g., associativity of composition has been identified as a danger point (cf. chapter 3), whereas most people would consider the latter an entirely harmless domestic assumption. Finally, this is not an isolated piece of semantic engineering, because relational transitions underlie such a vast range of computational processes. Arrow-style analysis is appropriate for relational algebra and propositional dynamic logic, but also for logics of belief revision or general process algebras. Indeed, the basic theory of labeled transition systems in computer science (cf. Stirling 1990, van Benthem and Bergstra 1995, van Benthem et al. 1994) is probably better viewed as a theory of states and arrows, with appropriate notions of two- sorted bisimulation for process equivalences (cf. van Benthem 1994). (One might even consider the introduction of computation 'paths' or 'branches' as a third kind of independent semantic object, thereby reflecting an earlier strand in our discussion.) Finally, there is no reason for staying inside the realm of computation, once our general points about modeling have been adopted. For instance, concrete geometrical arrows also reflect human thinking about their preferences, both scientific and domestic, and thus switching to arrow semantics might also throw some new light on preference logic (cf. the first exploration made in van Benthem et al. 1993) — and more generally, on various theories of individual and collective choice in the social sciences.

Generalized Predicate Logic

The same points apply to generalized predicate logic in the earlier algebraic spirit, which may be viewed as a theory of dynamic state domains allowing certain atomic actions of variable update. This paradigm has been developed in a number of recent dissertations, including Marx 1995, Mikulás 1995 and Alechina 1995. Moreover, dependency models with restricted sets of assignments have emerged in linguistic semantics, too (cf. Meyer Viol 1995, van den Berg 1996). One additional feature of interest here illustrates one more general phenomenon in semantic remodeling. Over a broader model class, one can usually interpret a more expressive language, supporting distinctions that were still collapsed over standard models. Generalized assignment models distinguish iterated standard quantifiers from genuinely polyadic ones, and abstract dependency structure also support various new generalized quantifiers. (Other telling examples of this phenomenon have been observed in relevant or linear logic.) Thus, semantic lightness may also induce semantic wealth.

5 General Issues

This paper has raised some general issues concerning semantic modeling, with a special interest in the locus of complexity. We have recommended a systematic search for alternative logical modelings, with the search for 'minimal complexity of wrappings' as a moving force toward new conceptual schemes. Moreover, we have noted some general features and benefits of this enterprise, as well as some of its philosophical repercussions. All this has by no means exhausted the topic of logical complexity. For instance, other important technical approaches exist in our field to high-lighting 'core content'. Among these, one may mention the use of 'oracles' supplying extraneous information to be bracketed out — as happens in Cook-style completeness theorems in programming semantics (Cook 1978) and in various parts of Complexity Theory (Buhrmann 1993). Such strategies remain to be compared with the present proposals. Another well-known way of coping with complexity is that of the 'moderate drinker': 'ingest only small amounts'. For instance, second-order complexity might be avoided by using only small fragments of the full second-order language, and the same holds for first-order logic. (By contrast, we have advocated retaining the full language as the expressive medium, while lowering the over-all complexity of its logic.) Finally, there are various statistical approaches to average-case complexity of logical reasoning that take the sharpest edges off the usual worst-case problems. In this division of labour, the statistician takes care of the harsh realities of life, so that the logical theorist can live at peace. What these alternative strategies have in common is that they do not require rebuilding basic logical frameworks, but rather offer ways of accommodating ourselves to their complexity. Perhaps, then, our own proposals here have been unduly principled and calvinistic. Indeed, they might also run against the no-nonsense spirit of our times. In particular, they seem to clash with a recent trend in logical modeling which aims at replacing language-oriented semantic analysis by what is called 'direct mathematics'. The latter is illustrated by current developments in the Dutch branch of Process Algebra (cf. the program outlined in Baeten 1992), but also by earlier 'set-theoretical' programs in the Philosophy of Science (Suppes 1960, Sneed 1971), whose proponents want to minimize encounters with the common syntactic idiosyncrasies of logical formalisms when describing their phenomena of interest. There are certainly good practical reasons for exploring such a line of investigation. Nevertheless, on the view of this paper, its guiding slogan may also be slightly misleading. 'Direct mathematics' with its seemingly innocent emphasis on concrete 'standard mathematical tools' carries many pre-wired decisions as to complexity, whereas our advice has been to take nothing for granted.

In all, this short essay has a very modest aim. We would be happy if our points were to contribute to a sustained probing discussion of received views in semantic modeling.

References

Alechina, N. 1995. *Modal Quantifiers*. ILLC Dissertation Series 1995–20, Institute for Logic, Language and Computation, University of Amsterdam.

Alechina, N., and J. van Benthem. 1993. Modal Quantification over Structured Domains. Report ML 93-02. Institute for Logic, Language and Computation, University of Amsterdam. To appear in M. de Rijke, ed., Advances in Intensional Logic, Kluwer Academic Publishers, Dordrecht.

Andréka, H. 1991. Complexity of the Equations Valid in Algebras of Relations. Thesis for D.Sc. (post–habilitation degree) with Math. Inst. Hungar. Ac. Sci. Budapest. A slightly updated version will appear in *Annals of Pure and Applied Logic*.

Bacon, J. 1985. Completeness of a Predicate-Functor Logic. *Journal of Symbolic Logic* 50:903–926.

Baeten, J. 1992. Informatica als Wetenschap (formele specificatie en wiskundige verificatie). *TUE Informatie* 35 (7/8):493–499.

de Bakker, J., W.P. de Roever, and G. Rozenberg (ed.). 1989. *Linear Time, Branching Time and Partial Order in Logics and Models for Concurrency*. Springer, Berlin.

Barwise, J. 1979. On Branching Quantifiers in English. *Journal of Philosophical Logic* 8:47–80.

Bealer, G. 1982. *Quality and Concept*. Oxford University Press, Oxford.

van Benthem, J. 1983. *The Logic of Time*. Reidel, Dordrecht.

van Benthem, J. 1984. Correspondence Theory. In Gabbay and Guenther 1984, 167–248.

van Benthem, J. 1985. *Modal Logic and Classical Logic*. Bibliopolis.

van Benthem, J. 1986. *Essays in Logical Semantics*. Dordrecht: Reidel.

van Benthem, J. 1989. Polyadic Quantifiers. *Linguistics and Philosophy* 12:4:437–464.
van Benthem, J. 1991. *Language in Action (Categories, Lambdas and Dynamic Logic)*. Studies in Logic, Vol. 130. North-Holland, Amsterdam.
van Benthem, J. 1994. A Note on Dynamic Arrow Logics. In *Logic and Information Flow*, ed. J. van Eijck and A. Visser. 15–29. MIT Press, Cambridge (Mass.).
van Benthem, J. to appear. Modal Foundations for Predicate Logic. In E. Orlowska, ed., *Memorial Volume for Elena Rasiowa*, Studia Logica Library, Kluwer, Dordrecht.
van Benthem, J., and J. Bergstra. 1995. Logic of Transition Systems. *Journal of Logic, Language and Information* 3:247–283.
van Benthem, J., and A. ter Meulen (eds.). to appear. *Handbook of Logic and Language*. Elsevier, Amsterdam.
van Benthem, J., J. van Eijck, and A. Frolova. 1993. Changing Preferences. Report CS-93-10. Centre for Mathematics and Computer Science, Amsterdam.
van Benthem, J., J. van Eijck, and V. Stebletsova. 1994. Modal Logic, Transition Systems and Processes. *Journal of Logic and Computation* 4:811–855.
van den Berg, M., 1996. *The Internal Structure of Discourse*. Doctoral dissertation. Institute for Logic, Language and Computation, University of Amsterdam.
Buhrmann, H. 1993. *Resource Bounded Reductions*. ILLC Dissertation Series 1993–2, Institute for Logic, Language and Computation, University of Amsterdam.
Burgess, J.P. 1984. Basic Tense Logic. In *Handbook of Philosophical Logic. Vol. 2*, ed. D. Gabbay and F. Guenthner. 89–133. Dordrecht: Reidel.
Cook, S. 1978. Soundness and Completeness of an Axiom System for Program Verification. *SIAM Journal of Computing* 7:70–90.
van der Does, J. 1992. *Applied Quantifier Logics*. Doctoral dissertation, Institute for Logic, Language and Computation, University of Amsterdam.
Enderton, H. 1972. *A Mathematical Introduction to Logic*. Academic Press, New York.
Etchemendy, J. 1990. *The Concept of Logical Consequence*. Harvard University Press, Cambridge.
Fine, K. 1985. *Reasoning with Arbitrary Objects*. Aristotelian Society Series, Vol. 3. Oxford: Basil Blackwell.
Gabbay, D.M., and F. Guenther (ed.). 1984. *Handbook of Philosohical Logic* Volume II. Dordrecht: Reidel.
Girard, J-Y. 1987. Linear Logic. *Theoretical Computer Science* 50:1–102.
Goldblatt, R. 1983. *Axiomatizing the Logic of Computer Programming*. Springer, Berlin.
Goldblatt, R. 1987. *Logics of Time and Computation*. Lecture Notes, Vol. 7. Stanford: CSLI Publications.
Groenendijk, J., and M. Stokhof. 1991. Dynamic Predicate Logic. *Linguistics and Philosophy* 14:39–100.

Gurevich, Y. 1985. Logic and the Challenge of Computer Science. Report CRL-TR-10-85. Computer Research Laboratory, University of Michigan.

Harel, D. 1984. Dynamic Logic. In Gabbay and Guenther 1984, 497–604.

Henkin, L. 1950. Completeness in the Theory of Types. *Journal of Symbolic Logic* 15:81–91.

Henkin, L., J. D. Monk, and A. Tarski. 1971 & 1985. *Cylindric Algebras, Parts I & II*. Amsterdam: North-Holland, Amsterdam.

Hintikka, J. 1979. Quantifiers in Natural Languages: Some Logical Problems. In *Game-Theoretical Semantics*, ed. E. Saarinen. 81–118. Reidel, Dordrecht.

Hoeksema, J. 1983. Plurality and Conjunction. In *Studies in Model-theoretic Semantics*, ed. A. ter Meulen. 63–83. Foris, Dordrecht.

Kamp, H. 1984. A Theory of Truth and Semantic Representation. In *Truth, Interpretation and Information*, ed. Th. Janssen J. Groenendijk and M. Stokhof. 1–41. Foris, Dordrecht.

Keenan, E., and D. Westerståhl. to appear. Generalized Quantifiers. In *Handbook of Logic and Language*, ed. by J. van Benthem and A. ter Meulen. Elsevier, Amsterdam.

Kneale, W., and M. Kneale. 1961. *The Development of Logic*. Clarendon Press, Oxford.

Kreisel, G. 1967. Informal Rigour and Completeness Proofs. In *Problems in the Philosophy of Mathematics*, ed. I. Lakatos. 138–186. North-Holland, Amsterdam.

Kugel, P. 1986. Thinking May Be More Than Computing. *Cognition* 22:137–198.

van Lambalgen, M. 1991. Natural Deduction for Generalized Quantifiers. In *Generalized Quantifiers; theory and applications*, ed. J. van der Does and J. van Eijck (eds.). 143–154. Amsterdam: Dutch PhD Network for Language, Logic and Information. To appear with CSLI Publications, Cambridge University Press.

Landman, F. 1989. Groups I & II. *Linguistics and Philosophy* 12:559–605 and 723–744.

Marx, M. 1995. *Algebraic Relativization and Arrow Logic*. Doctoral dissertation, Institute for Logic, Language and Computation, University of Amsterdam. ILLC Dissertation Series 1995-3.

McCarthy, J. 1980. Circumscription—A Form a Non-Monotonic Reasoning. *Artificial Intelligence* 13:27–39, 171–172.

Mey, D. 1992. Game-Theoretical Interpretation of a Logic Without Contraction. Technical Report. Department of Computer Science, Swiss Federal Institute of Technology, Zürich.

Meyer Viol, W. 1995. *Instantial Logic. An Investigation into Reasoning with Instances*. ILLC Dissertation Series 1995-11, Institute for Logic, Language and Computation, University of Amsterdam.

Mikulás, Sz. 1995. *Taming Logics*. Doctoral dissertation, Institute for Logic, Language and Computation, University of Amsterdam. ILLC Dissertation Series 95-12.

Morreau, M. 1985. Circumscription: A Sound and Complete Form of Non-Monotonic Reasoning. Technical report. Mathematical Institute, University of Amsterdam.

Moschovakis, Y. 1991. Sense and Reference as Algorithm and Value. Technical report. Department of Mathematics, University of California, Los Angeles.

Németi, I. 1991. Algebraizations of Quantifier Logics: an overview. *Studia Logica* L(3/4):485–569.

Németi, I. 1995. Decidability of Weakened Versions of First–Order logic. In *Logic Colloquium '92*, ed. L. Csirmaz, D. Gabbay, and M. de Rijke, 177–242. Studies in Logic, Language and Information, No. 1. Stanford. CSLI Publications.

de Rijke, M. 1993. *Extending Modal Logic*. Doctoral dissertation, Institute for Logic, Language and Computation, Universiteit van Amsterdam. ILLC Dissertation Series 1993-4.

Rodenburg, P.H. 1986. *Intuitionistic Correspondence Theory*. Doctoral dissertation, Mathematical Institute, University of Amsterdam.

Sanchez Valencia, V. 1991. *Studies on Natural Logic and Categorial Grammar*. Doctoral dissertation, Institute for Logic, Language and Computation, University of Amsterdam.

Simon, A., A. Kurucz, I. Németi, and I. Sain. 1993. Undecidability Issues of Some Boolean Algebras With Operators and Logics Related to Lambek Calculus. Workshop on Algebraization of Logic, Fifth European Summer School in Logic, Language and Information.

Sneed, J. 1971. *The Logical Structure of Mathematical Physics*. Reidel, Dordrecht.

Sommers, F. 1982. *The Logic of Natural Language*. Cambridge University Press, Cambridge.

Spaan, E. 1993. *Complexity of Modal Logics*. Doctoral dissertation, ILLC, University of Amsterdam.

Stirling, C. 1990. Modal and Temporal Logics. In *Handbook of Logic in Computer Science*, ed. S. Abramsky, D. Gabbay, and T. Maibaum. Oxford University Press.

Suppes, P. 1960. A Comparison of the Meaning and Uses of Models in Mathematics and the Empirical Sciences. *Synthese* 12:287–301.

Troelstra, A., and D. van Dalen. 1988. *Constructivism in Mathematics*. North-Holland, Amsterdam.

Turner, R. 1989. Two Issues in the Foundations of Semantics. In *Properties, Types and Meaning*, ed. G. Chierchia, B. Partee, and R. Turner. 63–84. Reidel, Dordrecht.

Veltman, F. to appear. Defaults in Update Semantics. *Journal of Philosophical Logic*.

Venema, Y. 1991. *Many–Dimensional Modal Logic*. Doctoral dissertation, Institute for Logic, Language and Computation, Universiteit van Amsterdam.

Zalta, E. 1993. A Philosophical Conception of Propositional Modal Logic. Technical report. Center for the Study of Language and Information, Stanford University.

10

A Fine-Structure Analysis of First-Order Logic

István Németi

1 Introduction

This work is partly motivated by the slogan in Sain 1979 and in van Benthem to appear. We quote van Benthem's wording:

"The complexity of any logical modeling reflects both the intrinsic structure of a logic described and the weight of the formal tools. Some of this weight seems inherent in even the most basic logical systems. Notably, standard predicate logic ..."

Indeed, predicate logic (or FOL for short) has some "expensive" properties which might not be needed for certain application areas. Such expensive features are: (i) undecidability of validities, (ii) non-axiomatizability of the finite variable fragments (i.e., to each n there is a 3-variable formula whose proof needs more than n variables), (iii) failure of the Craig interpolation property for the finite variable fragments (i.e., there are n-variable formulas whose interpolant needs more than n variables), (iv) non-finite axiomatizability of the valid schemas of formulas, etc.

These observations have lead to the project in which we decompose FOL into a hierarchy of well behaved but weaker logics such that the "expenses" (i)–(iv) listed above show up at different levels of the hierarchy. Such a hierarchy could, in principle, enable the user of FOL to decide which of the expenses (i)–(iv) are justified by the problem area he is working on. Then he can select the version or level of FOL which is strong enough, but not unnecessarily expensive, for the problem area in question.

In order to have a grasp on the fine-structure of FOL, we will reconstruct FOL as a propositional modal logic treating all extra-Boolean features of

---

Research supported by the Hungarian National Foundation for Scientific Research grants Nos. T16448, T7255, T7567.

Arrow Logic and Multi-Modal Logic
M. Marx, L. Pólos and M. Masuch, eds.
Copyright © 1996, CSLI Publications.

FOL as distinct modalities. Let Var be our set of individual variables and let $\mathfrak{M} = \langle M \ldots \rangle$ be a model. Then, according to the Tarskian tradition, ^{Var}M is the set of all evaluations (of the variables into $\mathfrak{M}$). We will treat the elements $k \in {^{Var}M}$ as states (or possible worlds) in the Kripke-frame $\langle {^{Var}M}, \ldots \rangle$. The various logical connectives, like $\varphi \mapsto \exists x \varphi$, and linguistic features, like $\varphi \mapsto \varphi[x/y]$, will re-appear here as modalities, like $\Diamond_x$, whose accessibility relations on ^{Var}M will determine their meanings. For the purposes of this introduction, let us select two "connectives" of FOL $\varphi \mapsto \exists x \varphi$ and $\varphi \mapsto \varphi[x/y]$, where $\varphi[x/y]$ is the formula obtained from φ by replacing all free occurrences of x with y (i.e., usual substitution). We will have more of these kinds of "linguistic devices" in the main text. The accessibility relation for the modality $\exists x$ $(= \Diamond_x)$ turns out to be $\stackrel{x}{\equiv}$ (where $k \stackrel{x}{\equiv} f$ iff $k \lceil (Var - \{x\}) = f \lceil (Var - \{x\})$). The accessibility relation S_y^x of the modality $[x/y]$ is the function $k \mapsto k(x/k(y))$, where $k(x/b) \in {^{Var}M}$ is the modified evaluation agreeing everywhere, except at x, with k such that $k(x/b)(x) = b$. (Depending on one's notational tradition, the accessibility relation of $[x/y]$ is either S_y^x or its inverse $(S_y^x)^{-1}$.) Both of the modalities ($\Diamond_x$ and $[x/y]$) are familiar from dynamic logic since, writing now exceptionally $[x/y]$ in front of the formula to which it applies, we obtain that, in the language of dynamic logic,

$$\text{``}[x/y]\varphi\text{''} = \text{``}[x := y]\varphi\text{''}$$
$$\text{``}\Diamond_x \varphi\text{''} = \text{``}[x :=?]\varphi\text{''}$$

(where "$x := y$" and "$x :=?$" are the "programs" of assignment and random assignment, respectively), cf. e.g. Harel 1984, van Benthem to appear. All this leaves us with the class of frames

$$\mathsf{K}_{qs} \stackrel{def}{=} \{\langle {^{Var}M}, \stackrel{x}{\equiv}, S_y^x \rangle_{x,y \in Var} : M \text{ is a set}\}.$$

The subscript "qs" refers to the presence of quantification "$\exists x$" and substitution "$[x/y]$". Let ML_{qs} be the propositional multi-modal logic determined by the class K_{qs} of frames. The modalities are "$\exists x$" and "$[x/y]$".

ML_{qs} is (a propositional modal logic which is) *equivalent* to FOL without equality, cf. Henkin et al. 1985, Németi 1991 RSC_α and SC_α in the last subsection of §5, Marx and Venema to appear, etc.[1]

---

[1] FOL is actually an alphabetic variant of ML_{qs}. In more detail, assume e.g. that $Var = \{v_0, \ldots, v_n\}$. Then the translation from ML_{qs} to FOL goes as follows: $p_i \mapsto R_i(v_0, \ldots, v_n)$, $S_y^x \varphi \mapsto \varphi[x/y]$ where $\varphi[x/y]$ is obtained from φ by replacing all free occurrences of x with y, $\Diamond_x \varphi \mapsto \exists x \varphi$. In the other direction, the only nontrivial step is translating formulas like $R(v_1 v_0)$ or $R(v_1 v_1)$ to ML_{qs}. For this we first bring FOL formulas to "normal form" by $R(v_1 v_1) \mapsto S_1^0 R(v_0 v_1)$, $R(v_1 v_0) \mapsto S_0^2 S_0^1 S_2^0 R(v_0 v_1)$ and then we apply the above indicated translation procedure (i.e. $\text{ML}_{qs} \mapsto FOL$) backwards. For more detail, and for the case of $|Var| \geq \omega$ the reader is referred to the above cited works.

The logic ML_{qs} is our starting point (in our analysis) but is not our destination. The modal algebras obtained from ML_{qs} the standard way form a distinguised brand of cylindric algebras called representable substitution (oriented) cylindric algebras, cf. e.g. Németi 1991, end of §5. Being equivalent with FOL, all the "expensive" properties (i)–(iv) listed way above are inherited by ML_{qs}. An implicit negative property of FOL surfaces, too, in the form of a theorem saying that ML_{qs} is not axiomatizable by any finite schema of axioms and rules, cf. Andréka 1992, Sain and Thompson 1991 (cf. properties (ii), (iv) of FOL listed above).[2]

Let us have a closer look at ML_{qs} trying to identify the causes of expenses (i)–(iv). What are the elements (possible states) of our frames? Well, they are evaluations $k \in {}^{Var}M$, and this seems to be natural (and will turn out to be harmless). However, the next requirement, saying that all frames are full Cartesian spaces (${}^{Var}M$, for some M), is not at all harmless, and is not so indispensable either for keeping our intuition concerning quantifier logics.[3] This requirement is equivalent with saying that if in some frame V evaluations k, f are available then all evaluations in ${}^{Var}(Rng(k) \cup Rng(f))$ are available in V, too (cf. Németi 1986). If we soften this "full space" requirement to requiring only that if k is available then all evaluations in ${}^{Var}Rng(k)$ are available, then we obtain the so called "locally full" or *locally square* frames which will turn out to cure almost all of our "expenses". E.g. if we extend ML_{qs} to locally square frames[4], then it becomes decidable, it will enjoy the finite model property, its admissible rules become decidable and so do the admissible sequents. All these are corollaries of Theorem 2.6 below. The remaining expenses in (i)–(iv) can be covered by fine tuning our approach in any one of the following two directions.

(I) Focusing on locally square frames is only one way of selecting nice kinds of frames of the form $V \subseteq {}^{Var}M$. Permitting many choices of $V \subseteq {}^{Var}M$ can serve purposes entirely *different* from being resource conscious in connection with expenses (i)–(iv). E.g. in the theory of generalized

---

[2]Let ML_s be obtained from ML_{qs} by omitting the modalities "$\exists x$". Then, ML_s is the modal logic of substitutions (cf. e.g. Venema 1995). A complete finite schema axiomatization of ML_s is given in Sági and Németi 1995, where it is also shown that this axiomatization is not strong enough for generating all admissible rules of ML_s. Decidability of validities of ML_s is proved in Sági and Németi 1995 by using the main result of the present paper. The Craig interpolation property seems to hold both for ML_s and for its finite variable fragments, cf. Madarász 1995d. These positive properties of ML_s point in the direction that the reasons of our expenses (i)–(iv) reside in the modality "$\exists x$" and perhaps in its interaction with "$[x/y]$" but not in the purely substitutional part.

[3]For the same conclusion (with sometimes different motivation) we refer to van Benthem to appear, Andréka et al. to appear, Andréka et al. 1995b, Németi 1986, Németi 1995, Alechina and van Lambalgen 1995, Alechina 1995.

[4]Or equivalently, dilute the semantics of ML_{qs} with the locally square frames.

quantifiers or dependent quantification the choice of V can reveal how quantification over say x and y might interact, cf. Alechina 1995, van Benthem and Alechina to appear, Alechina and van Lambalgen 1995. We will investigate the possible ways of selecting the right kinds of V's (frames) in the form of looking at the so called Crs_α-version, D_α-version, and G_α-version of our generalized semantics for FOL and of our algebras, too.

(II) By permitting frames like $V \subseteq {}^{Var}M$, some of the logical connectives and linguistic features which were derivable in ML_{qs} from $\exists x$ and $[x/y]$ will cease to be derivable. E.g. polyadic (or simultaneous) quantification ($\exists xy$) is no more derivable from $\exists x$ and $\exists y$, and the same applies to simultaneous substitutions like $[x/y, y/z]$ (not to mention $[x/y, y/x]$). Therefore, we will expand our modal logic ML_{qs} with (0) equality, (1) polyadic quantifiers, (2) simultaneous substitutions, and (3) atomic formulas of the form $R(zyx)$, $R(xxx)$, etc. with individual variables rearranged in arbitrary order.[5]

We will prove decidability and finite model property for this richer logic and also for admissible sequents, etc. as indicated above. The results extend to the "Crs_α-style", "D_α-style" and "G_α-style" choices of our frames $V \subseteq {}^{Var}M$.

For lack of space, we will only briefly and incompletely indicate statuses of the "expenses" in (i)–(iv) other than decidability. Although the approach taken in this paper is semantical, it can be formulated and pursued in a syntactical form, too. By this we mean to say that the approach of refining the notion of a model has a dual form, where we restrict the set of formulas of FOL (and leave the semantics unchanged). Indeed, this approach is discussed in Andréka et al. to appear, Andréka et al. 1995b, Németi 1986, Németi 1995.

Johan van Benthem formulated a question which in the present context can be reformulated as follows. "The results of the present approach seem to be too good. Namely, by Theorem 2.6 herein, both the locally square version of ML_{qs} and its expansions discussed above have the finite model property. This implies that infinity is not expressible in these logics. But infinity is expressible in FOL. So, we sacrificed some of the expressive power of FOL in order to get rid of expenses (i)–(iv). The question is whether this sacrifice was necessary." An answer to this question is the following. E.g. we can add cardinality quantifiers $\exists^2 x, \exists^3 x, \ldots, \exists^n x \ldots$ to ML_{qs} before "relativization" i.e., before relaxing the squareness condition $V = {}^{Var}M$ on the frames. (Here $\exists^2 x \varphi$ means that there are at least 2 different choices

---

[5] We will see that all these additions (0)–(3) can be handled in the framework of multi-modal logics, see section 3 (if a logic is proved equivalent with a quasi-variety of Boolean algebras with operators (BAO's), then this automatically means equivalence with a multi-modal logic because every quasivariety of BAO's is equivalent with a multi-modal (propositional) logic).

of x satisfying φ and similarly for $\exists^n x \varphi$.) After relativization, infinity (and related concepts) remains expressible while the new version of FOL becomes decidable, etc, cf. Mikulás and Németi 1995. (Of course, this logic does not have the finite model property.)[6]

2 First-Order Logic with Generalized Semantics

The syntax (set of formulas) will basically be the same as that of usual first-order logic (FOL). We will have equality in our language, but no constant or function symbols, only a set of (finitary) relation symbols. The only difference in syntax will be that we will explicitly introduce connectives that usually are treated as derived, defined ones in FOL. The semantics of our language will be a slight generalization of that of first-order logic: our models will differ from the usual ones only in that not arbitrary evaluations of variables are allowed, a model prescribes which evaluations of variables are "admissible" or allowed in that model.

We now begin to define our generalized semantics in detail.

Definition 2.1 (First-order logic with generalized semantics)

Syntax. Let $\mathcal{R}$ be a set, we will call it the set of *relation symbols*. Let $ar : \mathcal{R} \longrightarrow \omega$ be a function, where ω is the set of natural numbers. We will call $ar(R)$ the *arity*, or *rank*, of the relation symbol R. Let Var be an arbitrary nonempty set, disjoint from $\mathcal{R}$. We will call the elements of Var *variables*, thus Var is the set of variables.

The set of *atomic formulas* is

$$\{R(x_1 \ldots x_n) : R \in \mathcal{R}, \ n=ar(R), \ x_1,\ldots,x_n \in Var\} \cup \{x = y : x, y \in Var\}.$$

Our *logical connectives* are the Boolean ones "and" ($\wedge$), "not" ($\neg$), together with "polyadic quantifiers" ($\exists x_1 \ldots x_n$) and " simultaneous substitution" ($[x_1/y_1, \ldots, x_n/y_n]$). The set $Fm^p(\mathcal{R}, Var)$ is the smallest set Fm such that

each atomic formula is in Fm

$\varphi \wedge \psi, \ \neg \varphi, \ \exists x_1 \ldots x_n \ \varphi, \ \varphi[x_1/y_1, \ldots, x_n/y_n] \in Fm,$

whenever $\varphi, \psi \in Fm$ and $x_1, \ldots, x_n, y_1, \ldots, y_n \in Var$.

We will treat other Boolean connectives ($\vee, \longrightarrow, \leftrightarrow$, etc.) and the universal quantifiers ($\forall x_1 \ldots x_n$) as derived, defined connectives, e.g.

$\forall x_1 \ldots x_n \ \varphi$ denotes $\neg \exists x_1 \ldots x_n \neg \varphi.$

Semantics. Usual first-order models will be denoted by $\mathfrak{M}, \mathfrak{N}$, etc. If $\mathfrak{M}$ is a first-order model, then M (the Roman letter corresponding to the German one) denotes its universe (a nonempty set), and $R^{\mathfrak{M}}$ denotes the

---

[6] Apparently, this approach of turning negative properties into positive by relativization seems to work in the even more general case of $\exists^\kappa x \varphi$ where κ is an arbitrary but fixed cardinal.

concrete relation assigned by $\mathfrak{M}$ to the relation symbol $R \in \mathcal{R}$, i.e., if $ar(R) = n$, then $R^{\mathfrak{M}}$ is an n-ary relation over M. If $\mathfrak{M}$ is a first-order model, then our notation will be

$$\mathfrak{M} = \langle M, R^{\mathfrak{M}} \rangle_{R \in \mathcal{R}}.$$

$Mod(\mathcal{R})$, or Mod if $\mathcal{R}$ is understood from context, denotes the class of all models.

A *generalized model* is defined to be a pair $\langle \mathfrak{M}, V \rangle$, where $\mathfrak{M}$ is a first-order model, and V is a set of evaluations of the variables in M, i.e., functions mapping Var into M. $Mod^g(\mathcal{R}, Var)$ denotes the class of all generalized models. If

$$\underline{\mathfrak{M}} = \langle \mathfrak{M}, V \rangle$$

is a generalized model, then we call $\mathfrak{M}$ the *model part* of $\underline{\mathfrak{M}}$, we call M the *universe* of $\underline{\mathfrak{M}}$, V is the *evaluations part* of $\underline{\mathfrak{M}}$, and we call the elements of V *admissible evaluations*.

Truth of formulas in generalized models is defined very much the same way as in usual first-order logic. If $\underline{\mathfrak{M}}$ is a generalized model, k is an admissible evaluation of variables in $\underline{\mathfrak{M}}$, and φ is a formula, then $\underline{\mathfrak{M}} \models^g \varphi[k]$ denotes that "the formula φ is true in the model $\underline{\mathfrak{M}}$ under the evaluation k of variables". Its definition is as follows: Let $\underline{\mathfrak{M}} = \langle \mathfrak{M}, V \rangle \in Mod^g(\mathcal{R}, Var)$ and let $k \in V$. Then

$\underline{\mathfrak{M}} \models^g R(x_1 \ldots x_n)[k]$ iff $\langle k(x_1), \ldots, k(x_n) \rangle \in R^{\mathfrak{M}}$

$\underline{\mathfrak{M}} \models^g x = y[k]$ iff $k(x) = k(y)$

$\underline{\mathfrak{M}} \models^g (\varphi \wedge \psi)[k]$ iff $\underline{\mathfrak{M}} \models^g \varphi[k]$ and $\underline{\mathfrak{M}} \models^g \psi[k]$

$\underline{\mathfrak{M}} \models^g \neg\varphi[k]$ iff $\underline{\mathfrak{M}} \models^g \varphi[k]$ does not hold

$\underline{\mathfrak{M}} \models^g \exists x_1 \ldots x_n \varphi[k]$ iff $\underline{\mathfrak{M}} \models^g \varphi[k']$ for some $k' \in V$ which differs from k at most on $x_1, \ldots, x_n$, i.e., $(\forall x \in Var - \{x_1, \ldots, x_n\})\ k(x) = k'(x)$

$\underline{\mathfrak{M}} \models^g \varphi[x_1/y_1, \ldots, x_n/y_n][k]$ iff $\underline{\mathfrak{M}} \models^g \varphi[k']$ and $k' \in V$ for k' defined as $k'(x_i) = k(y_{\max\{j : x_j = x_i\}})$ for $1 \leq i \leq n$ and $k'(x) = k(x)$ for all $x \in Var - \{x_1, \ldots, x_n\}$.

We denote *first-order logic with generalized semantics* by $\mathrm{FOL}^g(\mathcal{R}, Var)$, in more detail

$$\mathrm{FOL}^g(\mathcal{R}, Var) \stackrel{\text{def}}{=} \langle Fm^p(\mathcal{R}, Var), Mod^g(\mathcal{R}, Var), \models^g \rangle.$$

We will omit $\mathcal{R}$, or Var, from the notation when they are understood from context, and we will e.g. write

$$\mathrm{FOL}^g = \langle Fm^p, Mod^g, \models^g \rangle,$$

or we will write FOL^g(*Var*), etc. By this, we have defined our first-order logic with generalized semantics. □

Before going on, we make some remarks on our notation and on connection with usual first-order logic (FOL).

Remark 2.2 (i) FOL^g is a generalization of FOL in the sense that FOL is equivalent to a sublogic of FOL^g. Namely, to a first-order model $\mathfrak{M}$ there corresponds the generalized model $\langle \mathfrak{M}, {}^{Var}M \rangle$ (where ${}^{Var}M$ denotes the set of all evaluations of the variables *Var* in the model $\mathfrak{M}$), in the following sense: To any first-order formula φ, any first-order model $\mathfrak{M}$ and evaluation k of the variables,

$$\mathfrak{M} \models \varphi[k] \quad \text{iff} \quad \langle \mathfrak{M}, {}^{Var}M \rangle \models^g \varphi[k].$$

(ii) The upper index "g" in FOL^g refers to "**g**eneralized". In some way, the truth is the same as that of FOL, except that it is "relativized" to the set of admissible evaluations. This will become clearer in our algebraic setting (in section 3). Relativizing is a general method for "taming a logic". More on this can be found e.g. in Mikulás 1995b, Marx 1995.

(iii) The upper index "p" in $Fm^p(\mathcal{R}, Var)$ refers to "**p**olyadic". We have polyadic quantifiers $\exists x_1 \ldots x_n$ in place of just "normal" quantifiers $\exists x$. In first-order logic $\exists x_1 \ldots x_n$ is expressible with $\exists x_1, \ldots, \exists x_n$, because for all formulas φ,

(1) $\quad \exists x_1 \ldots x_n \varphi$ is equivalent in FOL with $\exists x_1 \ldots \exists x_n \varphi$.

This is no longer true in FOL^g. (In the above, by (1) we mean that

$$\langle \mathfrak{M}, {}^{Var}M \rangle \models^g (\exists x_1 \ldots x_n \varphi \leftrightarrow \exists x_1 \ldots \exists x_n \varphi)[k],$$

for any formula φ.)

Also, substitution $[x_1/y_1, \ldots, x_n/y_n]$ is expressible in FOL (if *Var* is infinite), e.g. if x and y are different, then $\varphi[x/y]$ is equivalent in FOL with $\exists x(x = y \wedge \varphi)$. If some of $y_1, \ldots, y_n$ occurs in $\{x_1, \ldots, x_n\}$, then the corresponding formula will be somewhat more complicated, because we have to deal with "clash of variables", but still expressible without the "substitution" operation.

The expressibility (in FOL) of substitution by the quantifiers and equality was first noticed and used by Tarski 1965 to eliminate interdependence between the atomic formulas, which then made FOL into a (modal) propositional logic, and which also made real algebraization possible. For more on this see e.g. Henkin et al. 1971, 1985 section 4.3.

(iv) Our "connectives" $\exists x_1 \ldots x_n, [x_1/y_1, \ldots, x_n/y_n]$ are modalities in the sense that they distribute over disjunction, i.e.,

$$\exists x_1 \ldots x_n(\varphi \vee \psi) \leftrightarrow ((\exists x_1 \ldots x_n \varphi) \vee (\exists x_1 \ldots x_n \psi))$$
$$(\varphi \vee \psi)[x_1/y_1, \ldots, x_n/y_n] \leftrightarrow (\varphi[x_1/y_1, \ldots, x_n/y_n] \vee \psi[x_1/y_1, \ldots, x_n/y_n])$$

are true in each generalized model under all evaluations.

The substitution connective $\varphi[x_1/y_1, \ldots, x_n/y_n]$ is also introduced in van Benthem to appear, where it is denoted by $[x_1/y_1, \ldots, x_n/y_n]\varphi$ to highlight its modal operator character. □

Definition 2.3 (Validities, admissible rules, admissible sequents)

(i) Following the usual terminology (cf. e.g. in Andréka et al. 1995a, Andréka et al. 1994), if $\mathfrak{M} = \langle \mathfrak{M}, V \rangle$ is a generalized model and φ is a formula, then $\mathfrak{M} \models^g \varphi$ denotes that "φ is valid in the generalized model $\mathfrak{M}$", $\models^g \varphi$ denotes that "φ is valid in the generalized semantics", $Th \models^g \varphi$, if Th is a set of formulas, denotes that "φ is a consequence of Th", and $\mathsf{K} \models^g \varphi$ if K is a class of generalized models, denotes that "φ is valid in K". In more detail

$\mathfrak{M} \models^g \varphi$ iff $\mathfrak{M} \models^g \varphi[k]$ for all $k \in V$,

$\models^g \varphi$ iff $\mathfrak{M} \models^g \varphi$ for all $\mathfrak{M} \in Mod^g$,

$\mathfrak{M} \models^g Th$ iff $\mathfrak{M} \models^g \psi$ for all $\psi \in Th$,

$Th \models^g \varphi$ iff $\mathfrak{M} \models^g Th$ implies $\mathfrak{M} \models^g \varphi$ for all $\mathfrak{M} \in Mod^g$, and

$\mathsf{K} \models^g \varphi$ iff $\mathfrak{M} \models^g \varphi$ for all $\mathfrak{M} \in \mathsf{K}$.

(ii) Let $\varphi_1, \ldots, \varphi_n, \psi \in Fm^p$. We say that $\varphi_1, \ldots, \varphi_n/\psi$ is an *admissible* or *valid rule* of FOL^g, if for all models $\mathfrak{M} \in Mod^g$,

$$\mathfrak{M} \models^g \{\varphi_1, \ldots, \varphi_n\} \text{ implies } \mathfrak{M} \models^g \psi.$$

(If $n = 0$ in the above, then $/\psi$ means that ψ is valid, i.e., $\models^g \psi$.)
Let $\varphi_1, \ldots, \varphi_n, \psi_1, \ldots, \psi_k \in Fm^p$. $\varphi_1, \ldots, \varphi_n \longrightarrow \psi_1, \ldots, \psi_k$ is an *admissible* or *valid sequent* of FOL^g, if for all models $\mathfrak{M} \in Mod^g$,

$$\mathfrak{M} \models^g \{\varphi_1, \ldots, \varphi_n\} \text{ implies that } \mathfrak{M} \models^g \psi_i \text{ for some } 1 \leq i \leq k. \quad \square$$

Examples for formulas valid in the generalized semantics are $\exists xy\varphi \leftrightarrow \exists yx\varphi$ and $\varphi[x/y, x/z] \leftrightarrow \varphi[x/z]$. Examples for formulas not valid in the generalized semantics are $\exists x \exists y R(xy) \leftrightarrow \exists y \exists x R(xy)$ and $R(yx) \leftrightarrow R(xy)[x/y, y/x]$. In this paper we will give an algorithm for deciding about a formula whether it is valid in the generalized semantics or not (cf. Theorem 4.8). We will do the same for rules and sequents. We will also prove that if a formula is valid in all models with a finite universe, then it is valid in all models (see Theorem 2.6).

Remark 2.4 (More on the definition of generalized models.)
(i) Making restrictions on available evaluations of variables is familiar in the literature, e.g. this is what happens in many-sorted logic.
(ii) By selecting some set of evaluations and forbidding the rest, we can express dependence of variables from each other. E.g. if $V = \{k \in {}^{Var}M : k \text{ is one-to-one}\}$, then we are allowed only to choose a value for a variable which is different from the values of all the others. E.g. if $Var = \{v_i : i \in \omega\}$ and if we choose the value 1 to v_0, then we cannot choose the value 1 for v_1 any longer, but if we choose 0 for the value of v_0, then we can choose 1 as a value of v_1. For more on this feature, see Alechina and van Lambalgen 1995.
(iii) (Analogies with Henkin-type semantics for second order logic.) We can view our generalized semantics as adding some new models to the existing ones. We can say that a model $\mathfrak{M} = \langle \mathfrak{M}, V \rangle$ is "real" if $V = {}^{Var}M$, while it is "pseudo" if $V \subsetneq {}^{Var}M$. The point here is that "pseudo-models" are very similar to real ones in flavour. Adding "pseudo-models" to a logic in order to change some of its properties is also familiar from the literature. E.g. Henkin's generalized semantics for second (or higher) order logic does exactly the same, in order to make the set of validities recursively enumerable (from non-recursively-enumerable). See Henkin 1950. In that case we make restrictions on subsets of the universe, we "allow" some, these are the admissible subsets, and we forbid the others. A "real" model is one, where all subsets of the model are "allowed", and pseudo-models are the ones where not all the subsets of the universe are allowed.

The same thing is done in other branches of mathematics, e.g. in computer science. See Andréka et al. 1979, Andréka et al. 1981, Andréka et al. 1982. □

First, (in Theorem 2.5) we show that for checking validity of a formula φ in FOL^g, it is enough to consider evaluations of variables occurring in φ. This means the following. Let Var be the set of variables occurring in φ. Then for any set Var' such that $\varphi \in Fm^p(Var')$, φ is a validity of FOL$^g(Var')$ iff φ is a validity of FOL$^g(Var)$. (The same can be done w.r.t. "truth of φ in a model $\mathfrak{M}$", but we will not go into the details of that now.)

Theorem 2.5 Let $Var \subseteq Var'$ and $\varphi \in Fm^p(Var)$. Then
$$Mod^g(Var) \models^g \varphi \quad \text{iff} \quad Mod^g(Var') \models^g \varphi.$$

Proof. The proof is very similar to that of Theorem 4.13(i) in Németi 1995, we omit it. □

Theorem 2.5 above also states that the semantics of FOL^g behaves well w.r.t. "variable-fragmentation". I.e., it says that the set of validities of FOL^g(Var) is the set of validities of FOL^g(Var') intersected with Fm^p(Var).

Theorem 2.6 (Decidability of FOL^g) (i) *The sets of valid formulas, admissible rules and admissible sequents of* FOL^g *are all decidable.*

(ii) FOL^g *has the finite model property. I.e., if a formula fails in a model, then it also fails in a model with finite universe.*

Proof. The theorem will follow from statements proved in the next sections. In more detail, (i) follows from Lemma 3.3 and Theorem 4.1(i); and (ii) follows from Theorem 2.5, Lemma 3.3, Corollary 4.11 and Lemma 3.1. □

Remark 2.7 (Craig interpolation property for FOL^g and its fragments) We will briefly review this subject in Section 4. The reason for postponing it is that formulating the results in algebraic form will take less space. We will use the fact proved in Madarász 1995b, 1995c that a multi-modal logic $\mathcal{L}$ has the Craig interpolation property iff the class Alg($\mathcal{L}$) of modal algebras associated with $\mathcal{L}$ enjoys the so called super amalgamation property (SUPAP). For the normal case this was also established in Marx 1995. □

3 Polyadic Cylindric-Relativized Set Algebras and Connections with Generalized Semantics

In this section we define a class of algebras called Crs$^p_\alpha$, and show its connections with FOL^g($\mathcal{R}, Var$). Using this connections, we reduce decidability of FOL^g($\mathcal{R}, Var$) to decidability of the universal theory of Crs$^p_\alpha$, which we then will prove in the next section.

Polyadic Cylindric-Relativized Set Algebras

Let α and U be any sets. By a U-*termed* α-*sequence* we mean a function mapping α into U. If $s : \alpha \longrightarrow U$, then we also write $s = \langle s(i) : i \in \alpha \rangle$. By an α-*ary relation* (over U) we mean a set of (U-termed) α-sequences. $^\alpha U$ denotes the set of all functions mapping α into U. Thus R is an α-ary relation on U iff $R \subseteq {}^\alpha U$, and R is an α-ary relation iff $R \subseteq {}^\alpha U$ for some U. Clearly, the intersection and union of two α-ary relations are again α-ary relations. Now we define some other operations on α-ary relations. Let $\tau : \alpha \longrightarrow \alpha$ (i.e., $\tau \in {}^\alpha \alpha$), let $\Gamma \subseteq \alpha$ and let $s, z \in {}^\alpha U$. Then $s \circ \tau$, the *composition* of the functions s and τ, is again an α-sequence,

$$s \circ \tau = \langle s(\tau(i)) : i \in \alpha \rangle.$$

Thus $s \circ \tau$ is the sequence s "rearranged according to τ". We define the equivalence relation
$$\stackrel{\Gamma}{\equiv}$$
on α-sequences as follows:

$$s \stackrel{\Gamma}{\equiv} z \quad \text{iff} \quad (\forall i \in \alpha - \Gamma)\; s(i) = z(i).$$

Thus $s \stackrel{\Gamma}{\equiv} z$ denotes that s, z agree "off Γ", or that s and z are "Γ-neighbours". Let now V be an α-ary relation, let $X \subseteq V$, and let $i, j \in \alpha$. Then

$$S^V_{\tau,\Gamma}(X) \stackrel{\text{def}}{=} \{s \in V : (s \circ \tau) \stackrel{\Gamma}{\equiv} z \text{ for some } z \in X\}, \text{ and}$$

$$D^V_{ij} \stackrel{\text{def}}{=} \{s \in V : s(i) = s(j)\}.$$

Let A be a set of α-ary relations, such that A has a biggest element V, and A is closed under the Boolean operations, $S^V_{\tau,\Gamma}$ and D^V_{ij}. This means that for all $X, Y \in A$ and for all $\tau \in {}^\alpha \alpha$, $\Gamma \subseteq \alpha$ and $i, j \in \alpha$ we have that $X \cap Y$, $V - X$, $S^V_{\tau,\Gamma}(X)$ and D^V_{ij} are also in A. Then

$$\mathfrak{A} = \langle A, \cap, -, S^V_{\tau,\Gamma}, D^V_{ij} \rangle_{\tau \in {}^\alpha\alpha, \Gamma \subseteq \alpha, i,j \in \alpha}$$

is an algebra. (Actually, $\mathfrak{A}$ is an algebra whose elements are α-ary relations, and whose operations are our naturally defined operations on α-ary relations.) We will call such an algebra an (α-*dimensional*) *polyadic cylindric-relativized set algebra*, a Crs^p_α, and Crs^p_α denotes the class of all such algebras. I.e.,

$$\mathsf{Crs}^p_\alpha = \{\langle A, \cap, -, S^V_{\tau,\Gamma}, D^V_{ij}\rangle_{\tau \in {}^\alpha\alpha, \Gamma \subseteq \alpha, i,j \in \alpha} : A \text{ is a set of } \alpha\text{-ary}$$
$$\text{relations closed under the listed operations}\}.$$

Let $id : \alpha \longrightarrow \alpha$ be the identity mapping. Then

$$S^V_\tau(X) \stackrel{\text{def}}{=} S^V_{\tau,\emptyset}(X) = \{s \in V : s \circ \tau \in X\}, \text{ and}$$

$$C^V_{(\Gamma)}(X) \stackrel{\text{def}}{=} S^V_{id,\Gamma}(X) = \{s \in V : s \stackrel{\Gamma}{\equiv} z \text{ for some } z \in X\}.$$

In the literature S^V_τ and $C^V_{(\Gamma)}$ are usually called substitution operation and generalized cylindrification, respectively. Thus the operation $S^V_{\tau,\Gamma}$ "codes together" the substitutions and the cylindrifications. $S^V_{\tau,\Gamma}$ is not expressible from S^V_τ and $C^V_{(\Gamma)}$ in general, i.e., for any V. However, if V is a "straightenable unit", i.e., if $V \subseteq C^V_{(\{i\})} D^V_{ij}$ for all $i, j \in \alpha$, then

$$S^V_{\tau,\Gamma}(X) = S^V_\tau C^V_{(\Gamma)}(X), \quad \text{for all } X \subseteq V.$$

What we will prove in section 4 about Crs^p_α will be inherited by all generalized subreducts of Crs^p_α.

In section 4 we will investigate the class Crs^p_α of algebras from an algebraic point of view. E.g. we will prove that a universal (i.e., quantifier-free) formula is valid in Crs^p_α iff it is valid in finite members of Crs^p_α, and the set $Univ(\mathsf{Crs}^p_\alpha)$ of all universal formulas valid in Crs^p_α is decidable. In the next part of the present section we will show the connection of the semantics of FOL^g with Crs^p_α, cf. Lemma 3.3, and then we will use the theorem stating that $Univ(\mathsf{Crs}^p_\alpha)$ is decidable to prove that the set of validities, the set of admissible rules, and the set of admissible sequents of FOL^g are all decidable. Similarly, by the same methods, we will show that the fact that any finite Crs_α^p is isomorphic to one with a finite biggest element implies finite model property of FOL^g.

Connection between Crs_α^p and FOL^g

We now turn to working out the connections between FOL^g and Crs_α^p. Let φ be a formula in FOL^g, and let $\mathfrak{M} = \langle \mathcal{M}, V \rangle \in Mod^g$. Let $\alpha \stackrel{\text{def}}{=} Var$. We define the α-ary relation $\varphi^{\mathfrak{M}}$ over $\mathcal{M}$ as follows:

$$\varphi^{\mathfrak{M}} \stackrel{\text{def}}{=} \{s \in V : \mathfrak{M} \models^g \varphi[s]\}.$$

Let $id : \alpha \longrightarrow \alpha$ be the identity function, i.e., $id(x) = x$ for all $x \in Var$. For any $x_1, \ldots, x_n, y_1, \ldots, y_n$, let the function $[[x_1/y_1, \ldots, x_n/y_n]] : \alpha \longrightarrow \alpha$ be defined as:

$$[[x_1/y_1, \ldots, x_n/y_n]](x_i) \stackrel{\text{def}}{=} y_{\max\{j : x_j = x_i\}} \quad \text{for all } 1 \leq i \leq n, \text{ and}$$

$$[[x_1/y_1, \ldots, x_n/y_n]](x) \stackrel{\text{def}}{=} x \quad \text{for all } x \in Var - \{x_1, \ldots, x_n\}.$$

Now the following are not difficult to check, for any $\tau \in {}^\alpha\alpha$.

$$R(\tau(x_1) \ldots \tau(x_n))^{\mathfrak{M}} \cap S^V_{\tau, \alpha - \{x_1, \ldots, x_n\}}(V) =$$
$$S^V_{\tau, \alpha - \{x_1, \ldots, x_n\}}(R(x_1 \ldots x_n)^{\mathfrak{M}}),$$
$$(x = y)^{\mathfrak{M}} = D^V_{xy},$$
$$(\varphi \wedge \psi)^{\mathfrak{M}} = \varphi^{\mathfrak{M}} \cap \psi^{\mathfrak{M}},$$
$$(\neg\varphi)^{\mathfrak{M}} = V - \varphi^{\mathfrak{M}},$$
$$(\exists x_1 \ldots x_n \varphi)^{\mathfrak{M}} = S^V_{id, \{x_1, \ldots, x_n\}}(\varphi^{\mathfrak{M}}),$$
$$(\varphi[x_1/y_1, \ldots, x_n/y_n])^{\mathfrak{M}} = S^V_{[[x_1/y_1, \ldots, x_n/y_n]], \emptyset}(\varphi^{\mathfrak{M}}).$$

Next, let $\mathfrak{A}$ be an arbitrary Crs_α^p with biggest element V and with universe A. Let

$$X \stackrel{\text{def}}{=} \{R(x_1 \ldots x_n) : R \in \mathcal{R}, \, ar(R) = n, \, x_1, \ldots, x_n \in Var\}.$$

We will use X as a set of variables ranging over elements of algebras. Let $k : X \longrightarrow A$ be arbitrary. If s is an α-sequence, then $Rng(s)$ denotes the range of s, i.e., $Rng(s) = \{s(i) : i \in \alpha\}$. Let

$$M \stackrel{\text{def}}{=} \cup \{Rng(s) : s \in V\},$$
$$R^{\mathfrak{M}} \stackrel{\text{def}}{=} \{\langle s(x_1), \ldots, s(x_n)\rangle : s \in k(R(x_1 \ldots x_n)), \, x_1, \ldots, x_n \in Var\}$$
where $n = ar(R)$,
$$\mathcal{M} \stackrel{\text{def}}{=} \langle M, R^{\mathfrak{M}} \rangle_{R \in \mathcal{R}}, \quad \mathfrak{M} \stackrel{\text{def}}{=} \langle \mathcal{M}, V \rangle.$$

We will call $\mathfrak{M}$ the model corresponding to the pair $\langle \mathfrak{A}, k \rangle$, in notation $\mathfrak{M} = \mathfrak{M}(\mathfrak{A}, k)$.

Lemma 3.1 *Let $\mathfrak{A} \in \mathsf{Crs}_\alpha^p$ with biggest element V, and let $k : X \longrightarrow \mathfrak{A}$ satisfy*

$$k(R(\tau(x_1) \ldots \tau(x_n))) \cap S^V_{\tau, \alpha - \{x_1, \ldots, x_n\}}(V) =$$

(2) $\quad S^V_{\tau,\alpha-\{x_1,\ldots,x_n\}}(k(R(x_1\ldots x_n)))$,

for all $R \in \mathcal{R}$, $x_1,\ldots,x_n \in Var$, $\tau \in {}^\alpha\alpha$, $n = ar(R)$.

Then $R(x_1\ldots x_n)^{\underline{\mathfrak{M}(\mathfrak{A},k)}} = k(R(x_1\ldots x_n))$, *for all* $R(x_1\ldots x_n) \in X$.

Proof. Let $\mathfrak{A}$, V and k be as in the hypothesis. Let $\mathfrak{M} = \mathfrak{M}(\mathfrak{A},k)$. In this proof we will write $\bar{x}$ for $\langle x_1,\ldots,x_n\rangle$, $\tau(\bar{x})$ for $\langle\tau(x_1),\ldots,\tau(x_n)\rangle$, $s(\bar{x})$ for $\langle s(x_1),\ldots,s(x_n)\rangle$, etc. Assume $s \in k(R(\bar{x}))$. Then $s(\bar{x}) \in R^{\underline{\mathfrak{M}}}$ by the definition of $R^{\underline{\mathfrak{M}}}$, hence $\mathfrak{M} \models^g R(\bar{x})[s]$, i.e., $s \in R(\bar{x})^{\underline{\mathfrak{M}}}$. This shows $k(R(\bar{x})) \subseteq R(\bar{x})^{\underline{\mathfrak{M}}}$.

To show the other direction, let $s \in R(\bar{x})^{\underline{\mathfrak{M}}}$, i.e., $s(\bar{x}) \in R^{\underline{\mathfrak{M}}}$. Then $s(\bar{x}) = z(\bar{y})$ for some $\bar{y}$ and $z \in k(R(\bar{y}))$, by the definition of $R^{\underline{\mathfrak{M}}}$. Let $\tau : \alpha \longrightarrow \alpha$ be defined by

$$\tau(y_i) \stackrel{\text{def}}{=} x_{\min\{j:z(y_j)=z(y_i)\}}, \text{ and } \tau(x) \stackrel{\text{def}}{=} x \text{ for } x \in Var - \{y_1,\ldots,y_n\}.$$

Then $s \circ \tau \stackrel{\Gamma}{=} z$ for $\Gamma = \alpha - \{y_1,\ldots,y_n\}$, because for $1 \leq i \leq n$, $s \circ \tau(y_i) = s(\tau(y_i)) = s(x_j) = z(y_j) = z(y_i)$ by $s(\bar{x}) = z(\bar{y})$, where $j = \min\{p : y_p = y_i\}$. Then $s \in S^V_{\tau,\alpha-\{y_1,\ldots,y_n\}}(k(R(\bar{y}))) \subseteq k(R(\tau(\bar{y})))$. Let $\sigma : \alpha \longrightarrow \alpha$ be defined by

$$\sigma(x_i) \stackrel{\text{def}}{=} \tau(y_i), \text{ for all } 1 \leq i \leq n,$$

and

$$\sigma(x) \stackrel{\text{def}}{=} x \text{ if } x \in Var - \{x_1,\ldots,x_n\}.$$

This is a function, because if $x_i = x_j$, then $s(x_i) = s(x_j)$, so $z(y_i) = z(y_j)$, so $\tau(y_i) = \tau(y_j)$. Then $\sigma(\bar{y}) = \tau(\bar{x})$, and $s = s \circ \sigma$ because $s(\sigma(x_i)) = s(\tau(y_i)) = z(y_i) = s(x_i)$ if $1 \leq i \leq n$, and if $x \in Var - \{x_1,\ldots,x_n\}$ then $s(\sigma(x)) = s(x)$. Thus

$$s \in k(R(\tau(\bar{y}))) = k(R(\sigma(\bar{x}))) \cap S^V_{\sigma,\alpha-\bar{x}}(V) = S^V_{\sigma,\alpha-\bar{x}}(k(R(\bar{x}))).$$

This means that $s = s \circ \sigma \stackrel{\alpha-\bar{x}}{\equiv} z \in k(R(\bar{x}))$, i.e., $s \in S^V_{id,\alpha-\bar{x}}(k(R(\bar{x}))) \subseteq k(R(id(\bar{x}))) = k(R(\bar{x}))$. □

Now, for any formula $\varphi \in Fm^p(\mathcal{R}, Var)$ with $\mathcal{R}$ and Var finite we will associate a universal formula $\eta(\varphi)$ in the language of Crs^p_{Var}, using the algebraic variables X. The operation symbols of the language of Crs^p_α will be $\cap, -, S_{\tau,\Gamma}, D_{ij}$, i.e., we just omit the upper index. Also, we will use 1 as a constant denoting the biggest element of a Crs^p_α.

Definition 3.2 First we define $\xi(\varphi)$ as follows.

$\xi(R(\bar{x})) \stackrel{\text{def}}{=} R(\bar{x}) \in X$

$\xi(x = y) \stackrel{\text{def}}{=} D_{xy}$

$\xi(\varphi \wedge \psi) \stackrel{\text{def}}{=} \xi(\varphi) \cap \xi(\psi), \quad \xi(\neg\varphi) \stackrel{\text{def}}{=} -\xi(\varphi)$

$\xi(\exists x_1\ldots x_n \varphi) \stackrel{\text{def}}{=} S_{id,\{x_1,\ldots,x_n\}}(\xi(\varphi))$

$$\xi(\varphi[x_1/y_1,\ldots,x_n/y_n]) \stackrel{\text{def}}{=} S_{[[x_1/y_1,\ldots,x_n/y_n]],\emptyset}(\xi(\varphi)).$$

Let $\varrho(\mathcal{R}, Var)$ be the formula

$$\bigwedge \{S_{\tau,\alpha-\{x_1,\ldots,x_n\}}(R(x_1\ldots x_n)) =$$
$$R(\tau(x_1))\ldots\tau(x_n)) \cap S_{\tau,\alpha-\{x_1,\ldots,x_n\}}(1) :$$
$$R \in \mathcal{R},\ n = ar(R),\ x_1,\ldots,x_n \in Var,\ \tau \in {}^{Var}Var\}.$$

Then we define $\eta(\varphi)$ as

$$\varrho(\mathcal{R}, Var) \longrightarrow \xi(\varphi) = 1.$$

Lemma 3.3 *Let $\mathcal{R}$, Var be finite and let $\varphi, \psi, \varphi_1, \ldots, \varphi_n, \psi_1, \ldots, \psi_m \in Fm^p(\mathcal{R}, Var)$. Then (i)–(iii) below hold.*

(i) $\models^g \varphi$ iff $\mathsf{Crs}^p_{Var} \models \eta(\varphi)$.

(ii) $\varphi_1, \ldots, \varphi_n/\psi$ is an admissible rule iff $\mathsf{Crs}^p_{Var} \models (\eta(\varphi_1) \wedge \cdots \wedge \eta(\varphi_n)) \longrightarrow \eta(\psi)$.

(iii) $\varphi_1, \ldots, \varphi_n \longrightarrow \psi_1, \ldots, \psi_m$ is an admissible sequent iff $\mathsf{Crs}^p_{Var} \models (\eta(\varphi_1) \wedge \cdots \wedge \eta(\varphi_n)) \longrightarrow (\eta(\psi_1) \vee \cdots \vee \eta(\psi_m))$.

Proof. Let $\alpha \stackrel{\text{def}}{=} Var$. For any $\mathfrak{M} = \langle M, V\rangle \in Mod^g$, let $\mathcal{P}(V) = \{X : X \subseteq V\}$, and

$$\mathfrak{C}(\mathfrak{M}) \stackrel{\text{def}}{=} \langle \mathcal{P}(V), \cap, -, S^V_{\tau,\Gamma}, D^V_{xy}\rangle_{\tau \in {}^\alpha\alpha,\ \Gamma \subseteq \alpha,\ i,j \in \alpha},$$

and let $k \stackrel{\text{def}}{=} k(\mathfrak{M}) : X \longrightarrow \mathcal{P}(V)$ be defined by $k(R(\bar{x})) \stackrel{\text{def}}{=} R(\bar{x})^{\mathfrak{M}}$. Let $\varrho \stackrel{\text{def}}{=} \varrho(\mathcal{R}, Var)$. Then, as we have already seen, $\mathfrak{C}(\mathfrak{M}) \models \varrho[k(\mathfrak{M})]$. Thus

$$\mathfrak{C}(\mathfrak{M}) \models \eta(\varphi)[k(\mathfrak{M})] \quad \text{iff} \quad \mathfrak{C}(\mathfrak{M}) \models (\xi(\varphi) = 1)[k(\mathfrak{M})].$$

Then it is not hard to see that the latter holds iff

$$\varphi^{\mathfrak{M}} = V \quad \text{iff} \quad \mathfrak{M} \models^g \varphi.$$

We have seen that

(3) $\qquad\qquad \mathfrak{M} \models^g \varphi \quad \text{iff} \quad \mathfrak{C}(\mathfrak{M}) \models \eta(\varphi)[k(\mathfrak{M})].$

Thus $\mathsf{Crs}^p_{Var} \models \eta(\varphi)$ implies $\models^g \varphi$, because $\mathfrak{C}(\mathfrak{M}) \in \mathsf{Crs}^p_{Var}$.

To show the other direction of (i), assume that $\mathsf{Crs}^p_{Var} \not\models \eta(\varphi)$, and we want to show $\not\models^g \varphi$. Let $\mathfrak{A} \in \mathsf{Crs}^p_{Var}$ and $k : X \longrightarrow A$ be such that $\mathfrak{A} \not\models \eta(\varphi)[k]$. Then $\mathfrak{A} \models \varrho[k]$ but $\mathfrak{A} \not\models \xi(\varphi) = 1[k]$. Let $\mathfrak{M} \stackrel{\text{def}}{=} \mathfrak{M}(\mathfrak{A}, k)$. Then $k(R(\bar{x})) = R(\bar{x})^{\mathfrak{M}}$ for all $R(\bar{x}) \in X$ by Lemma 3.1, since $\mathfrak{A} \models \varrho[k]$ means that (2) holds for k. Then $k(\xi(\varphi)) = \varphi^{\mathfrak{M}}$, so $\mathfrak{A} \not\models \xi(\varphi) = 1[k]$ means that $\varphi^{\mathfrak{M}} \neq V$, i.e., $\mathfrak{M} \not\models^g \varphi$.

The proofs of (ii), (iii) are completely analogous, we omit them. $\square$

We note that the connections between FOL^g and Crs^p_α can be exploited in many other ways, see e.g. Németi 1990.

Now that we know that a FOL^g-formula ϕ is FOL^g-valid iff some computable universal formula ϕ' is Crs$_\alpha^p$-valid, we can bring the FOL^g-decision problem to Crs$_\alpha^p$. In the next section we concentrate on this feature of Crs$_\alpha^p$.

4 Decidability of the Universal Theory of Polyadic Cylindric-Relativized Set Algebras

First we introduce the classes Crs$_\alpha$, D$_\alpha$, G$_\alpha$ of cylindric-relativized set algebras investigated in the literature (see e.g. Henkin et al. 1981, 1985, Németi 1985, Monk 1993, Marx 1995, Alechina 1995, Marx and Venema to appear). Let V be an α-ary relation and let $i \in \alpha$. Let, for any $X \subseteq V$,

$$C_i^V(X) \stackrel{\text{def}}{=} \{s \in V : s \stackrel{\{i\}}{\equiv} z \text{ for some } z \in X\} = C_{(\{i\})}^V(X) = S_{id,\{i\}}^V(X).$$

Then

$$\mathsf{Crs}_\alpha \stackrel{\text{def}}{=} \{\langle A, \cap, -, C_i^V, D_{ij}^V\rangle_{i,j\in\alpha} \ : \ A \text{ is a set of } \alpha\text{-ary relations}$$
$$\text{closed under the listed operations}\}.$$

For any $s \in {}^\alpha U$, $i \in \alpha$ and $u \in U$, let $s(i/u)$ denote the sequence we obtain from s by changing its value at i to u and leaving the values of s at the other places unchanged. Now

$$\mathsf{D}_\alpha \stackrel{\text{def}}{=} \{\mathfrak{A} \in \mathsf{Crs}_\alpha : (\forall s \in 1^{\mathfrak{A}})(\forall i, j \in \alpha) \ s(i/s_j) \in 1^{\mathfrak{A}}\},$$
$$\mathsf{G}_\alpha \stackrel{\text{def}}{=} \{\mathfrak{A} \in \mathsf{Crs}_\alpha : (\forall s \in 1^{\mathfrak{A}})(\forall \tau \in {}^\alpha\alpha) \ s \circ \tau \in 1^{\mathfrak{A}}\}.$$

(In the above, $1^{\mathfrak{A}}$ is the greatest element of $\mathfrak{A}$.)

Let α be finite. In this section we will show that the universal theories of Crs$_\alpha^p$, and (hence) those of Crs$_\alpha$, D$_\alpha$, G$_\alpha$ are all decidable. For Crs$_\alpha$, D$_\alpha$, G$_\alpha$ this was asked as an open problem in Németi 1986, and in 1995. For D$_\alpha$ this problem was solved in Marx 1995 Thm.3.3.1, and Marx and Venema to appear contains a solution for G$_\alpha$; both use the method of filtration. In more detail: It was proved in Németi 1986, 1995 Theorem 4.3 that the equational theories of Crs$_\alpha$, D$_\alpha$, G$_\alpha$ are decidable, and it was asked there (see Németi 1995 Remark 4.15(ii)) whether the quasi-equational theories of these classes are decidable, and whether the equational theories of these classes are the same as those of their finite members. Theorem 4.1 below gives an affirmative answer for all these questions.

The notion of "strong decidability" was introduced in Németi 1986, 1995 section 2.7, and then it was asked whether Crs$_\alpha$, D$_\alpha$, G$_\alpha$ are strongly decidable. Theorem 4.1 below also gives an affirmative answer to this question. We note that if K is strongly decidable, then the universal theory of K is decidable, the universal theory of K is the same as that of the finite members of K, and it is decidable about a finite algebra whether it belongs to K or not. In this paper we show that the "mosaic method", which was invented in Németi 1986, is suitable for proving strong decidability. The method of "filtration", often used in modal logic, also proves the above listed corollaries of strong decidability. It would be interesting to investigate the connections between these two methods for proving decidability, i.e., the connections between the "mosaic method" and filtration.

Theorem 4.1 *Let α be finite. Then (i)–(v) below hold.*

(i) *The set of universal formulas valid in Crs_α^p is decidable. Hence, the set of quasi-equations, and the set of equations valid in Crs_α^p are decidable.*

(ii) *There is an algorithm deciding about a finite algebra whether it is isomorphic to a Crs_α^p or not.*

(iii) *A universal formula fails in Crs_α^p iff it fails in some finite Crs_α^p. Moreover, a bound on the size of this finite Crs_α^p can be computed from the formula.*

(iv) *Any finite substructure of a Crs_α^p is a substructure of a finite Crs_α^p. I.e., if $\mathfrak{B} \in \mathsf{Crs}_\alpha^p$, $X \subseteq B$, X is finite, then there is a finite $\mathfrak{A}$, isomorphic to a Crs_α^p, such that $\mathfrak{B} \lceil X = \mathfrak{A} \lceil X$. Here $\mathfrak{B} \lceil X$ denotes the partial algebra we get from $\mathfrak{B}$ by restricting it to X.*

(v) *Statements (i)–(iv) above hold for Crs_α, D_α, G_α in place of Crs_α^p.*

Proof. First we will show that all the statements of the theorem follow from (iii), and then we will concentrate on proving (iii).

(iii)$\Longrightarrow$(i) is straightforward. (i)$\Longrightarrow$(ii): Let $\mathfrak{A}$ be a finite algebra, let $\Delta(\bar{x})$ be the diagram of $\mathfrak{A}$, and let φ be $\forall \bar{x} \neg \Delta(\bar{x})$. Then ($\mathsf{Crs}_\alpha^p \models \varphi$ iff $\mathfrak{A}$ is not isomorphic to a Crs_α^p), and φ is a universal formula. (iii)$\Longrightarrow$(iv): Let $\Delta(\bar{x})$ be the diagram of $\mathfrak{B} \lceil X$, and let φ be $\forall \bar{x} \neg \Delta(\bar{x})$. Then φ fails in $\mathfrak{B} \in \mathsf{Crs}_\alpha^p$, hence by (iii), φ fails in a finite $\mathfrak{A}' \in \mathsf{Crs}_\alpha^p$. The failure of φ in $\mathfrak{A}'$ gives us a subset X' of A' such that $\mathfrak{A}' \lceil X'$ is isomorphic to $\mathfrak{B} \lceil X$. (iii)$\Longrightarrow$(v): It is enough to show that (iii) holds for Crs_α, D_α, G_α, because the arguments given for (iii)$\Longrightarrow$(i),(ii),(iv) work for any class K of algebras (in place of Crs_α^p). The idea here is that if (iii) holds for K, then (iii) holds for the class of any generalized subreducts of K, and Crs_α is a generalized subreduct of Crs_α^p. Instead of defining generalized subreducts in general, we tell the argument for the special case of Crs_α. Let φ be a universal formula in the language of Crs_α and assume that φ fails in some $\mathfrak{A} \in \mathsf{Crs}_\alpha$. Let $V = 1^{\mathfrak{A}}$ and let $\mathfrak{A}' = \langle \mathcal{P}(V), \cap, -, S_{\tau, \Gamma}^V, D_{ij}^V \rangle_{\tau \in {}^\alpha \alpha, \Gamma \subseteq \alpha, i, j \in \alpha}$. Then φ fails in $\mathfrak{A}' \in \mathsf{Crs}_\alpha^p$, because $\mathfrak{A}$ is a subalgebra of a generalized reduct of $\mathfrak{A}'$. Thus φ fails in a finite $\mathfrak{B}' \in \mathsf{Crs}_\alpha^p$. Let $\mathfrak{B} = \langle B, \cap, -, C_{(\{i\})}, D_{ij} \rangle_{i,j \in \alpha}$. Then $\mathfrak{B} \in \mathsf{Crs}_\alpha$ and φ fails in $\mathfrak{B}$ (because φ is in the language of Crs_α). So (iii) holds for Crs_α. To show (iii) for D_α and G_α, notice first that for any $\mathfrak{A} \in \mathsf{Crs}_\alpha$, ($\mathfrak{A} \in \mathsf{D}_\alpha$ iff $\mathfrak{A} \models \bigwedge_{i,j \in \alpha} C_i D_{ij} = 1$), and ($\mathfrak{A} \in \mathsf{G}_\alpha$ iff $\bigwedge_{\tau \in {}^\alpha \alpha} S_\tau 1^{\mathfrak{A}} = 1^{\mathfrak{A}}$). Now, if φ fails in D_α, then $\bigwedge_{i,j \in \alpha} C_i D_{ij} = 1 \longrightarrow \varphi$ fails in Crs_α^p, then $\bigwedge_{i,j \in \alpha} C_i D_{ij} = 1 \longrightarrow \varphi$ fails in a finite Crs_α, and then φ fails in a finite D_α. Similarly, if φ fails in a G_α, then $\bigwedge_{\tau \in {}^\alpha \alpha} S_\tau 1 = 1 \longrightarrow \varphi$

fails in Crs^p_α, hence it fails in a finite Crs^p_α, whose reduct is a finite G_α in which φ fails.

To prove Theorem 4.1(iii), we will prove other statements, which are interesting in their own right.

Definition 4.2 (Pattern)

(i) By a *pattern* $\mathfrak{P}$ we understand a finite partial algebra in the similarity type of Crs^p_α, together with a binary relation on this algebra. So $\mathfrak{P}$ is a pattern iff

$$\mathfrak{P} = \langle\langle X, \cap^{\mathfrak{P}}, -^{\mathfrak{P}}, S^{\mathfrak{P}}_{\tau,\Gamma}, D^{\mathfrak{P}}_{ij}\rangle_{\tau\in{}^\alpha\alpha, \Gamma\subseteq\alpha, i,j\in\alpha}, N\rangle,$$

where X is a finite set, $\cap^{\mathfrak{P}}$ is a binary partial function on X, $-^{\mathfrak{P}}$, $S^{\mathfrak{P}}_{\tau,\Gamma}$ are unary partial functions on X, and $D^{\mathfrak{P}}_{ij}$ are partial constants on X (i.e., constants either defined or not defined). Further, N is a binary relation on X.

(ii) Let the pattern $\mathfrak{P}$ be as in (i). By a *realization of* $\mathfrak{P}$ *in* $\mathfrak{A}$ we mean a homomorphism $h : \mathfrak{P} \longrightarrow \mathfrak{A}$ such that $h(x) \neq h(y)$ for all $(x,y) \in N$. We say that $\mathfrak{P}$ is *realizable in* $\mathfrak{A}$ if there is a realization $h : \mathfrak{P} \longrightarrow \mathfrak{A}$.

Lemma 4.3 *(i) To any universal formula φ one can construct patterns $\mathfrak{P}_1, \ldots, \mathfrak{P}_n$ such that for any algebra $\mathfrak{A}$, φ fails in $\mathfrak{A}$ iff one of $\mathfrak{P}_1, \ldots, \mathfrak{P}_n$ is realizable in $\mathfrak{A}$.*

(ii) Let $\mathfrak{A}$ be a finite algebra and let $\mathfrak{P} = \langle \mathfrak{A}, \{(a,b) : a,b \in A, a \neq b\}\rangle$. Then $\mathfrak{A}$ is isomorphic to a Crs^p_α iff $\mathfrak{P}$ is realizable in a Crs^p_α.

Proof. (of Lemma 4.3) (ii) is straightforward. To show (i), one first brings φ into a conjunctive normal form $\bigwedge \varphi_i$. Thus φ fails in $\mathfrak{A}$ iff one of φ_i fails in $\mathfrak{A}$. Therefore it is enough to construct a pattern $\mathfrak{P}$ to any disjunction $\psi = \bigvee_{j\in J} \tau_j = \sigma_j \vee \bigvee_{i\in I} \tau_i \neq \sigma_i$ of equations and negations of equations. Let X be the set of all subterms occurring in ψ, let the partial operations on X be as the terms themselves indicate (i.e., if $\tau, \sigma \in X$, then $\tau \cap \sigma$ is defined in X iff $\tau \cap \sigma \in X$, etc.). Let R be the congruence relation on $\langle X, \cap, -, \ldots\rangle$ generated by $\{(\tau_j, \sigma_j) : j \in J\}$, let $N = \{(\tau_i, \sigma_i) : i \in I\}$, and let $\mathfrak{P} = \langle\langle X, \cap, -, \ldots\rangle/R, N/R\rangle$. Now ψ fails in $\mathfrak{A}$ iff $\bigwedge_{j\in J} \tau_j = \sigma_j \wedge \bigwedge_{i\in I} \tau_i \neq \sigma_i$ is satisfiable in $\mathfrak{A}$, which then is equivalent to realizability of $\mathfrak{P}$ in $\mathfrak{A}$. □

We will show (in Theorem 4.8) that a pattern is realizable in a Crs^p_α iff it is realizable in a finite Crs^p_α with a computable size. This then will imply Theorem 4.1(iii) by Lemma 4.3(i).

From now on let us be given a pattern $\mathfrak{P} = \langle\langle X, \cap^{\mathfrak{P}}, \ldots\rangle, N\rangle$. We say that $\mathfrak{P}$ is *good* iff for all $x \in X$ and τ, Γ, if $S_{\tau,\Gamma}x$ is defined in $\mathfrak{P}$, then $C_{(\Gamma)}x$ is also defined in $\mathfrak{P}$. (Recall that $C_{(\Gamma)}x = S_{id,\Gamma}x$.)

Claim 4.4 *For any pattern $\mathfrak{P}$ we can construct a good pattern $\mathfrak{P}'$ such that for any algebra $\mathfrak{A}$, $\mathfrak{P}$ is realizable in $\mathfrak{A}$ iff $\mathfrak{P}'$ is realizable in $\mathfrak{A}$.*

Proof.(of Claim 4.4) Let

$$X' \stackrel{\text{def}}{=} X \cup \{\langle \Gamma, x \rangle : S_{\tau,\Gamma}x \text{ is defined in } \mathfrak{P} \text{ for some } \tau$$
$$\text{and } C_{(\Gamma)}x \text{ is not defined in } \mathfrak{P}\}.$$

(Here we assume that $\langle \Gamma, x \rangle \notin X$ for any $x \in X$ and for any Γ.) Define the operations of $\mathfrak{P}'$ such that they agree with the operations of $\mathfrak{P}$ except that $C_{(\Gamma)}^{\mathfrak{P}'}x = \langle \Gamma, x \rangle$ for any $\langle \Gamma, x \rangle \in X' - X$. It is easy to check that $\mathfrak{P}'$ satisfies the requirements. □

Because of Claim 4.4, from now on we assume that our pattern $\mathfrak{P}$ is good.

Definition 4.5 ($\mathfrak{P}$-mosaic)

(i) Let V be an α-ary relation. Then $\mathfrak{G}(V) \stackrel{\text{def}}{=} \{s \circ \tau : s \in V, \tau \in {}^{\alpha}\alpha\}$.
(ii) A $\mathfrak{P}$-*mosaic* is a pair (V, ℓ), where $V \subseteq {}^{\alpha}U$ for some U, $\ell : \mathfrak{G}(V) \longrightarrow \mathcal{P}(X)$ (where $\mathcal{P}(X) = \{Y : Y \subseteq X\}$ is the powerset of X) such that the following hold for all $s \in V$, $x, y \in X$, $i, j \in \alpha$, $\tau \in {}^{\alpha}\alpha$ and $\Gamma \subseteq \alpha$:
  - $x \cap y \in \ell(s)$ iff $x, y \in \ell(s)$, whenever $x \cap y$ is defined in $\mathfrak{P}$
  - $-x \in \ell(s)$ iff $x \notin \ell(s)$, whenever $-x$ is defined in $\mathfrak{P}$
  - $D_{ij} \in \ell(s)$ iff $s_i = s_j$, whenever D_{ij} is defined in $\mathfrak{P}$
  - $S_{\tau,\Gamma}x \in \ell(s)$ if $s \circ \tau \stackrel{\Gamma}{\equiv} z$ for some $z \in V$ with $x \in \ell(z)$, whenever $S_{\tau,\Gamma}x$ is defined in $\mathfrak{P}$
  - for all $s, z \in \mathfrak{G}(V)$, $\tau \in {}^{\alpha}\alpha$ and $\Gamma \subseteq \alpha$:
    $$S_{\tau,\Gamma}x \in \ell(s) \implies C_{(\Gamma)}x \in \ell(s \circ \tau)$$
    $$s \stackrel{\Gamma}{\equiv} z \implies [C_{(\Gamma)}x \in \ell(s) \text{ iff } C_{(\Gamma)}x \in \ell(z)].$$

Definition 4.6 Let (V, ℓ), (V', ℓ') be $\mathfrak{P}$-mosaics, $s \in \mathfrak{G}(V)$, $s' \in \mathfrak{G}(V')$.

(i) For any α-ary relation V, $base(V) \stackrel{\text{def}}{=} \bigcup \{Rng(s) : s \in V\}$.
(ii) Let $h : base(V) \longrightarrow base(V')$ be a bijection. We say that h is an isomorphism between $\langle (V, \ell), s \rangle$ and $\langle (V', \ell'), s' \rangle$, in symbols $(V, \ell), s \stackrel{h}{\cong} (V', \ell'), s'$, iff $V' = \{h \circ z : z \in V\}$, $s' = h \circ s$, and for all $z \in \mathfrak{G}(V)$, $\ell'(h \circ z) = \{x \in X : x \in \ell(z)\}$. We say that $\langle (V, \ell), s \rangle$ and $\langle (V', \ell'), s' \rangle$, or (V, ℓ) and (V', ℓ') are *isomorphic*, in symbols $(V, \ell), s \cong (V', \ell'), s'$ or $(V, \ell) \cong (V', \ell')$, if there is such an isomorphism h.
(iii) Let H be a set. Then $(V, \ell) \lceil H \stackrel{\text{def}}{=} (V \cap {}^{\alpha}H, \ell \lceil (\mathfrak{G}(V) \cap {}^{\alpha}H))$, this is called the *restriction* of (V, ℓ) to H.
(iv) (V, ℓ) is *compatible* with (V', ℓ'), in symbols $(V, \ell) \sim (V', \ell')$, iff $[(V, \ell) \lceil H = (V', \ell') \lceil H$, where $H = base(V) \cap base(V')$, and $(\exists s \in \mathfrak{G}(V)) C_{(\alpha)}x \in \ell(s)$ iff $(\exists s' \in \mathfrak{G}(V')) C_{(\alpha)}x \in \ell'(s')$, for all $x \in X$ such that $C_{(\alpha)}x$ is defined in $\mathfrak{P}$].

(v) Let $\Gamma \subseteq \alpha$. We say that $\langle (V, \ell), s \rangle$ is Γ-*close* to $\langle (V', \ell'), s' \rangle$, in symbols $(V, \ell), s \stackrel{\Gamma}{\equiv} (V', \ell'), s'$, iff there is an isomorphic copy $(V'', \ell''), s''$ of $(V', \ell'), s'$ such that $(V'', \ell'') \sim (V, \ell)$ and $s'' \stackrel{\Gamma}{\equiv} s$.

(vi) (V, ℓ) is *small* iff $V \subseteq {}^\alpha \alpha$.

Definition 4.7 Let $\mathcal{M}$ be a set of $\mathfrak{P}$-mosaics.

(i) $\mathcal{M}$ is *complete* iff for all $\Gamma \subseteq \alpha$, $x \in X$, $(V, \ell) \in \mathcal{M}$ and $s \in \mathfrak{G}(V)$ such that $C_{(\Gamma)} x \in \ell(s)$ there are $(V', \ell') \in \mathcal{M}$ and $s' \in V'$ such that $x \in \ell'(s')$ and
$$(V, \ell), s \stackrel{\Gamma}{\equiv} (V', \ell'), s'.$$

(ii) $\mathcal{M}$ *satisfies* N iff for all $(x, y) \in N$ there are $(V, \ell) \in \mathcal{M}$, $s \in V$ such that $(x \in \ell(s) \not\Leftrightarrow y \in \ell(s))$.

(iii) $\mathcal{M}$ is a *realization* of $\mathfrak{P}$ iff $\mathcal{M}$ is complete, $\mathcal{M}$ satisfies N, and the elements of $\mathcal{M}$ are pairwise compatible up to isomorphism, i.e., to any $m, m' \in \mathcal{M}$ there is an isomorphic copy m'' of m' such that $m \sim m''$.

Theorem 4.8 *(i)–(iii) below are equivalent.*

(i) $\mathfrak{P}$ *is realizable in a* Crs_α^p.

(ii) $\mathfrak{P}$ *is realizable in a finite* Crs_α^p, *the size of the finite* Crs_α^p *being computable from* $\mathfrak{P}$.

(iii) *There is a set of small $\mathfrak{P}$-mosaics which is a realization of $\mathfrak{P}$.*

Proof. We will prove (i)$\Longrightarrow$(iii)$\Longrightarrow$(ii). Since (ii)$\Longrightarrow$(i) holds trivially, this will prove the theorem.

(i)$\Longrightarrow$(iii): Assume that $h : \mathfrak{P} \longrightarrow \mathfrak{A}$ is a realization. Let $W \stackrel{\text{def}}{=} 1^\mathfrak{A}$. Define $\ell : \mathfrak{G}(W) \longrightarrow \mathcal{P}(X)$ as

$\ell(s) = \{x \in X : s \in h(x)\}$, if $s \in W$, and

$\ell(s) = \{C_{(\Gamma)} x \in X : s \stackrel{\Gamma}{\equiv} z \in h(C_{(\Gamma)} x) \text{ for some } z\}$, if $s \in \mathfrak{G}(W) - W$.

It can be checked that (W, ℓ) is a $\mathfrak{P}$-mosaic. Let

$$\mathcal{M} \stackrel{\text{def}}{=} \{(V, \ell') : V \subseteq {}^\alpha \alpha, \ (V, \ell') \cong (W, \ell) \lceil H \text{ for some } H\}.$$

It can be checked that $\mathcal{M}$ is a set of small $\mathfrak{P}$-mosaics which is a realization of $\mathfrak{P}$.

(iii)$\Longrightarrow$(ii): Assume that $\mathcal{M}$ is a set of small $\mathfrak{P}$-mosaics and $\mathcal{M}$ is a realization of $\mathfrak{P}$. We may assume that $\mathcal{M}$ is restriction-closed, i.e., that for all $(V, \ell) \in \mathcal{M}$ and set H, also $(V, \ell) \lceil H \in \mathcal{M}$.

Some Notation. $(V, \ell) \lceil s \stackrel{\text{def}}{=} (V, \ell) \lceil Rng(s)$, $\mathbf{I}\mathcal{M}$ denotes the class of all $\mathfrak{P}$-mosaics isomorphic to a member of $\mathcal{M}$, and

$$\bar{\mathcal{M}} \stackrel{\text{def}}{=} \{\langle (V, \ell), s \rangle : (V, \ell) \in \mathcal{M}, \ s \in V\}.$$

First, we want to build a (big) $\mathfrak{P}$-mosaic (W, ℓ) with the following properties:

(gi) $(\forall s \in W)\, (W, \ell) \lceil s \in \mathbf{I}\mathcal{M}$
(gii) $(\forall m \in \mathcal{M})(\exists H)\, m \cong (W, \ell) \lceil H$
(giii) $(\forall s \in \mathfrak{G}(W),\, (m, z) \in \bar{\mathcal{M}},\, \Gamma \subsetneq \alpha)$

$$[(W, \ell)\lceil s, s \stackrel{\Gamma}{\equiv} m, z \implies (\exists s' \in W)(s' \stackrel{\Gamma}{\equiv} s \text{ and } (W, \ell)\lceil s', s' \cong m, z].$$

The basic tool for achieving this will be:

Lemma 4.9 *If $(V, \ell) \sim (V', \ell')$, then $m \stackrel{\text{def}}{=} (V \cup V', \ell \cup \ell')$ is a $\mathfrak{P}$-mosaic and $(V, \ell) = m\lceil \text{base}(V)$.*

Proof.(of Lemma 4.9) Clearly, $\ell \cup \ell' : \mathfrak{G}(V \cup V') \longrightarrow \mathcal{P}(X)$. The conditions for $\cap, -, D_{ij}$ are easily seen to hold.

Assume that $S_{\tau,\Gamma} x$ is defined in $\mathfrak{P}$, $z \in V$, $x \in \ell(z)$, and

$$s \circ \tau \stackrel{\Gamma}{\equiv} z$$

for some $s \in V \cup V'$. We have to show $S_{\tau,\Gamma} x \in \ell''(s)$, where $\ell'' = \ell \cup \ell'$. If $s \in V$ then $S_{\tau,\Gamma} x \in \ell(s) = \ell''(s)$ and we are done. Assume that $s \in V'$. By $x \in \ell(z)$ and since $C_{(\Gamma)} x$ is defined, $C_{(\Gamma)} x \in \ell(z)$. If $\Gamma = \alpha$, then $C_{(\Gamma)} x \in \ell(s \circ \tau)$, because $(V, \ell) \sim (V', \ell')$. Assume $\Gamma \neq \alpha$, say $i \in \alpha - \Gamma$. Let

$$s' \stackrel{\text{def}}{=} s \circ \tau.$$

Then $z_i = s'_i$. Let $\sigma : \alpha \longrightarrow \alpha$ be such that σ is identity on $\alpha - \Gamma$, and σ takes every element of Γ to i. Then

$$z \stackrel{\Gamma}{\equiv} z \circ \sigma = s' \circ \sigma \stackrel{\Gamma}{\equiv} s',$$

since σ is identity on $\alpha - \Gamma$, and since $z \stackrel{\Gamma}{\equiv} s'$ and $z_i = s'_i$. Now $C_{(\Gamma)} x \in \ell(z)$ $\implies C_{(\Gamma)} x \in \ell(z \circ \sigma) = \ell'(s \circ \sigma) \implies C_{(\Gamma)} x \in \ell'(s) = \ell''(s)$.

The first statement in the last condition is easily seen to hold. So assume

$$s, z \in V \cup V',\; s \stackrel{\Gamma}{\equiv} z \text{ and } C_{(\Gamma)} x \in \ell''(s).$$

We want to show $C_{(\Gamma)} x \in \ell''(z)$. If $\Gamma = \alpha$ or $s, z \in V$ or $s, z \in V'$ then $C_{(\Gamma)} x \in \ell''(z)$ is easy to show. So assume $i \in \alpha - \Gamma$ and $s \in V$, $z \in V'$. Let σ be as in the previous case. Then $C_{(\Gamma)} x \in \ell''(s) \implies C_{(\Gamma)} x \in \ell(s) \implies C_{(\Gamma)} x \in \ell(s \circ \sigma) = \ell'(z \circ \sigma) \implies C_{(\Gamma)} x \in \ell'(z) = \ell''(z)$. $(V, \ell) = m\lceil \text{base}(V)$ is easy to see. $\square$

Notation. If $m = (W, \ell)$, $m' = (W', \ell')$, then $m \cup m' \stackrel{\text{def}}{=} (W \cup W', \ell \cup \ell')$.

Now, by using Lemma 4.9, we will "put together" a $\mathfrak{P}$-mosaic with properties (gi)–(giii) by induction along ω. Let $m_0 = (W_0, \ell_0)$ be the disjoint union of all members of $\mathcal{M}$. Then m_0 is finite, and (gi) is satisfied.

A FINE-STRUCTURE ANALYSIS OF FIRST-ORDER LOGIC / 241

Assume that $m_n = (W_n, \ell_n)$ is given such that m_n is finite. We will define $m_{n+1} = (W_{n+1}, \ell_{n+1})$ which will have the following properties:

(i) m_{n+1} is an extension of m_n, i.e., $m_n = m_{n+1}\lceil base(W_n)$.

(ii) $(\forall s \in \mathfrak{G}(W_n), (m,z) \in \bar{\mathcal{M}}, \Gamma \subseteq \alpha)\ [(W_n, \ell_n)\lceil s, s \stackrel{\Gamma}{\equiv} m, z \implies (\exists s' \in W_{n+1})(s' \stackrel{\Gamma}{\equiv} s$ and $(W_{n+1}, \ell_{n+1})\lceil s', s' \cong m, z)]$.

Let

$$\mathcal{N} \stackrel{\text{def}}{=} \{(s,(m,z),\Gamma) : s \in \mathfrak{G}(W_n),\ (m,z) \in \bar{\mathcal{M}},\ m_n\lceil s, s \stackrel{\Gamma}{\equiv} m, z\} = \{(s_1,(m_1,z_1),\Gamma_1),\ldots,(s_k,(m_k,z_k),\Gamma_k)\}.$$

We will define $m_n^0, m_n^1, \ldots, m_n^k$:

$m_n^0 \stackrel{\text{def}}{=} m_n$. Let $0 < p < k$ and assume that m_n^{p-1} is given. We know that

$$m_n\lceil s_p, s_p \stackrel{\Gamma_p}{\equiv} m_p, z_p.$$

Let $m'_p, z'_p \cong m_p, z_p$ be such that

$$z'_p \stackrel{\Gamma_p}{\equiv} s_p,\ m_n\lceil s_p \sim m'_p$$

and further, $base(m'_p) - Rng(s_p)$ is disjoint from $base(m_n^{p-1})$. Then $m_n^{p-1} \sim m'_p$. Let

$$m_n^p \stackrel{\text{def}}{=} m_n^{p-1} \cup m'_p.$$

Let $m_{n+1} \stackrel{\text{def}}{=} m_n^k$. Then m_{n+1} is finite, and satisfies (i), (ii) above.

Let $m \stackrel{\text{def}}{=} (W, \ell)$ be the union of all the m_n's, i.e.,

$$W \stackrel{\text{def}}{=} \bigcup\{W_n : n \in \omega\} \quad \text{and} \quad \ell \stackrel{\text{def}}{=} \bigcup\{\ell_n : n \in \omega\}.$$

Now it is not difficult to check that (W, ℓ) satisfies (gi)–(giii).

So assume that $m = (W, \ell)$ is a $\mathfrak{P}$-mosaic which satisfies (gi)–(giii). We define the equivalence relation $\equiv$ on W as follows: For all $s, z \in W$,

$$s \equiv z \quad \text{iff} \quad m\lceil s \cong m\lceil z.$$

By (gii), $\equiv$ has at most $|\bar{\mathcal{M}}|$-many blocks. For any $s \in W$, let $s/\equiv \stackrel{\text{def}}{=} \{z \in W : z \equiv s\}$. Let

$$At \stackrel{\text{def}}{=} \{s/\equiv\ :\ s \in W\}$$

$$A \stackrel{\text{def}}{=} \{\cup Y : Y \subseteq At\}$$

$$h(x) \stackrel{\text{def}}{=} \{s \in W : x \in \ell(s)\},\ \text{for any}\ x \in X.$$

Claim 4.10 (i) A is closed under $S_{\tau,\Gamma}^W$ and $D_{ij}^W \in A$ for all $\tau \in {}^\alpha\alpha$, $\Gamma \subseteq \alpha$ and $i, j \in \alpha$.

(ii) A is finite, in fact $|A| \leq 2^{|\bar{\mathcal{M}}|}$.

(iii) h is a realization of $\mathfrak{P}$ in $\langle A, \cap, -, S_{\tau,\Gamma}^W, D_{ij}^W \rangle_{\tau \in {}^\alpha\alpha,\ \Gamma \subseteq \alpha,\ i,j \in \alpha}$.

Proof.(of Claim 4.10) **(i):** By the definition of $\equiv$, it is clear that if $s \equiv z$, then $s_i = s_j$ iff $z_i = z_j$. Thus $D_{ij}^W = \bigcup\{s/\equiv \; : s \in W, \; s_i = s_j\}$. To show that A is closed under $S_{\tau,\Gamma}^W$, it is enough to show that $S_{\tau,\Gamma}^W(s/\equiv) \in A$ for all $s \in W$, and in order to show this, it is enough to show that if $z' \equiv z$ and $z \in S_{\tau,\Gamma}^W(s/\equiv)$, then $z' \in S_{\tau,\Gamma}^W(s/\equiv)$. Now, $z \in S_{\tau,\Gamma}^W(s/\equiv)$ means that $z \circ \tau \stackrel{\Gamma}{\equiv} s' \in W$ for some $s' \equiv s$. Then $z' \circ \tau \in \mathfrak{G}(W)$ and $z' \circ \tau \equiv z \circ \tau$. Let $m' \stackrel{\text{def}}{=} m\lceil s'$. Then $m' \in \mathbf{I}\mathcal{M}$ by (gii), and $m\lceil z \circ \tau, z \circ \tau \stackrel{\Gamma}{\equiv} m', s'$. By $z' \circ \tau \equiv z \circ \tau$, then $m\lceil z' \circ \tau, z' \circ \tau \stackrel{\Gamma}{\equiv} m', s'$. By (giii), then there is $s'' \in W$ such that $s'' \stackrel{\Gamma}{\equiv} z' \circ \tau$ and $m\lceil s'' \cong m'$. But then $s'' \equiv s' \equiv s$, i.e., $z' \in S_{\tau,\Gamma}^W(s/\equiv)$.

(ii): As we noted earlier, $\equiv$ has at most $|\bar{\mathcal{M}}|$-many blocks. (ii) follows from this, by the definition of A.

(iii): h is easily seen to be a homomorphism w.r.t. $\cap, -, D_{ij}$, by using the first three conditions in the definition of a $\mathfrak{P}$-mosaic. Assume now that $S_{\tau,\Gamma}x$ is defined in $\mathfrak{P}$, and we want to show $h(S_{\tau,\Gamma}x) = S_{\tau,\Gamma}^W h(x)$. $S_{\tau,\Gamma}^W h(x) \subseteq h(S_{\tau,\Gamma}x)$ holds by the last but one condition in the definition of a $\mathfrak{P}$-mosaic: Assume $s \in S_{\tau,\Gamma}^W h(x)$. Then $s \circ \tau \stackrel{\Gamma}{\equiv} z$ for some $z \in h(x)$, so $S_{\tau,\Gamma}x \in \ell(s)$, i.e., $s \in h(S_{\tau,\Gamma}x)$. The other inclusion, $h(S_{\tau,\Gamma}x) \subseteq S_{\tau,\Gamma}^W h(x)$ holds by the last condition in the definition of a $\mathfrak{P}$-mosaic, by completeness of $\mathcal{M}$, and by (gii)–(giii): Assume $s \in h(S_{\tau,\Gamma}x)$. Then $C_{(\Gamma)}x \in \ell(s \circ \tau)$. By (gii), then $m\lceil s \circ \tau \in \mathbf{I}\mathcal{M}$, and by completeness of $\mathcal{M}$, then there is $(m', z') \in \bar{\mathcal{M}}$ such that $m\lceil s \circ \tau, s \circ \tau \stackrel{\Gamma}{\equiv} m', z'$ and $x \in \ell'(z')$ (where $m' = (W', \ell')$). By (giii), then there is $s' \in W$ such that $s \circ \tau \stackrel{\Gamma}{\equiv} s'$ and $x \in \ell(s')$, i.e., $s' \in h(x)$. Then $s \in S_{\tau,\Gamma}^W h(x)$. We have seen that h is a homomorphism.

Let $(x, y) \in N$. Then $h(x) \neq h(y)$, because $\mathcal{M}$ satisfies N and because (gi) holds. $\square$

Now it is easy to give a bound on $|\bar{\mathcal{M}}|$, by using the fact that the elements of $\mathcal{M}$ are small $\mathfrak{P}$-mosaics. Thus, by Claim 4.10(ii), that gives a bound on $|A|$. By this, we have proved Theorem 4.1. $\square$

Corollary 4.11 *Let α be finite. Then Crs_α^p has the finite base property, i.e., any finite Crs_α^p is isomophic to one with a finite base (or unit).*

Proof. Let $\mathfrak{A} \in \mathsf{Crs}_\alpha^p$ be finite. We define a pattern $\mathfrak{P}$ as follows:

$$A' \stackrel{\text{def}}{=} \{\langle -, a\rangle, \langle \tau, \Gamma, a\rangle, \langle i, j\rangle, \langle 1\rangle \; : \; a \in A, \; \tau \in {}^\alpha\alpha, \; \Gamma \subseteq \alpha, \; i, j \in \alpha\}.$$

We may assume that A' is disjoint from A. Let

$$X \stackrel{\text{def}}{=} A \cup A'.$$

We define the partial operations $\cap^{\mathfrak{P}}, -^{\mathfrak{P}}, \ldots$ as follows. The operations are defined exactly in the cases listed below: Let $V \stackrel{\text{def}}{=} 1^{\mathfrak{A}}$, and let $a, b \in A$, $\tau \in {}^{\alpha}\alpha$, $\Gamma \subseteq \alpha$, $i, j \in \alpha$.

$$a \cap^{\mathfrak{P}} b = a \cap b$$
$$\langle -, a \rangle \cap^{\mathfrak{P}} V = -a, \quad \langle \tau, \Gamma, a \rangle \cap^{\mathfrak{P}} V = S_{\tau, \Gamma}^{V} a, \quad \langle i, j \rangle \cap^{\mathfrak{P}} V = D_{ij}^{V}$$
$$-^{\mathfrak{P}} a = \langle -, a \rangle$$
$$S_{\tau, \Gamma}^{\mathfrak{P}} a = \langle \tau, \Gamma, a \rangle, \quad S_{\tau}^{\mathfrak{P}} \langle 1 \rangle = \langle 1 \rangle$$
$$D_{ij}^{\mathfrak{P}} = \langle i, j \rangle$$
$$1^{\mathfrak{P}} = \langle 1 \rangle.$$

Let $N \stackrel{\text{def}}{=} \{(a, b) : a, b \in A,\ a \neq b\}$, and $\mathfrak{P} \stackrel{\text{def}}{=} \langle \langle X, \cap^{\mathfrak{P}}, -^{\mathfrak{P}}, \ldots \rangle, N \rangle$. Then $\mathfrak{P}$ is realizable in a $\mathsf{Crs}_{\alpha}^{p}$, namely it is realizable in

$$\langle \mathcal{P}(W), \cap, -, S_{\tau, \Gamma}^{W}, D_{ij}^{W} \rangle_{\tau \in {}^{\alpha}\alpha,\ \Gamma \subseteq \alpha,\ i, j \in \alpha} \text{ with } W = \mathfrak{G}(V).$$

Thus, by Theorem 4.8, $\mathfrak{P}$ is realizable in a finite $\mathfrak{B} \in \mathsf{Crs}_{\alpha}^{p}$. It is proved in Andréka and Németi 1995 that $\{\mathfrak{A} \in \mathsf{Crs}_{\alpha}^{p} : 1^{\mathfrak{A}} = \mathfrak{G}(1^{\mathfrak{A}})\}$ has the finite base property, therefore $\mathfrak{B}$ is isomorphic to some $\mathfrak{B}' \in \mathsf{Crs}_{\alpha}^{p}$ with $1^{\mathfrak{B}'}$ finite. Let $h : \mathfrak{P} \longrightarrow \mathfrak{B}'$ be a realization, and let $W \stackrel{\text{def}}{=} h(1^{\mathfrak{A}}) = h(V)$. Then W is finite, and $\mathfrak{A}$ is isomorphic to a $\mathsf{Crs}_{\alpha}^{p}$ with unit W (actually, h gives this isomorphism). □

The class of all weakly associative relation algebras has been investigated in many works, e.g. Marx 1995, Mikulás 1995b. This class is the same as that of representable relation algebras relativized to symmetric and reflexive relations (and subalgebras of the so obtained algebras). In Németi 1986 it was proved that the universal theory of weakly associated relation algebras is decidable. The following corollary is an improvement of that result.

Corollary 4.12 *The class of all weakly associative relation algebras has the finite base property, i.e. every finite weakly associative relation algebra is representable with a greatest element a finite symmetric and reflexive relation.*

Proof. This follows from Corollary 4.11, as follows. Let $\mathfrak{A}$ be a finite weakly associative relation algebra with greatest element W. Then W is a symmetric, reflexive binary relation. Let $V = \bigcup \{{}^{3}\{a, b\} : (a, b) \in W\}$, and for all $R \subseteq W$ let $h(R) = \{s \in V : (s_0, s_1) \in R\}$. Let $\mathfrak{C}$ be the full $\mathsf{Crs}_{\alpha}^{p}$ with unit V, i.e. $\mathfrak{C} \in \mathsf{Crs}_{\alpha}^{p}$ and the universe of $\mathfrak{C}$ is $\mathcal{P}(V)$. Then h is an embedding of $\mathfrak{A}$ into the relation algebraic reduct of $\mathfrak{C}$ because V is closed under substitutions, i.e. $\mathfrak{G}(V) = V$, where the relation algebraic reduct is taken in the usual sense, c.f. e.g. in Henkin et al. 1985. It is easy to check that $\{h(a) : a \in A\}$ generates a finite subalgebra $\mathfrak{C}'$ of $\mathfrak{C}$. Let $\mathfrak{C}''$ be a $\mathsf{Crs}_{\alpha}^{p}$ with finite unit V'' and isomorphic to $\mathfrak{C}'$. Then V'' is closed under substitutions. Then $\mathfrak{A}$ is isomorphic to a subalgebra of the relation algebraic reduct of $\mathfrak{C}''$, which gives us the desired weakly associative relation algebra with finite greatest element to which $\mathfrak{A}$ is isomorphic to. □

The logical counterpart of weakly associative relation algebras is the strongest well behaved arrow logic, which is the arrow logic of symmetric and reflexive pair frames.

Corollary 4.13 *The arrow logic of symmetric and reflexive pair frames has the finite model property, i.e. the finite pair-frame property.*

Remark 4.14 (The case of infinite α and open problems) Let α be infinite. Then Theorem 4.1 is true for Crs_α by Németi 1995, Theorem 4.13(i). Also, most likely the argument given for Theorem 4.13(i) in Németi 1995 proves that Theorem 4.1(iii) is true for G_α, hence Theorem 4.1(i)–(iv) would be true for G_α.

It was asked in Németi 1986, 1995 as an open problem whether the equational theory of D_α is decidable for infinite α. This question is still open.

As a related open problem we mention that we do not know whether G_α for $\alpha \geq \omega$ is equationally definable or not. (For finite α it is, as was stated in Németi 1995 below Def.4.2. As a misprint, in Németi 1995 the condition $\alpha < \omega$ was left out from this statement.) We note that Marx and Venema to appear contains an axiomatization of the expansion of G_ω with the "permutation operators".

To mention another related open problem, we still do not know whether the free Crs_α for $\alpha \geq 2$ is atomic or not (see Németi 1986). □

Remark 4.15 (On the interpolation and definability properties of our logics) It was proved in Madarász 1995b that a multi-modal logic $\mathcal{L}$ has the Craig interpolation property iff the associated class $\mathsf{Alg}(\mathcal{L})$ of modal algebras has SUPAP, cf. also Marx 1995. Therefore it is enough to discuss which of our classes of algebras has SUPAP.

By Németi 1985, Crs_α has SUPAP for all α. Since $\mathsf{ID}_\alpha \subset \mathsf{ICrs}_\alpha$, $\mathsf{ID}_\alpha \neq \mathsf{ICrs}_\alpha$ is definable over Crs_α by an equation $C_i(D_{ij}) = 1$ involving no variables, D_α inherits SUPAP from Crs_α. Madarász 1995d proved that G_α fails to have not only SUPAP but even some epimorphisms are not surjective in G_α and therefore (by e.g. Henkin et al. 1985 or by Sain 1990) both Beth's definability property and Craig's interpolation fail for the logic corresponding to G_α, for $\alpha > 1$. However, if we add the modalities $[x/y, y/x]$ for all $x, y \in \alpha$, then the so obtained enrichment G_α^+ of G_α has SUPAP, hence the corresponding logic has Craig's interpolation property, as it is proved in op. cit., cf. also Marx and Venema to appear. The proofs in Madarász 1995d, Madarász 1995b prove SUPAP for Crs_α^p (for all α), too.

Since Crs_α^p has SUPAP, this property is inherited by its subclasses D_α^p, G_α^p [7], where an $\mathfrak{A} \in \mathsf{Crs}_\alpha^p$ is in D_α^p (or G_α^p) iff it has a reduct in D_α (or G_α, respectively).

Finally, we note that stronger versions of Craig's interpolation property received algebraic investigations and sufficient conditions in Madarász 1995a, analogously to the investigations of Craig's original property in Németi 1985, Sain 1990, Marx 1995, Madarász 1995b, 1995c. □

Acknowledgment. The present work is part of an ongoing joint project with H. Andréka and J. van Benthem, cf. e.g. Andréka et al. to appear, 1995c, 1995b, Németi 1986, 1995, Mikulás 1995b, 1995a.

References

Alechina, N. 1995. *Modal Quantifiers*. Doctoral dissertation, University of Amsterdam.

Alechina, N., and M. van Lambalgen. 1995. Correspondence and Completeness for Generalized Quantifiers. *Bulletin of the IGPL* 3:167–190.

Andréka, H. 1992. Complexity of Equations Valid in Algebras of Relations. Doctoral Dissertation with the Hungarian Academy of Sciences, Budapest. To appear in *Annals of Pure and Applied Logic*.

Andréka, H., Á. Kurucz, I. Németi, and I. Sain. 1994. Applying algebraic logic; A general methodology. In *Algebraic Logic and the Methodology of Applying it*, ed. H. Andréka, I. Németi, and I. Sain, 1–72. A shortened version appeared as *Applying Algebraic Logic to Logic* in 'Algebraic Methodology and Software Technology (AMAST'93)', (eds: M.Nivat, C.Rattray, T.Rus and G.Scollo), in series *Workshops in Computing*, Springer-Verlag, 1994, 7–28.

Andréka, H., and I. Németi. 1995. Finite-base Property of Cylindric-relativized Set Algebras. Preprint. Math. Inst. Hungar. Acad. Sci., Budapest.

Andréka, H., I. Németi, and I. Sain. 1979. Henkin-type Semantics for Program Schemes to Turn Negative Results to Positive. In *Fundamentals of Computation Theory'79 (Proc. Conf. Berlin 1979)*, ed. L. Budach, 18–24. Berlin. Academie Verlag.

Andréka, H., I. Németi, and I. Sain. 1981. A Characterization of Floyd-provable Programs. In *Math. Foundations of Comp. Sci.'81 (Proc. Conf. Strbské Pleso, Czechoslovakia 1981)*, ed. J. Gruska and M. Chytil, 162–171. Lecture Notes in Computer Science Vol 118. Berlin. Springer Verlag.

Andréka, H., I. Németi, and I. Sain. 1982. A Complete Logic for Reasoning About Programs via Nonstandard Model Theory. *Theoretical Computer Science* 17.

Andréka, H., I. Németi, I. Sain, and Á. Kurucz. 1995a. General Algebraic Logic Including Algebraic Model Theory: An Overview. In *Logic Colloquium '92*, ed. L. Csirmaz, D. Gabbay, and M. de Rijke, 1–60. Studies in Logic, Language and Information. Stanford. CSLI Publications.

---

[7] Because these subclasses can be defined over Crs_α^p by equations involving *no* variables.

Andréka, H., J. van Benthem, and I. Németi. 1995b. Modal Logic and the Finite Variable Hierarchy. Manuscript. Institute for Logic, Language and Computation and Math. Inst. Hungar. Acad. Sci., Budapest.

Andréka, H., J. van Benthem, and I. Németi. 1995c. Submodel Preservation Theorems in Finite Variable Fragments. In *Modal Logic and Process Algebra. A Bisimulation Perspective*, ed. A. Ponse, M. de Rijke, and Y. Venema, 1–11. CSLI Lecture Notes Vol 53. CSLI Publications.

Andréka, H., J. van Benthem, and I. Németi. to appear. Back and Forth between Modal Logic and Classical Logic. *Journal of Philosophical Logic*.

van Benthem, J. n.d. Applications of Cylindric-relativized Set Algebras in Arrow Logics. Lecture at the Workshop: *Designing Logics*, CCSOM, University of Amsterdam, Sept.11, 1995.

van Benthem, J. to appear. Modal foundations for predicate logic. *Studia Logica*.

van Benthem, J., and N. Alechina. to appear. Modal Quantification over Structured Domains. In *Advances in Intensional Logic*, ed. M. de Rijke. Kluwer.

Harel, D. 1984. Dynamic Logic. In *Handbook of Philosophical Logic*, ed. D.M. Gabbay and F. Guenther. 497–604. Dordrecht: Reidel.

Henkin, L. 1950. Completeness in the Theory of Types. *Journal of Symbolic Logic* 15:81–91.

Henkin, L., J. D. Monk, and A. Tarski. 1971. *Cylindric Algebras, Part I*. North-Holland, Amsterdam.

Henkin, L., J. D. Monk, and A. Tarski. 1985. *Cylindric Algebras, Part II*. North-Holland, Amsterdam.

Henkin, L., J. D. Monk, A. Tarski, H. Andréka, and I. Németi. 1981. *Cylindric Set Algebras*. Lecture Notes in Mathematics, No. 883. Springer-Verlag, Berlin.

Madarász, J. 1995a. Characterization of A Very Strong Interpolation Property. Manuscript. Math. Inst. Hungar. Acad. Sci., Budapest.

Madarász, J. 1995b. Craig Interpolation in Algebraizable Logics and Meaningful Generalizations of Modal Logic. Preprint. Math. Inst. Hungar. Acad. Sci., Budapest.

Madarász, J. 1995c. The Craig Interpolation Theorem in Multimodal Logics. *Bulletin Section of Logic* 24(3):147–154.

Madarász, J. 1995d. Variants of Cylindric-relativized Set Algebras and Craig's Interpolation Property. Manuscript. Math. Inst. Hungar. Acad. Sci., Budapest.

Marx, M. 1995. *Algebraic Relativization and Arrow Logic*. Doctoral dissertation, Institute for Logic, Language and Computation, University of Amsterdam. ILLC Dissertation Series 1995–3.

Marx, M., and Y. Venema. to appear. *Multi-Dimensional Modal Logic*. Kluwer Academic Publisher.

Mikulás, Sz. 1995a. Taming first-order logic. In: *Proceedings of ACCOLADE'95*, editors: S. Fischer and M. Trautwein, Dutch Graduate School in Logic, University of Amsterdam, pp.91-104.

Mikulás, Sz. 1995b. *Taming Logics*. Doctoral dissertation, Institute for Logic, Language and Computation, University of Amsterdam. ILLC Dissertation Series 95-12.

Mikulás, Sz., and I. Németi. 1995. Relativized Algebras of Relations with Generalized Quantifiers (Positive Results). Manuscript. Math. Inst. Hungar. Acad. Sci., Budapest.

Monk, J.D., 1993. Lectures on cylindric set algebras, In: *Algebraic Methods in Logic and in Computer Science*, Banach Center Publications Vol. 28, Institute of Mathematics, Polish Academy of Sciences, Warsaw, pp.253-290.

Németi, I. 1985. Cylindric–Relativised Set Algebras Have Strong Amalgamation. *Journal of Symbolic Logic* 50(3):689–700.

Németi, I. 1986. Free Algebras and Decidability in Algebraic Logic. (In Hungarian), Doctoral Dissertation with the Hungarian Academy of Sciences, Budapest.

Németi, I. 1990. On Cylindric Algebraic Model Theory. In *Algebraic logic and universal algebra in computer science*, ed. C. H. Bergman, R. D. Maddux, and D. L. Pigozzi. Lecture Notes in Computer Science, Vol. 425, xi+292 p. Springer-Verlag, Berlin.

Németi, I. 1991. Algebraizations of quantifier logics: an overview (version 11.2). Preprint. Math. Inst. Hungar. Acad. Sci., Budapest. A short version without proofs appeared in *Studia Logica*, 50(3/4):485–569, 1991.

Németi, I. 1995. Decidable Versions of First-order Logic and Cylindric-relativized Set Algebras. In *Logic Colloquium '92*, ed. L. Csirmaz, D. Gabbay, and M. de Rijke, 177–242. Studies in Logic, Language and Information. Stanford. CSLI Publications.

Sági, G., and I. Németi. 1995. The Modal Logic of Substitutions (A Complete Axiomatization). Manuscript. Math. Inst. Hungar. Acad. Sci., Budapest.

Sain, I. 1979. There Are General Rules for Specifying Semantics: Observations on Abstract Model Theory. *CL and CL (Computational Linguistics and Computer Languages)* 13:195–250.

Sain, I. 1990. Beth's and Craig's properties via epimorphisms and amalgamation in algebraic logic. In *Algebraic logic and universal algebra in computer science*, ed. C. H. Bergman, R. D. Maddux, and D. L. Pigozzi. Lecture Notes in Computer Science, Vol. 425, 209–226. Springer-Verlag, Berlin.

Sain, I., and R. J. Thompson. 1991. Strictly finite schema axiomatization of quasi-polyadic algebras. In *Algebraic Logic (Proc. Conf. Budapest 1988)*, ed. H. Andréka, J. D. Monk, and I. Németi, 539–571. Amsterdam. Colloq. Math. Soc. J. Bolyai, North-Holland, Amsterdam.

Tarski, A. 1965. A Simplified Formulation of Predicate Logic with Identity. *Arch. Math. Logik Grundlagenforsch* 7:61–79.

Venema, Y. 1995. A Modal Logic of Quantification and Substitution. In *Logic Colloquium '92*, ed. L. Csirmaz, D. Gabbay, and M. de Rijke, 293–309. Studies in Logic, Language and Information. Stanford. CSLI Publications.